Advances in Inverse Problems for Partial Differential Equations

CONTEMPORARY MATHEMATICS

784

Advances in Inverse Problems for Partial Differential Equations

AMS Special Session
Recent Developments on Analysis and Computation for Inverse Problems for PDEs
March 13–14, 2021, Virtual

AMS Special Session
Recent Advances in Inverse Problems for PDEs
October 23–23, 2021, Virtual

Dinh-Liem Nguyen
Loc Hoang Nguyen
Thi-Phong Nguyen
Editors

AMERICAN MATHEMATICAL SOCIETY

2020 *Mathematics Subject Classification.* Primary 35J05, 35R30, 49N45, 65M32, 65N21, 65R20, 78A46, 78M22, 80A23, 86A22.

Library of Congress Cataloging-in-Publication Data

Names: Nguyen, Dinh-Liem, 1986- editor.

Title: Advances in inverse problems for partial differential equations : virtual AMS special session on recent developments on analysis and computation for inverse problems for PDEs, March 13-14, 2021; virtual AMS sectional meeting on recent advances in inverse problems for PDEs, October 23-24, 2021 / Dinh-Liem Nguyen, Loc Hoang Nguyen, Thi-Phong Nguyen, editors.

Description: Providence, Rhode Island : American Mathematical Society, [2023] | Contemporary mathematics, 0271-4132 ; volume 784 | Includes bibliographical references.

Identifiers: LCCN 2022046723 | ISBN 9781470469689 (paperback)

Subjects: Partial differential equations – Elliptic equations and systems – Laplacian operator, reduced wave equation (Helmholtz equation), Poisson equation. | Partial differential equations – Miscellaneous topics – Inverse problems. | Calculus of variations and optimal control; optimization – Miscellaneous topics – Inverse problems. | Numerical analysis – Partial differential equations, initial value and time-dependent initial-boundary value problems – Inverse problems. | Numerical analysis – Partial differential equations, boundary value problems – Inverse problems. | Numerical analysis – Integral equations, integral transforms – Integral equations. | Optics, electromagnetic theory – General – Inverse scattering problems. | Optics, electromagnetic theory – Basic methods – Spectral methods. | Classical thermodynamics, heat transfer – Thermodynamics and heat transfer – Inverse problems. | Geophysics – Geophysics – Inverse problems.

Classification: LCC QA374 .A24 2023

LC record available at https://lccn.loc.gov/2022046723

Contemporary Mathematics ISSN: 0271-4132 (print); ISSN: 1098-3627 (online)

DOI: https://doi.org/10.1090/conm/784

Color graphic policy. Any graphics created in color will be rendered in grayscale for the printed version unless color printing is authorized by the Publisher. In general, color graphics will appear in color in the online version.

∞ The paper used in this book is acid-free and falls within the guidelines
established to ensure permanence and durability.
Visit the AMS home page at `https://www.ams.org/`

10 9 8 7 6 5 4 3 2 1   28 27 26 25 24 23

Contents

Preface

An inverse problem for a partial differential equation (PDE) typically aims to determine some information about the PDE such as a coefficient, an initial condition, and geometry of a domain, from some knowledge about the solution of the PDE. Inverse problems for PDEs arise in a variety of applications including non-destructive testing, radar, medical imaging, geophysical exploration, image processing, finance, etc. These inverse problems are ill-posed problems where the solution usually does not depend continuously on the measured data. Also, in many situations, the dependence of the solution on the data is nonlinear. These features cause substantial challenges in computing numerical solutions to inverse problems for PDEs. During the past three decades, this area of research has been one of the fastest-growing areas in applied mathematics with significant developments in both numerical methods and theoretical analysis.

This volume focuses on new developments in numerical methods for solving inverse problems for PDEs. It contains eleven papers contributed by speakers of two special sessions of the American Mathematical Society Sectional Meeting: "Recent developments on Analysis and Computation for Inverse Problems for PDEs", virtually held from March 13-14, 2021, and "Recent Advances in Inverse Problems for Partial Differential Equations", virtually held from October 23-24, 2021. These papers are concerned with various research topics such as electrical impedance tomography, inverse scattering in radar and optics, reconstruction of initial conditions, control of acoustic fields, and stock price forecasting. The authors studied iterative and non-iterative approaches such as optimization-based, globally convergent, sampling, and machine learning-based methods.

We truly thank all speakers in the special sections and the authors of the papers in this volume for their excellent contributions. We are especially thankful to the administration and staff members of the American Mathematical Society for everything they did to guarantee the success of the meetings and this volume. We hope you enjoy this volume.

Dinh-Liem Nguyen
Loc Hoang Nguyen
Thi-Phong Nguyen

Contemporary Mathematics
Volume **784**, 2023
https://doi.org/10.1090/conm/784/15750

Discretization and convergence of the EIT optimal control problem in Sobolev spaces with dominating mixed smoothness

Ugur G. Abdulla and Saleheh Seif

Abstract. We consider the Inverse Electrical Impedance Tomography (EIT) problem on recovering electrical conductivity and potential in the body based on the measurement of the boundary voltages on the m electrodes for a given electrode current. The variational formulation is introduced as a PDE constrained coefficient optimal control problem in Sobolev spaces with dominating mixed smoothness. Electrical conductivity and boundary voltages are control parameters, and the cost functional is the L_2-norm declinations of the boundary electrode current from the given current pattern and boundary electrode voltages from the measurements. EIT optimal control problem is fully discretized using the method of finite differences. The existence of the optimal control and the convergence of the sequence of finite-dimensional optimal control problems to EIT coefficient optimal control problem is proved both with respect to functional and control in 2- and 3-dimensional domains.

1. Introduction

We consider inverse EIT problem of estimating an unknown conductivity inside the body based on voltage measurements on the surface of the body when electric currents are applied through a set of contact electrodes. Let $Q \in \mathbb{R}^n$ be an open and bounded set representing body, and assume $\sigma(x) : Q \to \mathbb{R}_+$ be an electrical conductivity function. Electrodes, $(E_l)_{l=1}^m$, with contact impedances vector $Z := (Z_l)_{l=1}^m \in \mathbb{R}_+^m$ are attached to the periphery of the body, ∂Q. Electric current vector $I := (I_l)_{l=1}^m \in \mathbb{R}^m$ is applied to the electrodes. Vector I is called *current pattern* if it satisfies conservation of charge condition

$$\sum_{l=1}^{m} I_l = 0. \tag{1}$$

2020 *Mathematics Subject Classification.* 35R30, 35Q93, 49M25, 49M05, 65M06, 65N21.
Key words and phrases. Electrical Impedance Tomography, PDE constrained optimal control, method of finite differences, Sobolev spaces with dominating mixed smoothness, Sobolev embedding, convergence in functional, convergence in control.

The induced constant voltage on electrodes is denoted by $U := (U_l)_{l=1}^m \in \mathbb{R}^m$. By specifying ground or zero potential it is assumed that

$$\sum_{l=1}^{m} U_l = 0. \tag{2}$$

Let $u : Q \to \mathbb{R}$ be an electrostatic potential. Mathematical model of EIT is described through the following mixed boundary-value problem for second order elliptic PDE:

$$-\nabla \cdot (\sigma(x)\nabla u(x)) = 0, \qquad x \in Q \tag{3}$$

$$\sigma(x)\frac{\partial u(x)}{\partial \nu} = 0, \qquad x \in \partial Q - \bigcup_{l=1}^{m} E_l \tag{4}$$

$$u(x) + Z_l\sigma(x)\frac{\partial u(x)}{\partial \nu} = U_l, \quad x \in E_l, \ l = \overline{1,m} \tag{5}$$

$$\int_{E_l} \sigma\frac{\partial u}{\partial \nu} ds = I_l, \qquad l = \overline{1,m} \tag{6}$$

where ν is the outward normal at $x \in \partial Q$.

Forward EIT Problem: *Given electrical conductivity map σ, electrode contact impedance vector Z, and electrode current pattern I it is required to find electrostatic potential u and electrode voltages U satisfying* (3)–(6):

$$(\sigma, Z, I) \longrightarrow (u, U) \tag{7}$$

Mathematical model (3)–(6) for the EIT, referred to as complete electrode model, was suggested in [**45**]. This model suggests the replacement of the complete potential measurements along the boundary with measurements of constant potential along the electrodes with contact impedances. In [**45**] it was demonstrated that the complete electrode model is physically more relevant, and it is capable of predicting the experimentally measured voltages to within 0.1 percent. Existence and uniqueness of the solution to the **forward EIT problem** (7) was proved in [**45**].

The goal of the paper is to prove the convergence of the method of finite differences for the variational formulation of the following inverse EIT problem of determining conductivity map σ from the measurements of the boundary voltages U^*.

Inverse EIT Problem: *Given electrode contact impedance vector Z, electrode current pattern I and boundary electrode measurement U^*, it is required to find electrostatic potential u and electrical conductivity map σ satisfying* (3)–(6) with $U = U^*$.

Inverse EIT problem has many important applications in medicine, industry, geophysics and material science [**26**]. We are especially motivated by medical applications in the detection of cancerous tumors from breast tissue or other parts of the body. The relevance of the inverse EIT problem for cancer detection is based on the fact that the conductivity of the cancerous tumor is higher than the conductivity of normal tissues [**37**]. Inverse EIT Problem is an ill-posed problem and belongs to the class of so-called Calderon type inverse problems, due to celebrated work [**19**], where well-posedness of the inverse problem for the identification of the conductivity coefficient σ of the second-order elliptic PDE (3) through Dirichlet-to-Neumann or Neumann-to-Dirichlet boundary maps is presented. Significant development in Calderon's inverse problem in the class of smooth conductivity function with spatial dimension $n \geq 3$, concerning questions on uniqueness, stability, reconstruction procedure, reconstruction with partial data was achieved in [**11, 16, 30, 31, 33, 34, 40, 46**]. Global uniqueness in spatial dimension $n = 2$ and reconstruction procedure through scattering transform and employment of the so-called D-bar method was presented in [**41**].

Inverse EIT Problem is more difficult than Calderon's problem due to the fact that the infinite-dimensional conductivity function σ and finite-dimensional voltage vector U must be identified based on the finitely many boundary electrode voltage measurements. Hence the input data is a finite-dimensional current vector, while in Calderon's problem input data is given via infinite-dimensional boundary operator "Dirichlet-to-Neuman" or "Neuman-to-Dirichlet". Therefore, the inverse EIT problem is under-determined, and powerful regularization methods are required for its solution. It is essential to note that the size of the input current vector is limited to the number of electrodes, and there is no flexibility to increase its size. It would be natural to suggest that multiple data sets - input currents can be implemented for the identification of the same conductivity function. However, note that besides the unknown conductivity function, there is an unknown boundary voltage vector with a size directly proportional to the size of the input current vector. Accordingly, multiple experiments with "current-to-voltage" measurements are not reducing the under-determinacy of the inverse problem. One can prove uniqueness and stability results by restricting conductivity to the finite-dimensional subset of piecewise analytic functions provided that the number of electrodes is large enough [**24, 38**]. Within the last three decades, many methods developed for the numerical solution of the ill-posed inverse EIT problem. Without any ambition to present a full review we refer to some significant developments such as the recovery of small inclusions from boundary measurements [**13, 35**]; hybrid conductivity imaging methods [**14, 44, 48**]; multi-frequency EIT imaging methods [**15, 43**]; finite element and adaptive finite element method [**29, 39**]; imaging algorithms based on the sparsity reconstruction [**15, 28**]; globally convergent method for shape reconstruction in EIT [**25**]; D-bar method, diction reconstruction method, recovering boundary shape and imaging the anisotropic electrical conductivity [**12, 20, 22, 23, 27**]; globally convergent regularization method using Carleman weight function [**32**].

In a recent paper, [**9**] a new variational formulation of the inverse EIT problem as a PDE constrained coefficient optimal control problem is introduced. The novelty of the new control-theoretic model is its adaptation to the clinical situation when additional "voltage-to-current" measurements can increase the size of the input data from the number of electrodes m up to $m!$ while keeping the size of the unknown parameters fixed. We explain this key novelty of the variational approach below in Remark 1 of Section 1.2. In [**9**] the existence of the optimal control and the Fréchet differentiability in the Besov space setting, formula for the Fréchet gradient, optimality condition, and numerical solution via gradient descent method in the 2D model example is addressed.

The goal of this paper is to prove the convergence of the method of finite differences for the EIT optimal control problem. We pursue discretization of the EIT optimal control problem in Sobolev-Hilbert spaces with dominating mixed smoothness. We prove the existence of the optimal control and the convergence of the sequence of finite-dimensional optimal control problems to EIT optimal control problem both with respect to functional and control in 2D and 3D domains. The setting of the EIT optimal control problem in a Sobolev-Hilbert space with dominating mixed smoothness is a key factor in the proof of the convergence of the finite difference method (see Remark 1 of Section 1.2 and Remark 2 of Section 1.3).

The organization of the paper is as follows. In Section 1.1 we introduce the notations of the functional spaces. In Section 1.2 we introduce Inverse EIT Problem as PDE constrained optimal control problem.In Section 1.3 we pursue discretization via finite differences, and introduce approximating sequence of finite-dimensional discrete optimal control problems. Section 2 formulates the main results. Various key preliminary results are proved in Section 3. Proofs of the main results are completed in Section 4. Finally, in Section 5 we outline the main conclusions.

1.1. Notations. Although the main results of the paper are established when number of spatial variables is 2 and 3, for technical reasons we will describe general notations in space of n independent variables.

Let Q be a bounded domain in $\mathbb{R}^n$ with Lipschitz bondary $S := \partial\Omega$; $B_r(x) = \{y \in \mathbb{R}^n : |y - x| < r\}$; $m_d(\cdot)$ - d-dimensional Lebesgue measure; We use the standard notation for Banach spaces $C^k(\overline{Q})$, $k \in \mathbb{Z}_* := \{0\} \cup \mathbb{Z}_+$ of k-times continuously differentiable functions on $\overline{Q}$, and we simply write $C(\overline{Q})$, if $k = 0$. The following standard notation will be used for Hölder spaces:

- For $k \in \mathbb{Z}_+, 0 < \gamma \leq 1$, Hölder space $C^{k,\gamma}(\overline{Q})$ is the Banach space of elements $u \in C^k(\overline{Q})$ with finite norm

$$\|u\|_{C^{k,\gamma}(\overline{Q})} := \sum_{|\alpha|\leq k} \|D^\alpha u\|_{C(\overline{Q})} + \sum_{|\alpha|=k} [D^\alpha u]_{C^{0,\gamma}(\overline{Q})}$$

where

$$[v]_{C^{0,\gamma}(\overline{Q})} := \sup_{\substack{x,x'\in Q \\ x\neq x'}} \frac{|v(x) - v(x')|}{|x - x'|^\gamma}.$$

Throughout the paper we use standard notations for $L_p(Q), 1 \leq p \leq \infty$ spaces; the following standard notations are used for Sobolev spaces [**17, 18**]:

- For $s \in \mathbb{Z}_+, 1 \leq p < \infty$, Sobolev space $W_p^s(Q)$ is the Banach space of measurable functions on Q with finite norm

$$\|u\|_{W_p^s(Q)} := \begin{cases} \Big(\int\limits_Q \sum\limits_{|\alpha|\leq s} |D^\alpha u(x)|^p dx\Big)^{\frac{1}{p}}, & \text{if } 1 \leq p < \infty, \\ \sum\limits_{|\alpha|\leq s} \|D^\alpha u(x)\|_{L_\infty(Q)}, & \text{if } p = \infty, \end{cases}$$

where $\alpha = (\alpha_1, \dots, \alpha_n) \in \mathbb{Z}_+^n$, $|\alpha| = \alpha_1 + \dots + \alpha_n$, $D_k = \frac{\partial}{\partial x_k}$, $D^\alpha = D_1^{\alpha_1} \dots D_n^{\alpha_n}$. In particular if $p = 2$, $H^s(Q) := W_2^s(Q)$ is a Hilbert space with inner product

$$(f, g)_{H^s(Q)} = \sum_{|\alpha|\leq s} (D^\alpha f(x), D^\alpha g(x))_{L_2(Q)}$$

- Equivalent inner product and norm in $H^1(Q)$ are given as

$$((f, g))_{H^1(Q)} := \int\limits_Q Df \cdot Dg \, dx + \sum_{l=1}^m \int\limits_{E_l} fg \, dS, \quad |||f|||_{H^1(Q)} := ((f, f))^{\frac{1}{2}}.$$

- For $k \in \mathbb{N}, 1 \leq p \leq \infty$, Sobolev space $S_p^k(\mathbb{R}^n)$ with dominating mixed derivatives is a Banach space of measurable functions on $\mathbb{R}^n$ with finite norm
$$\|u\|_{S_p^k(\mathbb{R}^n)} := \sum_{\alpha \in \mathbb{Z}_+^n, \alpha_i \leq k} \left\| \frac{\partial^{|\alpha|} u(x)}{\partial x_1^{\alpha_1} \cdots \partial x_n^{\alpha_n}} \right\|_{L_p(\mathbb{R}^n)}.$$
- Let $Q \subset \mathbb{R}^n$ be a bounded domain. For $k \in \mathbb{N}, 1 \leq p \leq \infty$, Sobolev space $S_p^k(Q)$ with dominating mixed derivatives is defined as
$$S_p^k(Q) = \{f \in \mathcal{D}'(Q) : \exists g \in S_p^k(\mathbb{R}^n) \text{ with } g|_Q = f\}$$
and with
$$\|f\|_{S_p^k(Q)} := \inf \|g\|_{S_p^k(\mathbb{R}^n)},$$
where the infimum is taken over all $g \in S_p^k(\mathbb{R}^n)$ such that its restriction $g|_Q$ to Q coincides with f in the space of distributions $\mathcal{D}'(Q)$. If $p = 2$, we use a notation $\tilde{H}(Q) := S_2^1(Q)$.

Standard notation will be employed for embedding of Banach spaces:

- $B_1 \hookrightarrow B_2$ means bounded embedding of B_1 into B_2, i.e. $B_1 \subset B_2$, and
$$\|u\|_{B_2} \leq C\|u\|_{B_1}, \ \forall u \in B_1, \text{ for some constant } C.$$
- $B_1 \Subset B_2$ denotes compact embedding of B_1 into B_2, meaning that $B_1 \hookrightarrow B_2$, and every bounded subset of B_1 is precompact in B_2.

1.2. EIT Optimal Control Problem. Consider the optimal control problem on the minimization of the cost functional

$$\mathcal{J}(v) = \sum_{l=1}^{m} \left| \int_{E_l} \frac{U_l - u(x)}{Z_l} ds - I_l \right|^2 + \beta |U - U^*|^2 \tag{8}$$

on the control set

$$\mathscr{F}^R = \Big\{ v = (\sigma, U) \in \tilde{H}^1(Q) \times \mathbb{R}^m \ \Big| \ \sum_{l=1}^{m} U_l = 0,$$
$$\|\sigma\|_{\tilde{H}^1}^2 + |U|^2 \leq R^2, \ \sigma(x) \geq \sigma_0 > 0, \ \forall x \in Q \Big\} \tag{9}$$

where $\beta > 0, R > 0$, and $u = u(\cdot; v) \in H^1(Q)$ is a weak solution of the elliptic problem (3)–(5), i.e.

$$\int_Q \sigma \nabla u \cdot \nabla \eta dx + \sum_{l=1}^{m} \frac{1}{Z_l} \int_{E_l} u \eta ds = \sum_{l=1}^{m} \frac{U_l}{Z_l} \int_{E_l} \eta \, ds, \quad \forall \eta \in H^1(Q). \tag{10}$$

This optimal control problem will be called Problem $\mathcal{E}$. Note that the first term in the cost functional $\mathcal{J}(v)$ characterizes the mismatch of the condition (6) in light of the Robin condition (5).

REMARK 1. Inverse EIT problem on the identification of the electrical conductivity function σ with m input data $(I_l)_{l=1}^m$ is highly ill-posed. Optimal control Problem $\mathcal{E}$ inherits ill-posedness of the inverse EIT problem. We can modify an optimal control problem to adapt to the situation when the size of the input data can be increased through additional measurements while keeping the size of the unknown parameters fixed. Let $I^1 := I$ is a current pattern input, and $U^1 = (U_1, \ldots, U_m)$

is a corresponding boundary electrode voltage measurements. Consider $m-1$ new permutations of boundary voltages

$$U^j = (U_j, \dots, U_m, U_1, \dots, U_{j-1}),\ j = 2, \dots, m \tag{11}$$

applied to electrodes $E_1, E_2, \dots, E_m$ respectively. Assume that the "voltage–to–current" measurement allows us to measure associated currents $I^j = (I_1^j, \cdots, I_m^j)$. By setting $U^1 = U^*$ and having a new set of m^2 input data $(I^j)_{j=1}^m$, we now consider optimal control problem on the minimization of the new cost functional

$$\mathcal{K}(v) = \sum_{j=1}^m \sum_{l=1}^m \Big| \int_{E_l} \frac{U_l^j - u^j(x)}{Z_l} ds - I_l^j \Big|^2 + \beta |U - U^*|^2 \tag{12}$$

on a control set $\mathscr{F}^R$, where each function $u^j(\cdot; A, U^j), j = 1, \dots, m$, solves elliptic PDE problem (10) with U replaced by U^j. Let us call this modified optimal control problem a Problem $\mathcal{K}$. Note that the number of input currents in the Problem $\mathcal{K}$ has increased from m to m^2. However, the size of unknown control vector is unchanged, and in particular there are only m unknown voltages $U_1, \cdots, U_m$, whereas all vectors $U^j, j = 2, \dots, m$ are formed by their permutation as in (11). The price we pay for this gain is the increase of the number of PDE constraints, which has increased from 1 to m. It should be noted that similar approach can be pursued to increase the size of input data up to $m!$ by adding possible permutations of U in (11). In a recent paper **[9]** existence of the optimal control, Fréchet differentiability in the Besov space setting, formula for the Fréchet gradient, optimality condition are proved for both problems $\mathcal{E}$ and $\mathcal{K}$ when the control space for conductivity function is $H^\epsilon(Q) \cap L_\infty(Q)$. Numerical solution via gradient descent method in 2D model example demonstrated that the accuracy of the optimal conductivity vector σ is significantly improved in a Problem $\mathcal{K}$ with input data increased from $m = 16$ to $m^2 = 256$. Hence, the increase of the input data with additional measurements without changing the size of the unknown control parameters is demonstrated to be an effective regularization tool for solving highly ill-posed EIT optimal control problem.

Since $\tilde{H}(Q) \subset H^\epsilon(Q) \cap L_\infty(Q)$, first we extend the well-posedness result of the paper **[9]** by proving the existence of the optimal control for the Problem $\mathcal{E}$ (Theorem 4, Section 2). A similar proof can be applied for the proof of the existence of the optimal control for the Problem $\mathcal{K}$. The results of the paper **[9]** on the Fréchet differentiability, formula for the Fréchet gradient and optimality condition are applicable for both problems $\mathcal{E}$ and $\mathcal{K}$ in a new setting with identical proofs. The aim of this paper is to prove the convergence of the method of finite differences for the EIT optimal control problem with the control parameter σ in Sobolev-Hilbert space $\tilde{H}(Q)$ with dominating mixed smoothness (Theorem 5, Section 2). In Section 1.3 we pursue discretization of the Problem $\mathcal{E}$. In Sections 3 and 4 we prove the convergence of the sequence of finite-dimensional optimal control problems to EIT optimal control problem both with respect to functional and control in 2D and 3D domains. The setting of the EIT optimal control problem in a Sobolev-Hilbert space with dominating mixed smoothness is a key factor in the proof of the convergence of the finite difference method (see Remark 2 of Section 1.3).

Since the proof of the convergence of the method of finite differences for the Problem $\mathcal{K}$ is similar to the proof for the Problem $\mathcal{E}$, to avoid technical complexity we will focus on the proof of convergence for the Problem $\mathcal{E}$.

1.3. Discrete Optimal Control Problem. To discretize optimal control problems $\mathcal{E}$ we pursue finite difference method following the framework introduced in [**8**]. Let $h > 0$ and cut $\mathbb{R}^n$ by the planes

$$x_i = k_i h, \; i = 1, \ldots, n \;\; \forall\, k_i \in \mathbb{Z}.$$

into a collection of elementary cells with length h in each x_i-direction. For every $h > 0$ and multi-index $\alpha = (k_1, \ldots, k_n)$ we define a cell C_h^α as

$$C_h^\alpha = \{x \in \mathbb{R}^n |\; k_i h \leq x_i \leq (k_i + 1)h, \; i = 1, \ldots, n\}, \tag{13}$$

and consider the collection of cells which have non-empty intersection with Q:

$$\mathscr{C}_h^Q = \{C_h^\alpha |\; C_h^\alpha \cap Q \neq \emptyset\}. \tag{14}$$

We now introduce the exterior approximation of $\overline{Q}$ as follows:

$$Q_h = \bigcup_{C_h^\alpha \in \mathscr{C}_h^Q} C_h^\alpha. \tag{15}$$

Obviously, we have $\overline{Q} \subset Q_h$. Let $S_h = \partial Q_h$. The vertex of the prism C_h^α whose coordinates are the smallest relative to the other vertices, is called its *natural corner.* We are going to identify each prism (cell) by its natural corner. With slight abuse of notation we denote as C_h^z, a cell in $\mathbb{R}^n$ of side length h and with a natural corner at z. Hence, C_h^α and $C_h^{\alpha h}$ are identical. Consider a lattice

$$\mathscr{L} = \Big\{x \in \mathbb{R}^n \mid \exists \alpha \in \mathbb{Z}^n \text{ s.t. } x_i = k_i h, \; i = 1, \ldots, n\Big\}.$$

We will write $x_\alpha = (k_1 h, \ldots, k_n h)$. Bijection $\alpha \mapsto x_\alpha$ will henceforth be referred as natural. Given a set X which is in natural bijection with a subset of the set of multi-indexes α, we write $\mathscr{A}(X)$ as the indexing set. Moreover, if $X \subset \mathbb{R}^n$, then $\mathscr{L}(X) := \mathscr{L} \cap X$. When $X = \mathscr{L}(Y) \subset \mathbb{R}^n$, we'll agree to write $\mathscr{A}(Y)$ instead of $\mathscr{A}(\mathscr{L}(Y))$. These indices are also in natural bijection with the natural corners of these prisms. In particular, some of the corresponding lattice points may fall on the boundary S_h. We contrast this set to the set $\mathscr{A}(Q_h')$ of indexes in natural bijection to the lattice points that lie strictly in the interior of Q_h, and to the set $\mathscr{A}(Q_h)$, of all indices which are in natural bijection with the lattice points that lie in Q_h. We will write

$$\sum_{\mathscr{A}(X)} \quad \text{instead of} \quad \sum_{\alpha \in \mathscr{A}(X)},$$

and likewise for other expressions requiring subscripts. We adopt the notation

$$\alpha \pm e_i := (k_1, \ldots, k_i \pm 1, \ldots, k_n).$$

To discretize optimal control problem $\mathscr{E}$, we need to introduce some refined subsets of grid points of Q_h.

$$Q_h^+ = \{x_\alpha \in Q_h : C_h^\alpha \cap Q \neq \emptyset\}$$

be a subset of natural corners of the cells in Q_h. We denote as

$$Q_h^{(i)} = \{x_\alpha \in Q_h : x_{\alpha + e_i} \in Q_h\}$$

the subset of all grid points x_α in Q_h such that the edge $[x_\alpha, x_{\alpha+e_i}] \subset Q_h$, and similarly

$$Q_h^{(i,j)} = \{x_\alpha \in Q_h : x_\alpha + e_i + e_j \in Q_h\}, \quad i, j = 1, 2, 3. \tag{16}$$

Subset of natural corners x_α of cells in Q_h which intersect the boundary S is denoted as

$$\hat{S}_h = \{x_\alpha \in Q_h : C_h^\alpha \cap S \neq \emptyset\} \tag{17}$$

and

$$\hat{E}_{lh} = \{\ x_\alpha \in Q_h :\ E_{l\alpha} := C_h^\alpha \cap E_l \neq \emptyset\ \},\quad l = 1,\dots,m$$

is a collection of grid pints which are natural corners of C_h^α containing portion $E_{l\alpha}$ of the boundary manifold E_l. Let $\Gamma_{l\alpha} = m_{n-1}(E_{l\alpha})$, $l = 1,\dots,m$ is an $n-1$-dimensional Lebesgue measure of $E_{l\alpha}$. Note that according to the definition of the space $\tilde{H}(Q)$ any control vector σ has an extension to $\mathbb{R}^n$ as an element of $\tilde{H}(\mathbb{R}^n)$. We introduce discrete grid function by discretizing σ through Steklov average:

$$\sigma_\alpha = \frac{1}{h^n}\int_{x_1}^{x_1+h}\cdots\int_{x_n}^{x_n+h}\sigma(y_1,\dots,y_n)\,dy_1\cdots dy_n,\quad \alpha\in\mathscr{A}(Q_h), \tag{18}$$

where x_i is the i-th coordinate of x_α. We use standard notation for finite differences of grid functions u_α, σ_α:

$$u_{\alpha x_i} = \frac{u_{\alpha+e_i}-u_\alpha}{h},\ u_{\alpha\bar{x}_i} = \frac{u_\alpha - u_{\alpha-e_i}}{h}, u_{\alpha x_i} = \frac{u_{\alpha+e_i}-u_\alpha}{h},\quad i = 1,\dots,n$$

$$\sigma_{\alpha x_i x_j} = \frac{\sigma_{(\alpha+e_j)x_i}-\sigma_{\alpha x_i}}{h} = \frac{\sigma_{\alpha+e_j+e_i}-\sigma_{\alpha+e_j}-\sigma_{\alpha+e_i}+\sigma_\alpha}{h^2},$$

$$\sigma_{\alpha x_1x_2x_3} = \frac{\sigma_{(\alpha+e_3)x_1x_2}-\sigma_{\alpha x_1x_2}}{h} = \frac{\sigma_{(\alpha+e_3+e_2)x_1}-\sigma_{(\alpha+e_3)x_1}-\sigma_{(\alpha+e_2)x_1}+\sigma_{\alpha x_1}}{h^2}$$

$$= \frac{\sigma_{\alpha+e_3+e_2+e_1}-\sigma_{\alpha+e_3+e_2}-\sigma_{\alpha+e_3+e_1}-\sigma_{\alpha+e_2+e_1}+\sigma_{\alpha+e_3}+\sigma_{\alpha+e_2}+\sigma_{\alpha+e_1}-\sigma_\alpha}{h^3}.$$

For a given discretization with step size h, we employ the notation

$$[\eta]_h := \{\eta_\alpha \in \mathbb{R} :\ \alpha\in\mathscr{A}(Q_h)\},$$

for the grid function. Next, we define the discrete $\mathscr{H}^1(Q_h)$, $\tilde{\mathscr{H}}^1(Q_h)$ and $L_\infty(Q_h)$ norms:

$$\|[u]_h\|^2_{\mathscr{H}^1(Q_h)} := \sum_{\mathscr{A}(Q_h)} h^n u_\alpha^2 + \sum_{i=1}^n \sum_{\mathscr{A}(Q_h^{(i)})} h^n u_{\alpha x_i}^2$$

$$|||[u]_h|||^2_{\mathscr{H}^1(Q_h)} := \sum_{i=1}^n h^n \sum_{\mathscr{A}(Q_h^{(i)})} u_{\alpha x_i}^2 + \sum_{l=1}^m \sum_{\mathscr{A}(\hat{E}_{lh})} \Gamma_{l\alpha} u_\alpha^2$$

$$\|[\sigma]_h\|^2_{\tilde{\mathscr{H}}^1(Q_h)} := \sum_{\mathscr{A}(Q_h)} h^2\sigma_\alpha^2 + \sum_{i=1}^2 \sum_{\mathscr{A}(Q_h^{(i)})} h^2\sigma_{\alpha x_i}^2 + \sum_{\mathscr{A}(Q_h^+)} h^2 \sigma_{\alpha x_1x_2}^2,\ Q\in\mathbb{R}^2$$

$$\|[\sigma]_h\|^2_{\tilde{\mathscr{H}}^1(Q_h)} := \sum_{\mathscr{A}(Q_h)} h^3\sigma_\alpha^2 + \sum_{i=1}^3 \sum_{\mathscr{A}(Q_h^{(i)})} h^3\sigma_{\alpha x_i}^2 + \sum_{\substack{i,j=1\\ i<j}}^3 \sum_{\mathscr{A}(Q_h^{(i,j)})} h^3\sigma_{\alpha x_i x_j}^2$$

$$+ \sum_{\mathscr{A}(Q_h^+)} h^3 \sigma_{\alpha x_1x_2x_3}^2,\ Q\in\mathbb{R}^3$$

$$\|[\sigma]_h\|_{L_\infty(Q_h)} := \max_{\alpha\in\mathscr{A}(Q_h)} |\sigma_\alpha|$$

For fixed $R > 0$, define the discrete control sets $\mathscr{F}_h^R$ as

$$\mathscr{F}_h^R := \Big\{ [v]_h = ([\sigma]_h, U) \Big| \sum_{l=1}^{m} U_l = 0, \ \|[\sigma]_h\|^2_{\tilde{\mathscr{H}}^1(Q_h)} + |U|^2_{\mathbb{R}^m} \leq R^2,$$

$$\sigma_\alpha \geq \sigma_0 > 0, \ \forall \alpha \in \mathscr{A}(Q_h) \Big\} \tag{19}$$

and the interpolating map $\mathscr{P}_h$ as

$$\mathscr{P}_h : \bigcup_R \mathscr{F}_h^R \to \bigcup_R \mathscr{F}^R, \qquad \mathscr{P}_h([v]_h) = (\mathscr{P}_h([\sigma]_h), U) = (\sigma'_h, U)$$

where σ'_h is a multilinear interpolation of $[\sigma]_h$, which assigns the value σ_α to each grid point of C_h^α, and it is a piecewise linear with respect to each variable x_i when the other variables are fixed. Precisely,

$$\begin{aligned} \sigma'_h(x) &= \sigma_\alpha + \sigma_{\alpha x_1}(x_1 - k_1 h) + \sigma_{\alpha x_2}(x_2 - k_2 h) \\ &+ \sigma_{\alpha x_1 x_2}(x_1 - k_1 h)(x_2 - k_2 h), x \in C_h^\alpha, \quad n = 2 \end{aligned} \tag{20}$$

$$\begin{aligned} \sigma'_h(x) = \sigma_\alpha + \sum_{i=1}^{3} \sigma_{\alpha x_i}(x_i - k_i h) + \sum_{\substack{i,j=1 \\ i<j}}^{3} \sigma_{\alpha x_i x_j}(x_i - k_i h)(x_j - k_j h) \\ + \sigma_{\alpha x_1 x_2 x_3} \prod_{1 \leq i \leq 3} (x_i - k_i h), x \in C_h^\alpha, \quad n = 3. \end{aligned} \tag{21}$$

We also define the discretizing map $\mathscr{Q}_h$ as

$$\mathscr{Q}_h : \bigcup_R \mathscr{F}^R \to \bigcup_R \mathscr{F}_h^R, \qquad \mathscr{Q}_h(v) = (\mathscr{Q}_h(\sigma), U) = ([\sigma]_h, U)$$

where $[\sigma]_h = \{\sigma_\alpha\}$, with σ_α given by (18) for each $\alpha \in \mathscr{A}(Q_h)$.

REMARK 2. One can easily check that the multilinear interpolation σ'_h is an element of $S_p^1(Q)$. Note that every partial derivative of σ'_h indicated in the definition of the space $S_p^1(Q)$ exists. Moreover, each of these derivatives is piecewise constant with respect to the differentiated variables, and piecewise linear with respect to the non-differentiated variables, when all the other variables are fixed. Therefore, none of the other weak derivatives of σ besides the ones included in the definition $S_p^1(Q)$ exist. Hence, the multilinear interpolations are typical elements of the Sobolev space $S_p^1(Q)$ of dominating mixed smoothness. In a recent paper [**10**] the compact embedding of the Sobolev spaces $S_p^1(Q)$ into the Hölder spaces $C^{1-\frac{1}{p}}(\overline{Q})$ is proved (see Lemma 14 of Section 3). Thus the Sobolev spaces $S_p^1(Q)$ with dominating mixed smoothness provide a compactness criterion for the family of multilinear interpolations of discrete control parameters in Hölder spaces. Due to these advantages, Sobolev spaces with dominating mixed smoothness are very well suited for the proof of the convergence of the method of finite differences in the EIT optimal control problem.

Next, we define a discrete state vector, which is a solution of the discretized elliptic problem (3)–(5).

DEFINITION 3. Given $[v]_h$, the grid function $[u([v]_h)]_h$ is called a discrete state vector of problem $\mathcal{E}$ if it satisfies

$$h^n \sum_{\mathscr{A}(Q_h^+)} \sigma_\alpha \sum_{i=1}^{n} u_{\alpha x_i}\eta_{\alpha x_i} + \sum_{l=1}^{m} \frac{1}{Z_l} \sum_{\mathscr{A}(\hat{E}_{lh})} \Gamma_{l\alpha} u_\alpha \eta_\alpha + J_h([u]_h, [\eta]_h)$$

$$= \sum_{l=1}^{m} \frac{U_l}{Z_l} \sum_{\mathscr{A}(\hat{E}_{lh})} \Gamma_{l\alpha}\eta_\alpha \tag{22}$$

for arbitrary grid function $[\eta]_h$, where

$$J_h([u]_h, [\eta]_h) = h^n \sum_{\mathscr{A}(S_h)} \sum_{i=1}^{n} \theta^i_\alpha u_{\alpha x_i}\eta_{\alpha x_i}, \ \theta^i_\alpha = \begin{cases} 1 & \text{if } \alpha \in \mathscr{A}(Q_h^{(i)} \setminus Q_h^+) \\ 0 & \text{otherwise} \end{cases}$$

The necessity of adding J_α to (22) is that some $u_{\alpha x_i}$ and $\eta_{\alpha x_i}$ values on S_h are not present in the first term of (22). Addition of these terms to (22) through J_α is essential for the proof of stability of our discrete scheme.

In Section 3, it will be proved that for a given $[v]_h \in \mathscr{F}_h^R$ there exists a unique discrete state vector of problem $\mathcal{E}$. Consider minimization of the discrete cost functional

$$\mathcal{J}_h([v]_h) = \sum_{l=1}^{m} \Big(\sum_{\mathscr{A}(\hat{E}_{lh})} \Gamma_{l\alpha} \frac{U_l - u_\alpha}{Z_l} - I_l \Big)^2 + \beta |U - U^*|^2 \tag{23}$$

on a control set $\mathscr{F}_h^R$, where u_α's are components of the discrete state vector $[u([v]_h)]_h$ of the Problem $\mathscr{E}$. The formulated discrete optimal control problem will be called Problem $\mathscr{E}_h$.

Next, we define three interpolations of the discrete state vector $[u]_h$. Piecewise constant interpolation $\tilde{u}_h : Q_h \to \mathbb{R}$ assigns to the interior of each cell in Q_h the value of u_α at its natural corner:

$$\tilde{u}_h\big|_{C_h^{\alpha'}} = u_\alpha, \qquad \forall \alpha \in \mathscr{A}(Q_h^+). \tag{24}$$

Piecewise constant interpolation of the discrete x_i-derivative $\tilde{u}_h^i : Q_h \to \mathbb{R}$, $i = 1, .., n$ assign to the interior of each cell in Q_h the value of the forward spatial difference at the natural corner:

$$\tilde{u}_h^i\big|_{C_h^{\alpha'}} = u_{\alpha x_i}, \qquad \forall \alpha \in \mathscr{A}(Q_h^+). \tag{25}$$

Multilinear interpolation $u_h' : Q_h \to \mathbb{R}$ assigns the value u_α to each grid point in $\mathscr{L}(Q_h)$, and it is a peicewise linear with respect to each variable x_i when the rest of variables are fixed.

2. Main Result

The following are the main results of the paper.

THEOREM 4. *The optimal control problem $\mathcal{E}$ has a solution, i.e.*

$$\mathscr{F}_* := \Big\{ v \in \mathscr{F}^R \ \Big| \ \mathscr{J}(v) = \mathscr{J}_* \Big\} \neq \emptyset$$

THEOREM 5. *The sequence of discrete optimal control problems $\mathcal{E}_h$ approximates the optimal control problem $\mathcal{E}$ with respect to functional, i.e.*

$$\lim_{h\to 0} \mathscr{J}_{h_*} = \mathscr{J}_*, \tag{26}$$

where

$$\mathscr{J}_{h_*} = \inf_{\mathscr{F}_h^R} \mathscr{J}_h([v]_h), \quad \mathscr{J}_* := \inf_{f\in\mathscr{F}^R} \mathscr{J}(v). \tag{27}$$

Furthermore, let $\{\epsilon_h\}$ *be a sequence of positive real numbers with* $\lim_{h\to 0}\epsilon_h = 0$. *If the sequence* $[v]_{h,\epsilon} = ([\sigma]_{h,\epsilon}, U^{h,\epsilon}) \in \mathscr{F}_h^R$ *is chosen so that*

$$\mathscr{J}_{h_*} \le \mathscr{J}_h([v]_{h,\epsilon}) \le \mathscr{J}_{h_*} + \epsilon_h, \tag{28}$$

then we have

$$\lim_{h\to 0} \mathscr{J}(\mathscr{P}_h([v]_{h,\epsilon})) = \mathscr{J}_*, \tag{29}$$

the sequence $\{(\mathscr{P}_h([\sigma]_{h,\epsilon}), U^{h,\epsilon})\}$

- *is precompact in Tikhonov topology of* $\tilde{H}^1(Q)\times\mathbb{R}^m$ *formed with the product of the weak topology of* $\tilde{H}^1(Q)$ *and Euclidean topology of* $\mathbb{R}^m$;
- *is precompact in Tikhonov topology of* $C^{0,\mu}(\overline{Q})\times\mathbb{R}^m$, $0<\mu<\frac{1}{2}$ *formed with the product of the strong topology of Hölder space* $C^{0,\mu}(\overline{Q})$ *and Euclidean topology of* $\mathbb{R}^m$;

and all the corresponding limit points $v_* = (\sigma_*, U_*)$ *are optimal controls of the problem* $\mathcal{E}$. *Moreover, if* $v_* = (\sigma_*, U_*)$ *is any such limit point, then there exists a subsequence* h' *such that the multilinear interpolations* $u'_{h'}$ *of the discrete state vectors* $[u([v]_{h',\epsilon})]_{h'}$ *converge to weak solution* $u = u(x; v_*)$ *of the elliptic problem* (3)-(5), *weakly in* $H^1(Q)$, *strongly in* $L_2(Q)$, *and almost everywhere on* Q.

3. Preliminary Results

The following lemma presents a key discrete energy estimate for the elliptic PDE problem:

LEMMA 6 (Discrete Energy Estimate). *For any* $[v]_h \in \mathscr{F}_h^R$, *discrete state vector* $[u([v]_h)]_h$ *satisfies the energy estimate:*

$$\|\|[u]_h\|\|_{\mathscr{H}^1(Q_h)} \le M|U|, \tag{30}$$

where M *depends on* σ_0, Z *and* Q.

Proof: We set $\eta_\alpha = u_\alpha$ in (22) to get

$$h^n \sum_{\mathscr{A}(Q_h^+)} \sigma_\alpha \sum_{i=1}^n u_{\alpha x_i}^2 + \sum_{l=1}^m \frac{1}{Z_l} \sum_{\mathscr{A}(\hat{E}_{lh})} \Gamma_{l\alpha} u_\alpha^2 + J_\alpha(u_\alpha, u_\alpha) = \sum_{l=1}^m \frac{U_l}{Z_l} \sum_{\mathscr{A}(\hat{E}_{lh})} \Gamma_{l\alpha} u_\alpha,$$

and by recalling the definition of J_α and the fact that $0<\sigma_0\le\sigma_\alpha$ we have

$$\mu\|\|[u([v]_h)]_h\|\|_{\mathscr{H}^1(Q_h)}^2 \le \sum_{l=1}^m Z_l^{-1} U_l \sum_{\mathscr{A}(\hat{E}_{lh})} \Gamma_{l\alpha} u_\alpha \tag{31}$$

where $\mu = \min\{\sigma_0, \min_l Z_l^{-1}\}$. Using Cauchy-Schwarz inequality we derive

$$\sum_{l=1}^m Z_l^{-1} U_l \sum_{\mathscr{A}(\hat{E}_{lh})} \Gamma_{l\alpha} u_\alpha \le m_{n-1}^{\frac{1}{2}}(\partial Q) \max_l Z_l^{-1} |U| \; \|\|[u([v]_h)]_h\|\|_{\mathscr{H}^1(Q_h)} \tag{32}$$

From (31) and (32), (30) follows with $M = \mu^{-1} m_{n-1}^{\frac{1}{2}}(\partial Q) \max_l Z_l^{-1}$. □

COROLLARY 7. *For any* $[v]_h \in \mathscr{F}_h^R$*, there exists a unique discrete state vector* $[u([v]_h)]_h$.

Assertion of the corollary follows from energy estimate with similar arguments as in [**36**]. By replacing u_{x_i}, η_{x_i} with respective difference quotients, from (22) it follows

$$\sum_{\mathscr{A}(Q_h)} \Big\{\mathscr{L}_\alpha \cdot [u]_h - \mathscr{G}_\alpha(U)\Big\}\eta_\alpha = 0, \tag{33}$$

where $\mathscr{L}_\alpha$ is a vector of the same size as $[u]_h$ and $\mathscr{G}_\alpha : \mathbb{R}^m \to \mathbb{R}$ is a linear functional. Since the values of η_α are independent, (33) is equivalent to the following system of linear algebraic equations (SLAE)

$$\mathscr{L}_\alpha \cdot [u]_h = \mathscr{G}_\alpha(U), \ \alpha \in \mathscr{A}(Q_h). \tag{34}$$

Note that the number of equations and the number of unknowns u_α in (34) are both equal to a number of vertices in a grid Q_h. Addition of the expression $J_\alpha(u_\alpha, \eta_\alpha)$ to the discrete identity (22) served exactly to this purpose. From the energy estimate (30) it easily follows that the corresponding homogeneous SLAE has only a zero solution. Therefore, a claim of the corollary is a consequence of the well-known result of linear algebra.

Another crucial consequence of the energy estimate (30) is uniform $H^1(Q)$-boundedness of the interpolations of the discrete state vector:

COROLLARY 8. *Multilinear interpolation* u_h' *of the discrete state vector is uniformly bounded in* $H^1(Q)$:

$$\sup_{[v]_h \in \mathscr{F}_h^R} \|u_h'\|_{H^1(Q)} \leq C, \tag{35}$$

where C *depends on* σ_0, Z, Q, R, n.

Indeed, first of all from [**8**] (formula (4.13)) it follows that

$$\int_Q |Du_h'|^2 dx \leq 2^{n-1} \sum_{i=1}^{n} \sum_{\mathscr{A}(Q_h^+)} h^n |u_{\alpha x_i}|^2. \tag{36}$$

Next, we establish that the sequences u_h' and $\tilde{u}_h$ are equivalent in strong topology of $L_2(E_l)$, as $h \to 0$. The proof is similar to the statement (d) of Theorem 14 in [**8**]. The following estimate is proved in [**8**] (estimate (4.23)):

$$|\tilde{u}_h(x) - u_h'(x)| \leq (2^n - 1)n \sum_{\text{edges of } C_h^\alpha} h^2 |u_{\alpha' x_j}|^2, \ x \in C_h^\alpha, \tag{37}$$

where the summation on the right-hand side is taken over all α' and j such that $\alpha' \in \mathscr{A}(C_h^\alpha)$ and $\alpha' + e_j \in \mathscr{A}(C_h^\alpha)$. Since, E_l is Lipschitz, we have $m_{n-1}(E_{l\alpha}) \leq Lh^{n-1}$, where L is a Lipschitz constant of the boundary S. Therefore, from (37) and (30) it follows that

$$\|\tilde{u}_h - u_h'\|^2_{L_2(E_l)} \leq L(2^n - 1)2^{n-1} n \sum_{i=1}^{n} \sum_{\mathscr{A}(Q_h^{(i)})} h^{n+1} u^2_{\alpha x_i} \to 0, \tag{38}$$

as $h \to 0$. Assuming that $h \leq 1$, from (36) and (38) it follows that

$$\||u_h'\||^2_{H^1(Q)} \leq C_1 \||[u]_h\||^2_{\mathscr{H}^1(Q_h)}, \tag{39}$$

where $C_1 = 2^{n-1}(2L(2^n-1)nm+1)$. Due to equivalency of the norms $\|\cdot\|$ and $|||\cdot|||$ in $H^1(Q)$ (see Lemma 5.1 in [**9**]), from (39), (35) follows.

Discrete energy estimate implies the following interpolation result:

LEMMA 9. *Let $R>0$ is fixed, and $\{[v]_h\}$ is a sequence of discrete control vectors such that $[v]_h \in \mathscr{F}_h^R$ for each h. Then the following statements hold:*

(a) The sequences $\{u'_h\}$ and $\{\tilde{u}_h\}$ are uniformly bounded in $L_2(Q_h)$.

(b) For each $i \in \{1,\dots,n\}$, the sequences $\{\tilde{u}^i_h\}$, $\{\frac{\partial u'_h}{\partial x_i}\}$ are uniformly bounded in $L_2(Q_h)$.

(c) the sequence $\{\tilde{u}_h - u'_h\}$ converges strongly to 0 in $L_2(Q)$ as $h\to 0$.

(d) For each $i \in \{1,\dots,n\}$, the sequences $\{\frac{\partial u'_h}{\partial x_i} - \tilde{u}^i_h\}$ converges weakly to zero in $L_2(Q)$ as $h \to 0$.

(e) the sequence $\{\tilde{u}_h - u'_h\}$ converges strongly to 0 in $L_2(S)$ as $h\to 0$.

The proof of the claims (a)-(d) coincides with the proofs of similar claims in Theorem 14 of [**8**]. Claim (e) is proved above in (38).

Next, we recall the necessary and sufficient condition for the convergence of the discrete optimal control problems $\mathscr{E}_h$, which is the suitable criteria to employ for the proof of method of finite differences for the optimal control problems with distributed parameters ([**1**]-[**8**]).

LEMMA 10. [**47**] *The sequence of discrete optimal control problems $\mathscr{E}_h$ approximates the continuous optimal control problem $\mathscr{E}$ with respect to the functional if and only if the following conditions are satisfied:*

(1) *For arbitrary sufficiently small $\epsilon>0$ there exists $h_1 = h_1(\epsilon)$ such that $\mathscr{Q}_h(v) \in \mathscr{F}_h^R$ for all $v \in \mathscr{F}^{R-\epsilon}$ and $h \le h_1$; Moreover, for any fixed $\epsilon>0$ and for all $v\in\mathscr{F}^{R-\epsilon}$ the following inequality is satisfied:*

$$\limsup_{h\to 0}(\mathscr{J}_h(\mathscr{Q}_h(v)) - \mathscr{J}(v)) \le 0. \tag{40}$$

(2) *For arbitrary sufficiently small $\epsilon>0$ there exists $h_2 = h_2(\epsilon)$ such that $\mathscr{P}_h([v]_h) \in \mathscr{F}^{R+\epsilon}$ for all $[v]_h \in \mathscr{F}_h^R$ and $h\le h_2$; moreover, for all $[v]_h \in \mathscr{F}_h^R$, the following inequality is satisfied:*

$$\limsup_{h\to 0}(\mathscr{J}(\mathscr{P}_h([v]_h)) - \mathscr{J}_h([v]_h)) \le 0. \tag{41}$$

(3) *For arbitrary sufficiently small $\epsilon>0$, the following inequalities are satisfied:*

$$\limsup_{\epsilon\to 0} \mathcal{J}_*(\epsilon) \ge \mathcal{J}_*, \ \liminf_{\epsilon\to 0} \mathcal{J}_*(-\epsilon) \le \mathcal{J}_*,$$

where $\mathcal{J}_*(\pm\epsilon) = \inf_{\mathscr{F}^{R\pm\epsilon}} \mathcal{J}(v)$.

Our next goal is to show that the mappings $\mathscr{P}_h$ and $\mathscr{Q}_h$ satisfy the conditions of Lemma 10. The following lemma plays a key role to prove this claim. The proof is similar to the proof of Proposition 11 in [**8**].

LEMMA 11. *Let $Q \in \mathbb{R}^2$. Then for $\forall\epsilon>0$, there exists $\delta>0$ such that*

$$\sum_{\mathscr{A}(Q_h^+)} h^2|\sigma_{\alpha x_1x_2}|^2 \le (1+\epsilon)\Big\|\frac{\partial^2\sigma}{\partial x_1\partial x_2}\Big\|^2_{L_2(Q_h)}$$

whenever $h<\delta$.

Proof: For each $h > 0$, define the function $\tilde{\sigma}_h^{12}$ as

$$\tilde{\sigma}_h^{12}\Big|_{C_h^{\alpha'}} = \sigma_{\alpha x_1 x_2}, \quad \forall \alpha \in \mathscr{A}(Q_h^+) \tag{42}$$

In the following we will prove that

$$\tilde{\sigma}_h^{12} \to \frac{\partial^2 \sigma}{\partial x_1 \partial x_2} \quad \text{strongly in } L_2(Q) \text{ as } h \to 0 \tag{43}$$

As an element of $\tilde{H}^1(Q)$, almost all restrictions of σ to lines parallel to the x_i direction are absolutely continuous, moreover, restrictions of $\frac{\partial \sigma}{\partial x_1}$ (or $\frac{\partial \sigma}{\partial x_2}$) to lines parallel to the x_2 (or x_1) direction are absolutely continuous [**10**]. Therefore, for almost every $z = (z_1, z_2) \in Q$ we have

$$\int\limits_{C_h^z} \frac{\partial^2 \sigma}{\partial x_1 \partial x_2}\, dy = \sigma(z + he_2 + he_1) - \sigma(z + he_2) - \sigma(z + he_1) + \sigma(z) \tag{44}$$

In the following transformation, we write simply $\mathscr{A}$ instead of summation index set $\mathscr{A}(Q_h^+)$. Using the definition of Steklov average (18), (44) and Cauchy-Schwartz inequality, we get

$$\Big\|\tilde{\sigma}_h^{12} - \frac{\partial^2 \sigma}{\partial x_1 \partial x_2}\Big\|^2_{L_2(Q_h)} = \sum_{\mathscr{A}} \int\limits_{C_h^\alpha} \Big|\sigma_{\alpha x_1 x_2} - \frac{\partial^2 \sigma(x)}{\partial x_1 \partial x_2}\Big|^2 dx$$

$$= \sum_{\mathscr{A}} \int\limits_{C_h^\alpha} \Big|\frac{1}{h^4}\Big[\int\limits_{C_h^{\alpha+e_1+e_2}} dz - \int\limits_{C_h^{\alpha+e_1}} dz - \int\limits_{C_h^{\alpha+e_2}} dz + \int\limits_{C_h^\alpha} dz \sigma(z)\Big] - \frac{\partial^2 \sigma(x)}{\partial x_1 \partial x_2}\Big|^2 dx$$

$$= \sum_{\mathscr{A}} \int\limits_{C_h^\alpha} \Big|\frac{1}{h^4}\int\limits_{C_h^\alpha} \Big[\sigma\big|_{z+he_1+he_2} - \sigma\big|_{z+he_2} - \sigma\big|_{z+he_1} + \sigma\big|_z\Big] dz - \frac{\partial^2 \sigma(x)}{\partial x_1 \partial x_2}\Big|^2 dx$$

$$= \sum_{\mathscr{A}} \frac{1}{h^8} \int\limits_{C_h^\alpha} \Big|\int\limits_{C_h^\alpha} \int\limits_{C_h^z} \Big[\frac{\partial^2 \sigma(y)}{\partial x_1 \partial x_2} - \frac{\partial^2 \sigma(x)}{\partial x_1 \partial x_2}\Big] dy\, dz\Big|^2 dx$$

$$\leq \sum_{\mathscr{A}} \frac{1}{h^4} \int\limits_{C_h^\alpha} \int\limits_{C_h^\alpha} \int\limits_{C_h^z} \Big|\frac{\partial^2 \sigma(y)}{\partial x_1 \partial x_2} - \frac{\partial^2 \sigma(x)}{\partial x_1 \partial x_2}\Big|^2 dy\, dz dx \tag{45}$$

Changing integration order with respect to y and z, we have

$$\int\limits_{C_h^\alpha} \int\limits_{C_h^z} \Big|\frac{\partial^2 \sigma(y)}{\partial x_1 \partial x_2} - \frac{\partial^2 \sigma(x)}{\partial x_1 \partial x_2}\Big|^2 dy\, dz = \Big(\int\limits_{C_h^\alpha} (y_1 - k_1 h)(y_2 - k_2 h)\, dy$$

$$+ \int\limits_{C_h^{\alpha+e_1}} ((k_1+2)h - y_1)(y_2 - k_2 h)\, dy + \int\limits_{C_h^{\alpha+e_2}} (y_1 - k_1 h)((k_2+2)h - y_2)\, dy$$

$$\int\limits_{C_h^{\alpha+e_1+e_2}} ((k_1+2)h - y_1)((k_2+2)h - y_2)\, dy\Big)\Big|\frac{\partial^2 \sigma(y)}{\partial x_1 \partial x_2} - \frac{\partial^2 \sigma(x)}{\partial x_1 \partial x_2}\Big|^2 \leq h^2$$

$$\times\Big(\int\limits_{C_h^\alpha} dy + \int\limits_{C_h^{\alpha+e_1}} dy + \int\limits_{C_h^{\alpha+e_2}} dy + \int\limits_{C_h^{\alpha+e_1+e_2}} dy\Big)\Big|\frac{\partial^2 \sigma(y)}{\partial x_1 \partial x_2} - \frac{\partial^2 \sigma(x)}{\partial x_1 \partial x_2}\Big|^2 \tag{46}$$

From (45), (46) it follows that

$$\Big\|\tilde{\sigma}_h^{12} - \frac{\partial^2 \sigma}{\partial x_1 \partial x_2}\Big\|^2_{L_2(Q_h)} \le \sum_{\mathscr{A}} \frac{1}{h^2} \int\limits_{C_h^\alpha} \Big(\int\limits_{C_h^{\alpha+e_1+e_2}} dz$$

$$+ \int\limits_{C_h^{\alpha+e_1}} dz + \int\limits_{C_h^{\alpha+e_2}} dz + \int\limits_{C_h^\alpha} dz \Big) \Big| \frac{\partial^2 \sigma(z)}{\partial x_1 \partial x_2} - \frac{\partial^2 \sigma(x)}{\partial x_1 \partial x_2}\Big|^2 dx \tag{47}$$

Let $\forall\ \epsilon > 0$ is fixed. Since $C^2(\overline{Q + B_1(0)})$ is dense in $\tilde{H}^1(Q + B_1(0))$ we can choose $g \in C^2(\overline{Q + B_1(0)})$ such that

$$\Big\|\frac{\partial^2 \sigma}{\partial x_1 \partial x_2} - \frac{\partial^2 g}{\partial x_1 \partial x_2}\Big\|^2_{L_2(Q+B_1(0))} < \frac{\epsilon}{24(1 + m_n(Q))}. \tag{48}$$

From (47) it follows

$$\Big\|\tilde{\sigma}_h^{12} - \frac{\partial^2 \sigma}{\partial x_1 \partial x_2}\Big\|^2_{L_2(Q_h)} \le \sum_{\mathscr{A}} \frac{3}{h^2} \int\limits_{C_h^\alpha} (I_1 + I_2 + I_3)\, dx, \tag{49}$$

where

$$I_1 = \Big(\int\limits_{C_h^{\alpha+e_1+e_2}} dz + \int\limits_{C_h^{\alpha+e_1}} dz + \int\limits_{C_h^{\alpha+e_2}} dz + \int\limits_{C_h^\alpha} dz \Big) \Big| \frac{\partial^2 \sigma(z)}{\partial x_1 \partial x_2} - \frac{\partial^2 g(z)}{\partial x_1 \partial x_2}\Big|^2,$$

$$I_2 = \Big(\int\limits_{C_h^{\alpha+e_1+e_2}} dz + \int\limits_{C_h^{\alpha+e_1}} dz + \int\limits_{C_h^{\alpha+e_2}} dz + \int\limits_{C_h^\alpha} dz \Big) \Big| \frac{\partial^2 g(z)}{\partial x_1 \partial x_2} - \frac{\partial^2 g(x)}{\partial x_1 \partial x_2}\Big|^2,$$

$$I_3 = \Big(\int\limits_{C_h^{\alpha+e_1+e_2}} dz + \int\limits_{C_h^{\alpha+e_1}} dz + \int\limits_{C_h^{\alpha+e_2}} dz + \int\limits_{C_h^\alpha} dz \Big) \Big| \frac{\partial^2 g(x)}{\partial x_1 \partial x_2} - \frac{\partial^2 \sigma(x)}{\partial x_1 \partial x_2}\Big|^2.$$

Since $\frac{\partial^2 g}{\partial x_1 \partial x_2}$ is uniformly continuous on $Q + B_1(0)$, there exists $\delta = \delta(\epsilon) > 0$ such that

$$\Big| \frac{\partial^2 g(z)}{\partial x_1 \partial x_2} - \frac{\partial^2 g(x)}{\partial x_1 \partial x_2}\Big|^2 < \frac{\epsilon}{24(1 + m_n(Q))} \tag{50}$$

whenever $|z - x| < \delta$. Let $h_\epsilon > 0$ satisfy

$$\sqrt{8}\, h_\epsilon < \delta, \tag{51}$$

Then (50) is satisfied for each $h < h_\epsilon$, any $\alpha \in \mathscr{A}$, and any $x, z \in C_h^{\alpha+e_1+e_2} \cup C_h^{\alpha+e_1} \cup C_h^{\alpha+e_2} \cup C_h^\alpha$. Assuming h is chosen so small that $m_n(Q_h) \le 2m_n(Q)$, from (48), (50) it follows

$$\sum_{\mathscr{A}} \frac{3}{h^2} \int\limits_{C_h^\alpha} I_1\, dx \le 12 \Big\|\frac{\partial^2 \sigma}{\partial x_1 \partial x_2} - \frac{\partial^2 g}{\partial x_1 \partial x_2}\Big\|^2_{L_2(Q+B_1(0))} < \frac{\epsilon}{2(1 + m_n(Q))}$$

$$\sum_{\mathscr{A}} \frac{3}{h^2} \int\limits_{C_h^\alpha} I_2\, dx < \frac{\epsilon m_n(Q)}{1 + m_n(Q)}$$

$$\sum_{\mathscr{A}} \frac{3}{h^2} \int\limits_{C_h^\alpha} I_3\, dx \le 12 \Big\|\frac{\partial^2 \sigma}{\partial x_1 \partial x_2} - \frac{\partial^2 g}{\partial x_1 \partial x_2}\Big\|^2_{L_2(Q+B_1(0))} < \frac{\epsilon}{2(1 + m_n(Q))}$$

From (49) we deduce

$$\left\|\tilde{\sigma}_h^{12} - \frac{\partial^2 \sigma}{\partial x_1 \partial x_2}\right\|_{L_2(Q_h)}^2 < \epsilon, \quad \forall h \leq h_\epsilon \tag{52}$$

Lemma is proved. □

The following lemma expresses similar result for 3D domains:

LEMMA 12. *Let $Q \in \mathbb{R}^3$. Then for $\forall\ \epsilon > 0$, there exists $\delta > 0$ such that*

$$\sum_{\mathscr{A}(Q_h^+)} h^3 |\sigma_{\alpha x_1 x_2 x_3}|^2 \leq (1+\epsilon)\left\|\frac{\partial^3 \sigma}{\partial x_1 \partial x_2 \partial x_3}\right\|_{L_2(Q_h)}^2$$

whenever $h < \delta$.

Although it is more technical, the proof of Lemma 12 is very similar to the proof of Lemma 11.

Lemmas 11 and 12 imply that the mappings $\mathscr{P}_h$ and $\mathscr{Q}_h$ satisfy the conditions of Lemma 10.

COROLLARY 13. *Assume $Q \in \mathbb{R}^2$ or $\mathbb{R}^3$. For arbitrary sufficiently small $\epsilon > 0$ there exists h_ϵ such that*

$$\mathscr{Q}_h(v) \in \mathscr{F}_h^R \quad \text{for all} \quad v \in \mathscr{F}^{R-\epsilon} \quad \text{and} \quad h \leq h_\epsilon, \tag{53}$$

$$\mathscr{P}_h([v]_h) \in \mathscr{F}^{R+\epsilon} \quad \text{for all} \quad [v]_h \in \mathscr{F}_h^R \quad \text{and} \quad h \leq h_\epsilon. \tag{54}$$

To prove (53), we first choose h'_ϵ such that for $\forall h < h'_\epsilon$ we have

$$\|\sigma\|_{\tilde{H}^1(Q_h)}^2 \leq \left(R - \frac{\epsilon}{2}\right)^2 \tag{55}$$

Then we apply Lemmas 11, 12, Proposition 11 in [**8**] with $\epsilon_1 = \left(\frac{R}{R-\frac{\epsilon}{2}}\right)^2 - 1$, and select $h_\epsilon < h'_\epsilon$ such that for $\forall\ h < h_\epsilon$

$$\|\mathscr{Q}_h(\sigma)\|_{\tilde{\mathscr{H}}^1(Q_h)}^2 \leq (1+\epsilon_1)\|\sigma\|_{\tilde{H}^1(Q_h)}^2 \leq R^2, \tag{56}$$

which proves (53). To prove (54), we derive the following estimation via straightforward calculation of the respective norm of multilinear interpolation σ'_h:

$$\|\mathscr{P}_h([\sigma]_h)\|_{\tilde{H}^1(Q)}^2 = \|\sigma'_h\|_{\tilde{H}^1(Q)}^2 \leq \|[\sigma]_h\|_{\tilde{\mathscr{H}}^1(Q_h)}^2 + Ch, \tag{57}$$

where C is independent of h. The latter easily imply (54).

Final statement of this section is the following embedding result of [**10**]:

LEMMA 14. [**10**] *If $Q \subset \mathbb{R}^n$ is a bouned domain, then*

$$\tilde{H}^1(Q) \hookrightarrow C^{0,\frac{1}{2}}(\overline{Q}); \quad \tilde{H}^1(Q) \Subset C^{0,\mu}(\overline{Q}), 0 < \mu < \frac{1}{2}. \tag{58}$$

4. Approximation Theorem and Convergence of the Discrete Optimal Control Problems

The following approximation theorem establishes the convergence of the discretized PDE problem:

THEOREM 15. *Let* $\{[v]_h\} = \{([\sigma]_h, U^h)\}$ *be a sequence of discrete control vectors such that there exists* $R > 0$ *for which* $[v]_h \in \mathscr{F}_h^R$ *for each* h*, and such that the sequence* $\{(\mathscr{P}_h([\sigma]_h), U^h)\}$ *converges to* $v = (\sigma, U)$ *in Tikhonov topology of* $\tilde{H}^1(Q) \times \mathbb{R}^m$ *formed with the product of the weak topology of* $\tilde{H}^1(Q)$ *and Euclidean topology of* $\mathbb{R}^m$*. Then the sequence of multilinear interpolations* $\{u'_h\}$ *of associated discrete state vectors* $\{[u]_h([v]_h)\}$ *converges to the solution* $u = u(x; v) \in H^1(Q)$ *of the elliptic problem* (3)–(5)*, weakly in* $H^1(Q)$*, strongly in* $L_2(Q)$*, strongly in* $L_2(S)$*, and almost everywhere on* Q*.*

Proof. By (35) of Corollary 8, sequence $\{u'_h\}$ is uniformly bounded in $H^1(Q)$. Consequently, $\{u'_h\}$ is weakly precompact in $H^1(Q)$. Let $u \in H^1(Q)$ be any weak limit point. By the Rellich-Kondrachev Theorem [**42**], it is known that there is a subsequence of $\{u'_h\}$ that converges to u, weakly in $H^1(Q)$, and strongly in $L_2(Q)$ and $L_2(S)$. By selecting further subsequence, if necessary, one can achieve that the convergence is almost everywhere on Q. We proceed to show that u satisfies the integral identity (10). Without loss of generality, we assume that the whole sequence $\{u'_h\}$ converges to $u \in H^1(Q)$. Let $Q' \subset \mathbb{R}^n$ be bounded open domain such that $\bar{Q} \subset Q'$ and choose arbitrary function $\eta \in C^1(\overline{Q}')$. We assume $h > 0$ is small enough that $Q_h \subset Q'$. We choose a grid function

$$[\eta]_h = \{\eta_\alpha : \ \eta_\alpha = \eta(x_\alpha), \alpha \in \mathscr{A}(Q_h)\}$$

in (22). Introducing standard interpolations $\tilde{\eta}_h$ and η_h^i as

$$\tilde{\eta}_h|_{C_h^\alpha} = \eta_\alpha, \quad \tilde{\eta}_h^i|_{C_h^\alpha} = \eta_{\alpha x_i}, \quad \forall \alpha \in \mathscr{A}(Q_h^+)$$

we write (22) in an equivalent form:

$$\sum_{i=1}^n \int_Q \tilde{\sigma}_h \tilde{u}_h^i \tilde{\eta}_h^i \, dx + \sum_{l=1}^m \frac{1}{Z_l} \int_{E_l} \tilde{u}_h \tilde{\eta}_h \, dS - \sum_{l=1}^m \frac{U_l^h}{Z_l} \int_{E_l} \tilde{\eta}_h \, dS$$

$$= -J_h([u]_h, [\eta]_h) - \sum_{i=1}^n \int_{Q_h \setminus Q} \tilde{\sigma}_h \tilde{u}_h^i \tilde{\eta}_h^i \, dx \tag{59}$$

Since, $\tilde{\sigma}_h$ converges to σ strongly in $L_2(Q)$, $\tilde{u}_h^i$ converges to $\frac{\partial u}{\partial x_i}$ weakly in $L_2(Q)$, $\tilde{u}_h$ converges to u strongly in $L_2(S)$, $\tilde{\eta}_h^i$ and $\tilde{\eta}_h$ converges to $\frac{\partial \eta}{\partial x_i}$ and η uniformly on $\overline{Q}$, the limit of three terms on the left hand side of (59) imply the respective terms of the integral identity (10) as $h \to 0$. Hence, it remains to prove that limit of the remaining terms in (59) vanishes. By applying Cauchy-Schwartz inequality we have

$$|J_h([u]_h, [\eta]_h)| \leq \Big(\sum_{i=1}^n \sum_{\mathscr{A}(Q_\Delta^{(i)})} h^n u_{\alpha x_i}^2 \Big)^{\frac{1}{2}} \|\eta\|_{C^1(\overline{Q'})} \sqrt{nh} \Big(\sum_{\mathscr{A}(S_h)} h^{n-1} \Big)^{\frac{1}{2}}. \tag{60}$$

Noting that every grid point on S_h belongs to cell (with 2^n vertices) which intersects S, and by recalling the definition of $\hat{S}$ we can estimate

$$\sum_{\mathscr{A}(S_h)} h^{n-1} \leq 2^n \sum_{\mathscr{A}(\hat{S}_h)} h^{n-1} \leq 2^n \sup_{h>0} \sum_{\mathscr{A}(\hat{S}_h)} h^{n-1} = 2^n \mathcal{H}^{n-1}(S), \tag{61}$$

where $\mathcal{H}^{n-1}(\cdot)$ is $n-1$-dimensional Hausdorff measure on $\mathbb{R}^n$. Since S is Lipschitz, $\mathcal{H}^{n-1}(S)$ coincides with the surface measure $m_{n-1}(S)$ [**21**]. Therefore, from (60),(61) and discrete energy estimate (30) it follows that

$$J_h([u]_h, [\eta]_h) = O(\sqrt{h}) \to 0, \text{ as } h \to 0. \tag{62}$$

Using Cauchy-Schwartz inequality for the second term in the right hand side of (59) we have

$$\begin{aligned} \Big|\sum_{i=1}^{n} \int_{Q_h\setminus Q} \tilde{\sigma}_h \tilde{u}^i_h \tilde{\eta}^i_h \, dx\Big| &\leq \sup_{Q_h\setminus Q} |\tilde{\sigma}_h| \sum_{i=1}^{n} \|\tilde{u}^i_h\|_{L_2(Q_h\setminus Q)} \|\tilde{\eta}^i_h\|_{L_2(Q_h\setminus Q)} \\ &\leq \|\sigma'_h\|_{C(Q_h)} \||[u]_h\||_{\mathscr{H}^1(Q_h)} \|D\eta\|_{C^1(\overline{Q}')} (m_n(Q_h\setminus Q))^{\frac{1}{2}} \end{aligned} \tag{63}$$

From the embedding result of Lemma 14 and (57) it follows that for sufficiently small h

$$\|\sigma'_h\|_{C(Q_h)} \leq C\|\sigma'_h\|_{\tilde{H}^1(Q_h)} \leq C\|[\sigma]_h\|^2_{\tilde{\mathscr{H}}^1(Q_h)} + 1 \leq CR + 1. \tag{64}$$

Since Lebesgue measure of $Q_h \setminus Q$ converges to zero as $h \to 0$, from the energy estimate (30) and (64) it follows that (63) converges to zero as $h \to 0$. Hence, passing to limit as $h \to 0$, from (59) it follows that the limit function u satisfies the integral identity (10). □

Approximation Theorem 15 imply the existence of the optimal control. The proof of the Theorem 4 is similar to the proof of existence Theorem 4.4 in [**9**].

In light of the approximation Theorem 15, to complete the proof of Theorem 5 it remains to prove that the conditions of Lemma 10 are satisfied. Proof of the condition (iii) of Lemma 10 coincide with the proof of similar fact from [**1, 5**]. Hence, it only remains to prove that the conditions (40) and (41) of Lemma 10 are satisfied (see Corollary 13).

Let $v \in \mathscr{F}^{(R-\epsilon)}$. By Corollary 13 we have $\mathscr{Q}_h(\sigma) = [\sigma]_h \in \mathscr{F}^R_h$. Applying Corollary 13 again, we deduce that $\mathscr{P}_h([\sigma]_h)$ belong to $\mathscr{F}^{R+\epsilon}$, and therefore it forms a weakly precompact sequence in $\tilde{H}^1(Q)$. From compact embedding result of Lemma 14 it follows that it forms a precompact sequence in a strong topology of $C^{0,\mu}(\overline{Q})$, $0 < \mu < \frac{1}{2}$. It easily follows that the whole sequence $\mathscr{P}_h([\sigma]_h)$ converges to σ weakly in $\tilde{H}^1(Q)$, and strongly in $C^{0,\mu}(\overline{Q})$. From Theorem 15 it follows that the sequence of multilinear interpolations $\{u'_h\}$ of associated discrete state vectors $\{[u([v]_h)]_h\}$ converges to the solution $u = u(x;v) \in H^1(Q)$ of the elliptic problem (3)–(5), weakly in $H^1(Q)$, strongly in $L_2(Q)$ and $L_2(S)$, and almost everywhere on Q. Claim (e) of Lemma 9 implies that the sequence $\tilde{u}_h$ converges to u strongly in

$L_2(S)$. Therefore, we have

$$\lim_{h\to 0} \mathscr{J}_h(\mathscr{Q}_h(v) = \lim_{h\to 0}\Big(\sum_{l=1}^m\Big(\int_{E_l}\frac{U_l-\tilde{u}_h}{Z_l}ds - I_l\Big)^2 + \beta|U-U^*|^2\Big)$$

$$= \sum_{l=1}^m\Big(\int_{E_l}\frac{U_l-u}{Z_l}ds - I_l\Big)^2 + \beta|U-U^*|^2 = \mathscr{J}(v)$$

which proves (40).

Let $\{[v]_h = ([\sigma]_h, U^h)\} \in \mathscr{F}_h^R$ be arbitrary sequence. From the Corollary 13 it follows that $(\mathscr{P}_h([\sigma]_h), U^h) \in \mathscr{F}^{R+1}$ for sufficiently small h, and therefore it is a precompact sequence in Tikhonov topology of $\tilde{H}^1(Q)\times\mathbb{R}^m$ formed as a product of weak topology of $\tilde{H}^1(Q)$ and Euclidean topology of $\mathbb{R}^m$. From compact embedding result of Lemma 14 it follows that $\{\mathscr{P}_h([\sigma]_h)\}$ is a precompact sequence in a strong topology of $C^{0,\mu}(\overline{Q})$, $0<\mu<\frac{1}{2}$. Without loss of generality assume that the whole sequence $(\mathscr{P}_h([\sigma]_h), U^h)$ converges to some limit $\tilde{v} = (\tilde{\sigma}, \tilde{U}) \in \tilde{H}^1(Q)\times\mathbb{R}^m$. We have

$$\mathscr{J}(\mathscr{P}_h([v]_h)) - \mathscr{J}_h([v]_h) = \mathscr{J}(\mathscr{P}_h([v]_h)) - \mathscr{J}(\tilde{v}) + \mathscr{J}(\tilde{v}) - \mathscr{J}_h([v]_h)$$

From Theorem 15 it follows that

$$\lim_{h\to 0}(\mathscr{J}(\mathscr{P}_h([v]_h)) - \mathscr{J}(\tilde{v})) = 0.$$

The proof of the limit

$$\lim_{h\to 0}(\mathscr{J}(\tilde{v}) - \mathscr{J}_h([v]_h)) = 0$$

is almost identical to the preceding proof of (40). Hence, (41) is proved and this completes the proof of the Theorem 5.

5. Conclusions

Inverse Electrical Impedance Tomography (EIT) problem on recovering electrical conductivity and potential in the body based on the measurement of the boundary voltages on the m electrodes for a given electrode current is considered. The variational formulation is introduced as a PDE constrained coefficient optimal control problem in Sobolev spaces with dominating mixed smoothness. Electrical conductivity and boundary voltages are control parameters, and the cost functional is the L_2-norm declinations of the boundary electrode current from the given current pattern and boundary electrode voltages from the measurements. The novelty of the control-theoretic model is its adaptation to the clinical situation when additional “voltage-to-current” measurements can increase the size of the input data from the number of electrodes m up to $m!$ while keeping the size of the unknown parameters fixed. In a recent paper [**9**], existence of the optimal control in Besov spaces framework, Fréchet differentiability, the formula for the Fréchet gradient, and optimality condition are proved. Numerical solution via gradient descent method in 2D model example demonstrated that the increase of the input data with additional measurements without changing the size of the unknown control parameters is an effective regularization tool for solving highly ill-posed EIT optimal control problem.

The setting of the EIT optimal control problem in the framework of the Sobolev spaces with dominating mixed smoothness is motivated by the proof of the convergence of the method of finite differences. In light of the recently established compact embedding into the Hölder spaces [**10**], Sobolev spaces with dominating mixed smoothness provide a suitable compactness criterion in Hölder spaces for the multilinear interpolations of the discrete grid functions on a lattice.

In this paper, the EIT optimal control problem in Sobolev spaces with dominating mixed smoothness is fully discretized using the method of finite differences. The existence of the optimal control and the convergence of the sequence of finite-dimensional optimal control problems to the EIT coefficient optimal control problem is proved both with respect to functional and control in 2- and 3-dimensional domains.

References

[1] Ugur G. Abdulla, *On the optimal control of the free boundary problems for the second order parabolic equations. I. Well-posedness and convergence of the method of lines*, Inverse Probl. Imaging **7** (2013), no. 2, 307–340, DOI 10.3934/ipi.2013.7.307. MR3063536

[2] Ugur G. Abdulla, *On the optimal control of the free boundary problems for the second order parabolic equations. II. Convergence of the method of finite differences*, Inverse Probl. Imaging **10** (2016), no. 4, 869–898, DOI 10.3934/ipi.2016025. MR3610744

[3] Ugur G. Abdulla and Jonathan M. Goldfarb, *Frechet differentiability in Besov spaces in the optimal control of parabolic free boundary problems*, J. Inverse Ill-Posed Probl. **26** (2018), no. 2, 211–227, DOI 10.1515/jiip-2017-0014. MR3778603

[4] Ugur G. Abdulla, Evan Cosgrove, and Jonathan Goldfarb, *On the Frechet differentiability in optimal control of coefficients in parabolic free boundary problems*, Evol. Equ. Control Theory **6** (2017), no. 3, 319–344, DOI 10.3934/eect.2017017. MR3669653

[5] Ugur G. Abdulla and Bruno Poggi, *Optimal control of the multiphase Stefan problem*, Appl. Math. Optim. **80** (2019), no. 2, 479–513, DOI 10.1007/s00245-017-9472-7. MR4008673

[6] Ugur G. Abdulla, Vladislav Bukshtynov, and Ali Hagverdiyev, *Gradient method in Hilbert-Besov spaces for the optimal control of parabolic free boundary problems*, J. Comput. Appl. Math. **346** (2019), 84–109, DOI 10.1016/j.cam.2018.06.043. MR3864148

[7] Ugur G. Abdulla, Jonathan Goldfarb, and Ali Hagverdiyev, *Optimal control of coefficients in parabolic free boundary problems modeling laser ablation*, J. Comput. Appl. Math. **372** (2020), 112736, 21, DOI 10.1016/j.cam.2020.112736. MR4055048

[8] Ugur G. Abdulla and Bruno Poggi, *Optimal Stefan problem*, Calc. Var. Partial Differential Equations **59** (2020), no. 2, Paper No. 61, 40, DOI 10.1007/s00526-020-1712-z. MR4073207

[9] Ugur G. Abdulla, Vladislav Bukshtynov, and Saleheh Seif, *Cancer detection through electrical impedance tomography and optimal control theory: theoretical and computational analysis*, Math. Biosci. Eng. **18** (2021), no. 4, 4834–4859, DOI 10.3934/mbe.2021246. MR4276124

[10] Ugur G. Abdulla, Generalized Newton-Leibniz Formula and the Embedding of the Sobolev Functions with Dominating Mixed Smoothness into Hölder Spaces, *preprint* (2020)

[11] Giovanni Alessandrini, *Stable determination of conductivity by boundary measurements*, Appl. Anal. **27** (1988), no. 1-3, 153–172, DOI 10.1080/00036818808839730. MR922775

[12] Melody Alsaker and Jennifer L. Mueller, *A D-bar algorithm with a priori information for 2-dimensional electrical impedance tomography*, SIAM J. Imaging Sci. **9** (2016), no. 4, 1619–1654, DOI 10.1137/15M1020137. MR3554880

[13] Habib Ammari and Hyeonbae Kang, *Reconstruction of small inhomogeneities from boundary measurements*, Lecture Notes in Mathematics, vol. 1846, Springer-Verlag, Berlin, 2004, DOI 10.1007/b98245. MR2168949

[14] Habib Ammari, Lingyun Qiu, Fadil Santosa, and Wenlong Zhang, *Determining anisotropic conductivity using diffusion tensor imaging data in magneto-acoustic tomography with magnetic induction*, Inverse Problems **33** (2017), no. 12, 125006, 15, DOI 10.1088/1361-6420/aa907e. MR3729795

[15] Giovanni S. Alberti, Habib Ammari, Bangti Jin, Jin-Keun Seo, and Wenlong Zhang, *The linearized inverse problem in multifrequency electrical impedance tomography*, SIAM J. Imaging Sci. **9** (2016), no. 4, 1525–1551, DOI 10.1137/16M1061564. MR3553924

[16] Kari Astala and Lassi Päivärinta, *Calderón's inverse conductivity problem in the plane*, Ann. of Math. (2) **163** (2006), no. 1, 265–299, DOI 10.4007/annals.2006.163.265. MR2195135

[17] Oleg V. Besov, Valentin P. Il′in, and Sergey M. Nikol′skiĭ, *Integral representations of functions and imbedding theorems. Vol. II*, Scripta Series in Mathematics, V. H. Winston & Sons, Washington, D.C.; Halsted Press [John Wiley & Sons], New York-Toronto, Ont.-London, 1979. Edited by Mitchell H. Taibleson. MR521808

[18] Oleg V. Besov, Valentin P. Il′in, and Sergey M. Nikol′skiĭ, *Integral representations of functions and imbedding theorems. Vol. II*, Scripta Series in Mathematics, V. H. Winston & Sons, Washington, D.C.; Halsted Press [John Wiley & Sons], New York-Toronto, Ont.-London, 1979. Edited by Mitchell H. Taibleson. MR521808

[19] Alberto-P. Calderón, *On an inverse boundary value problem*, Seminar on Numerical Analysis and its Applications to Continuum Physics (Rio de Janeiro, 1980), Soc. Brasil. Mat., Rio de Janeiro, 1980, pp. 65–73. MR590275

[20] Melody Dodd and Jennifer L. Mueller, *A real-time D-bar algorithm for 2-D electrical impedance tomography data*, Inverse Probl. Imaging **8** (2014), no. 4, 1013–1031, DOI 10.3934/ipi.2014.8.1013. MR3295956

[21] Lawrence C. Evans and Ronald F. Gariepy, *Measure theory and fine properties of functions*, Revised edition, Textbooks in Mathematics, CRC Press, Boca Raton, FL, 2015. MR3409135

[22] S. J. Hamilton, J. M. Reyes, S. Siltanen, and X. Zhang, *A hybrid segmentation and D-bar method for electrical impedance tomography*, SIAM J. Imaging Sci. **9** (2016), no. 2, 770–793, DOI 10.1137/15M1025992. MR3507090

[23] S. J. Hamilton, M. Lassas, and S. Siltanen, *A direct reconstruction method for anisotropic electrical impedance tomography*, Inverse Problems **30** (2014), no. 7, 075007, 33, DOI 10.1088/0266-5611/30/7/075007. MR3233020

[24] Bastian Harrach, *Uniqueness and Lipschitz stability in electrical impedance tomography with finitely many electrodes*, Inverse Problems **35** (2019), no. 2, 024005, 19, DOI 10.1088/1361-6420/aaf6fc. MR3900209

[25] Bastian Harrach and Mach Nguyet Minh, *Enhancing residual-based techniques with shape reconstruction features in electrical impedance tomography*, Inverse Problems **32** (2016), no. 12, 125002, 21, DOI 10.1088/0266-5611/32/12/125002. MR3628886

[26] D.S. Holder, *Electrical impedance tomography: methods, history and applications*, CRC Press, 2004.

[27] Nuutti Hyvönen, Lassi Päivärinta, and Janne P. Tamminen, *Enhancing D-bar reconstructions for electrical impedance tomography with conformal maps*, Inverse Probl. Imaging **12** (2018), no. 2, 373–400, DOI 10.3934/ipi.2018017. MR3810163

[28] Bangti Jin, Taufiquar Khan, and Peter Maass, *A reconstruction algorithm for electrical impedance tomography based on sparsity regularization*, Internat. J. Numer. Methods Engrg. **89** (2012), no. 3, 337–353, DOI 10.1002/nme.3247. MR2876564

[29] Bangti Jin, Yifeng Xu, and Jun Zou, *A convergent adaptive finite element method for electrical impedance tomography*, IMA J. Numer. Anal. **37** (2017), no. 3, 1520–1550, DOI 10.1093/imanum/drw045. MR3671504

[30] Carlos E. Kenig, Johannes Sjöstrand, and Gunther Uhlmann, *The Calderón problem with partial data*, Ann. of Math. (2) **165** (2007), no. 2, 567–591, DOI 10.4007/annals.2007.165.567. MR2299741

[31] Carlos Kenig and Mikko Salo, *The Calderón problem with partial data on manifolds and applications*, Anal. PDE **6** (2013), no. 8, 2003–2048, DOI 10.2140/apde.2013.6.2003. MR3198591

[32] Michael Klibanov, Jingzhi Li, and Wenlong Zhang, *Convexification of electrical impedance tomography with restricted Dirichlet-to-Neumann map data*, Inverse Problems **35** (2019), no. 3, 035005, 33, DOI 10.1088/1361-6420/aafecd. MR3916465

[33] Robert Kohn and Michael Vogelius, *Determining conductivity by boundary measurements*, Comm. Pure Appl. Math. **37** (1984), no. 3, 289–298, DOI 10.1002/cpa.3160370302. MR739921

[34] R. V. Kohn and M. Vogelius, *Determining conductivity by boundary measurements. II. Interior results*, Comm. Pure Appl. Math. **38** (1985), no. 5, 643–667, DOI 10.1002/cpa.3160380513. MR803253

[35] Ohin Kwon, Jin Keun Seo, and Jeong-Rock Yoon, *A real-time algorithm for the location search of discontinuous conductivities with one measurement*, Comm. Pure Appl. Math. **55** (2002), no. 1, 1–29, DOI 10.1002/cpa.3009. MR1857878

[36] O. A. Ladyzhenskaya, *Kraevye zadachi matematicheskoĭ fiziki* (Russian), Izdat. "Nauka", Moscow, 1973. MR0599579

[37] S. Laufer, A. Ivorra and V. Reuter, Electrical impedane characterization of normal and cancerous human hepatic tissue, *Physiological Measurements*, **31** (2010), 995–1009.

[38] Armin Lechleiter and Andreas Rieder, *Newton regularizations for impedance tomography: convergence by local injectivity*, Inverse Problems **24** (2008), no. 6, 065009, 18, DOI 10.1088/0266-5611/24/6/065009. MR2456956

[39] Matthias Gehre, Bangti Jin, and Xiliang Lu, *An analysis of finite element approximation in electrical impedance tomography*, Inverse Problems **30** (2014), no. 4, 045013, 24, DOI 10.1088/0266-5611/30/4/045013. MR3190315

[40] Adrian I. Nachman, *Reconstructions from boundary measurements*, Ann. of Math. (2) **128** (1988), no. 3, 531–576, DOI 10.2307/1971435. MR970610

[41] Adrian I. Nachman, *Global uniqueness for a two-dimensional inverse boundary value problem*, Ann. of Math. (2) **143** (1996), no. 1, 71–96, DOI 10.2307/2118653. MR1370758

[42] S.M. Nikol'skii, Approximation of Functions of Several Variables and Imbedding Theorems, Springer-Verlag, (1975).

[43] J.K. Seo, J. Lee, S.W. Kim, H. Zribi and E.J. Woo, Frequency-difference electrical impedance tomography: algorithm development and feasibility study, *Phys. Meas.*, **29** (2008), 929–941.

[44] Jin Keun Seo and Eung Je Woo, *Magnetic resonance electrical impedance tomography (MREIT)*, SIAM Rev. **53** (2011), no. 1, 40–68, DOI 10.1137/080742932. MR2785879

[45] Erkki Somersalo, Margaret Cheney, and David Isaacson, *Existence and uniqueness for electrode models for electric current computed tomography*, SIAM J. Appl. Math. **52** (1992), no. 4, 1023–1040, DOI 10.1137/0152060. MR1174044

[46] John Sylvester and Gunther Uhlmann, *A global uniqueness theorem for an inverse boundary value problem*, Ann. of Math. (2) **125** (1987), no. 1, 153–169, DOI 10.2307/1971291. MR873380

[47] F.P. Vasil'ev, Methods for Solving Extremal Problems. Minimization problems in function spaces, regularization, approximation, *Moscow, Nauka* (1981).

[48] Thomas Widlak and Otmar Scherzer, *Hybrid tomography for conductivity imaging*, Inverse Problems **28** (2012), no. 8, 084008, 28, DOI 10.1088/0266-5611/28/8/084008. MR2956564

Department of Mathematical Sciences, Florida Institute of Technology, Melbourne, Florida 32901, USA; and Analysis & PDE Unit, Okinawa Institute of Science and Technology, Okinawa, 904-0495 Japan

Email address: `Ugur.Abdulla@oist.jp`

Department of Mathematical Sciences, Florida Institute of Technology, Melbourne, Florida 32901, USA

Contemporary Mathematics
Volume **784**, 2023
https://doi.org/10.1090/conm/784/15751

Global reconstruction of initial conditions of nonlinear parabolic equations via the Carleman-contraction method

Thuy T. Le

ABSTRACT. We propose a global convergent numerical method to reconstruct the initial condition of a nonlinear parabolic equation from the measurement of both Dirichlet and Neumann data on the boundary of a bounded domain. The first step in our method is to derive, from the nonlinear governing parabolic equation, a nonlinear system of elliptic partial differential equations (PDEs) whose solution yields directly the solution of the inverse source problem. We then establish a contraction mapping-like iterative scheme to solve this system. The convergence of this iterative scheme is rigorously proved by employing a Carleman estimate and the argument in the proof of the traditional contraction mapping principle. This convergence is fast in both theoretical and numerical senses. Moreover, our method, unlike the methods based on optimization, does not require a good initial guess of the true solution. Numerical examples are presented to verify these results.

1. Introduction

The problem of recovering the initial condition for a nonlinear parabolic equation from the measurement of the Dirichlet and Neumann data in a bounded domain arises in many real-world applications, for example, detecting the pollution on the surface of rivers or lakes [**1**], reconstructing of the spatially distributed temperature inside a solid from the heat and the heat flux on the boundary in the time domain [**7**], effective monitoring heat and conduction processes in steel industries, glass and polymer-forming and nuclear power station [**28**]. Due to its realistic applications, this inverse source problem has been studied intensively. It is formulated as follows.

Let $d \geq 1$ be the spatial dimension, let Ω be an open and bounded domain in $\mathbb{R}^d$ with smooth boundary $\partial\Omega$ of the class C^1, and let T be a positive number. Let $F : \mathbb{R}^d \times \mathbb{R} \times \mathbb{R} \times \mathbb{R}^d \to \mathbb{R}$ and $c : \mathbb{R}^d \to [c_0, \infty)$, for some $c_0 > 0$, be functions of the class C^1. Denote by Q_T the set $\Omega \times (0, T)$.

2020 *Mathematics Subject Classification.* 35R30, 35K20.

Key words and phrases. Nonlinear equations, initial condition, inverse source problem, convergent numerical method, Carleman estimate, iteration, contraction mapping.

The work of Thuy Le was partially supported by US Army Research Laboratory and US Army Research Office grant W911NF-19-1-0044, by National Science Foundation grant DMS-2208159, and by funds provided by the Faculty Research Grant program at UNC Charlotte Fund No. 111272.

Consider the function $u(\mathbf{x}, t)$ is governed by the following problem:

$$\begin{cases} c(\mathbf{x})u_t(\mathbf{x}, t) = \Delta u(\mathbf{x}, t) + F(\mathbf{x}, t, u(\mathbf{x}, t), \nabla u(\mathbf{x}, t)) & (\mathbf{x}, t) \in \mathbb{R}^d \times (0, T) \\ u(\mathbf{x}, 0) = p(\mathbf{x}) & \mathbf{x} \in \mathbb{R}^d, \end{cases} \tag{1.1}$$

where $p \in C_c^1(\mathbb{R}^d)$ is a source function. Assume that p is compactly supported in Ω. Since the main aim of this paper concerns with inverse problem, not forward problem, the existence, uniqueness and regularity results for the solution to (1.1) are considered as assumptions. For the completeness, we provide here a set of conditions that guarantees that these assumptions hold true. Assume that $p(\mathbf{x})$ is in $H^{2+\beta}(\mathbb{R}^d)$ for some $\beta \in [0, 1 + 4/d]$. Assume further for all $\mathbf{x} \in \mathbb{R}^d$, $t \in [0, T]$, $s \in \mathbb{R}$, and $\mathbf{r} \in \mathbb{R}^d$,

$$|F(\mathbf{x}, t, s, \mathbf{r})| \leq C \max\left\{(1 + |\mathbf{r}|)^2, 1 + |s|\right\}$$

for some positive constant C. Then, due to Theorem 6.1 in [**21**, Chapter 5, §6] and Theorem 2.1 in [**21**, Chapter 5, §2], problem (1.1) has a unique solution with $|u(\mathbf{x}, t)| \leq M_1$, $|\nabla u(\mathbf{x}, t)| \leq M_2$ for some positive constants M_1, M_2, and $u(\mathbf{x}, t) \in H^{2+\beta, 1+\beta/2}(\mathbb{R}^d \times [0, T])$.

We are interested in the problem of determining the source function $p(\mathbf{x}), \mathbf{x} \in \Omega$, from the lateral Cauchy data. This problem is formulated as follows.

PROBLEM 1.1 (Inverse Source Problem). *Assume that $\|u\|_{C^1(\overline{Q_T})} \leq M$ for some known large number M. Given the lateral Cauchy data*

$$g(\mathbf{x}, t) = u(\mathbf{x}, t) \quad \textit{and} \quad q(\mathbf{x}, t) = \partial_\nu u(\mathbf{x}, t) \tag{1.2}$$

for $\mathbf{x} \in \partial\Omega$, $t \in [0, T]$, determine the function $u(\mathbf{x}, 0) = p(\mathbf{x}), \mathbf{x} \in \Omega$.

The uniqueness of Problem 1.1 is still an open question. We temporarily consider the uniqueness as an assumption. This result will be written in a separate paper. In the case when (1.1) is linear, the uniqueness of Problem 1.1 is well-known, see [**23**]. The logarithmic stability results in the linear case were rigorously proved in [**7, 28**].

A natural approach to solve the Problem 1.1 is based on least squares optimization. However, since our problem is highly nonlinear, the cost functional might have multiple minima and ravines. Thus, optimization-based methods might not deliver good solutions; especially, when a good initial guess is unavailable. An effective way to overcome the lack of a good initial guess is the convexification method, which first introduced in [**10**]. The main idea of the convexification method is to convexify the cost functional by employing a suitable Carleman weight function. We refer the reader to [**2, 6, 9, 11–13, 15, 17–19, 26, 37–40**] for the development of the convexification method. Although effective, the convexification method is time consuming. Therefore, another method should be investigated.

In this paper, we introduce an iterative method to solve Problem 1.1. The first step in our numerical method is to derive a system of quasi-linear elliptic equations whose solution yields directly the solution of Problem 1.1 by truncating a Fourier series with respect to the special basis in [**9**]. In the second step, we propose a fixed point-like iterative process to solve the system mentioned above. The iterative process can start from an arbitrarily function. The idea to design the iterative process is similar to [**24, 25, 33, 34**]. Especially, in [**25**], we study a similar problem to recover the initial condition of a nonlinear parabolic equation from the lateral Cauchy data. The main difference between [**25**] and this paper is that in this paper, we study a more complicated case at which the nonlinear term F depends on both $u(\mathbf{x}, t)$ and $\nabla u(\mathbf{x}, t)$. The convergence of our scheme is rigorously proved based on a Carleman estimate and the analogue contraction mapping principle. Our iterative scheme converges quickly to the true solution at the rate θ^n for some constant $\theta \in (0, 1)$ and n is the number of iterations. In particular, the convergence of our numerical method is rigorously proved in H^2. This result is a significant improvement in comparison to the main theorem in [**30, 33**] which shows a similar convergence in H^1 only. These results is verified rigorously by our numerical tests.

The paper is organized as follows. In Section 2, we recall some preliminaries which we employ directly in our numerical method, including a Carleman estimate and a special orthonormal basis. In Section 3, we introduce two steps to solve Problem 1.1. The first step is to derive a system of nonlinear PDEs whose solutions yields directly the solutions to Problem 1.1. The second step is to establish an iterative scheme to solve the above system of nonlinear PDEs. In Section 4, we prove the convergence of our iterative scheme to the true solution. In Section 5, we discuss the implementation of our method and present some numerical results. Section 6 is some concluding remarks.

2. Preliminaries

In this section, we recall a Carleman estimate established in [**25**]. This Carleman estimate plays an important role this paper. On the other hand, we recall a special orthonormal basis $\{\Phi_n\}_{n\geq 1}$ of $L^2(0, T)$, which will be used in the numerical implementation section, Section 5. This special basis was first introduced in [**9**].

2.1. A Carleman estimate. The Carleman estimate is a powerful tool in the field of PDEs which were first employed to prove the unique continuation principle, see [**5, 36**], the uniqueness of a long list of inverse problems, see [**4**] and in cloaking [**31**]. Here we recall a Carleman estimate which was established in [**25**]. The analysis of this paper is based on this Carleman estimate.

LEMMA 2.1 (Carleman estimate, see [**25**]). *Let $\mathbf{x}_0$ be a point in $\mathbb{R}^d \setminus \overline{\Omega}$ such that $r(\mathbf{x}) = |\mathbf{x} - \mathbf{x}_0| > 1$ for all $\mathbf{x} \in \Omega$. Let $b > \max_{\mathbf{x}\in\overline{\Omega}} r(\mathbf{x})$ be a fixed constant. There exists a positive constant β_0 depending only on b, $\mathbf{x}_0$, Ω and d such that for all function $v \in C^2(\overline{\Omega})$ satisfying*

$$v(\mathbf{x}) = \partial_\nu v(\mathbf{x}) = 0 \quad \textit{for all } \mathbf{x} \in \partial\Omega,$$

the following estimate holds true

$$\begin{aligned}(2.1)\quad \int_\Omega e^{2\lambda b^{-\beta} r^\beta(\mathbf{x})} |\Delta v(\mathbf{x})|^2 d\mathbf{x} \geq \frac{C}{\lambda \beta^{7/4} b^{-\beta}} \int_\Omega e^{2\lambda b^{-\beta} r^\beta(\mathbf{x})} r^{2\beta}(\mathbf{x}) |D^2 v(\mathbf{x})|^2 d\mathbf{x} \\ + C\lambda^3 \beta^4 b^{-3\beta} \int_\Omega r^{2\beta}(\mathbf{x}) e^{2\lambda b^{-\beta} r^\beta} |v(\mathbf{x})|^2 d\mathbf{x} \\ + C\lambda \beta^{1/2} b^{-\beta} \int_\Omega e^{2\lambda b^{-\beta} r^\beta(\mathbf{x})} |\nabla v(\mathbf{x})|^2 d\mathbf{x}\end{aligned}$$

for $\beta \geq \beta_0$ and $\lambda \geq \lambda_0$. Here, $D^2 v = (v_{x_i x_j})_{i,j=1}^d$ is the Hessian matrix of v, $\lambda_0 = \lambda_0(b, \Omega, d, \mathbf{x}_0) > 1$ is a positive number with $\lambda_0 b^{-\beta} \gg 1$ and $C = C(b, \Omega, d, \mathbf{x}_0) > 1$ is a constant. These numbers depend only on listed parameters.

COROLLARY 2.1. *Recall β_0 and λ_0 as in Lemma 2.1. Fix $\beta \geq \beta_0$ and let the constant C depend on $\mathbf{x}_0$, Ω, d and β. There exists a constant λ_0 depending only on $\mathbf{x}_0$, Ω, d and β such that for all function $v \in H^2(\Omega)$ with*

$$v(\mathbf{x}) = \partial_\nu v(\mathbf{x}) = 0 \quad \text{on } \partial\Omega,$$

we have

$$\begin{aligned}(2.2)\quad \int_\Omega e^{2\lambda b^{-\beta} r^\beta(\mathbf{x})} |\Delta v(\mathbf{x})|^2 d\mathbf{x} &\geq C\lambda^{-1} \int_\Omega e^{2\lambda b^{-\beta} r^\beta(\mathbf{x})} |D^2 v(\mathbf{x})|^2 d\mathbf{x} \\ &+ C\lambda^3 \int_\Omega e^{2\lambda b^{-\beta} r^\beta} |v(\mathbf{x})|^2 d\mathbf{x} + C\lambda \int_\Omega e^{2\lambda b^{-\beta} r^\beta(\mathbf{x})} |\nabla v(\mathbf{x})|^2 d\mathbf{x}\end{aligned}$$

for all $\lambda \geq \lambda_0$.

We refer the reader to [**25**, Theorem 3.1] for the proof of Lemma 2.1. We also refer the readers to [**29, 35**] for some versions of Carleman estimates for parabolic operators, and to [**3, 4, 16**] for several other versions of Carleman estimates for a variety kinds of differential operators and their applications in inverse problems.

2.2. A special orthonormal basis. In this section, we recall a special basis of $L^2(0,T)$. This special basis will be employed in our numerical study. Let $\phi_n(t) = (t - T/2)^{n-1} e^{t-T/2}$ for $n \in \mathbb{N}$. The set $\{\phi_n\}_{n=1}^\infty$ is complete in $L^2(0,T)$. Applying the GramSchmidt orthonormalization process to this complete set gives a basis of $L^2(0,T)$, named as $\{\Psi_n\}_{n=1}^\infty$. The following proposition holds true.

PROPOSITION 2.1 (see [**9**]). *The basis $\{\Psi_n\}_{n=1}^\infty$ satisfies the following properties:*

(i) *Ψ_n' is not identically zero for all $n \geq 1$*
(ii) *For all $m, n \geq 1$*

$$(2.3)\qquad s_{mn} = \int_0^T \Psi_n'(t)\Psi_m(t) = \begin{cases} 1 \text{ if } m = n, \\ 0 \text{ if } n < m. \end{cases}$$

As a result, for all integer $N > 1$, the matrix $S = (s_{mn})_{m,n=1}^N$ is invertible.

This basis was first introduced to solve the electrical impedance tomography problem with partial data in [**9**]. Afterward, it is widely used in our research group to solve variety kinds of inverse problems, including ill-posed inverse source problems for elliptic equations [**35**], parabolic equations [**25**] [**29**] and hyperbolic equations [**27**], nonlinear coefficient inverse problems for elliptic equations [**38**], and parabolic equations [**20, 26, 32, 39, 40**], transport equations [**14**] and full transfer equations [**37**].

3. A numerical method to solve Problem 1.1

In this section, we present our numerical method to solve Problem 1.1. Our method consists of two (2) main steps. Firstly, we derive a nonlinear system of elliptic equations by cutting of a Fourier series with respect to an orthonormal basis of $L^2(0,T)$. Solution of this system directly leads to that of Problem 1.1. Secondly, we propose a fixed point-like iterative scheme to solve the nonlinear system mentioned in step 1.

3.1. A system of nonlinear elliptic equations. Let $\{\Psi_n\}_{n=1}^{\infty}$ be an orthonormal basis of $L^2(0,T)$. For all $(\mathbf{x},t) \in Q_T$, we can rewrite $u(\mathbf{x},t)$ as follows.

$$u(\mathbf{x},t) = \sum_{n=1}^{\infty} u_n(\mathbf{x})\Psi_n(t) \tag{3.1}$$

where

$$u_n(\mathbf{x}) = \int_0^T u(\mathbf{x},t)\Psi_n(t)dt \quad \text{for all } n \geq 1. \tag{3.2}$$

We now derive our approximation model. For some cut-off number N, chosen later in Section 5, we approximate the function $u(\mathbf{x},t)$ by truncating the series in (3.1) as

$$u(\mathbf{x},t) \approx u^N(\mathbf{x},t) := \sum_{n=1}^{N} u_n(\mathbf{x})\Psi_n(t) \quad \text{for all } (\mathbf{x},t) \in Q_T \tag{3.3}$$

We also approximate

$$u_t(\mathbf{x},t) \approx u_t^N(\mathbf{x},t) := \sum_{n=1}^{N} u_n(\mathbf{x})\Psi_n'(t) \quad \text{for all } (\mathbf{x},t) \in Q_T \tag{3.4}$$

REMARK 3.1. *The approximation in* (3.3) *is numerically verified by Figure* 1.

Plugging (3.3) and (3.4) into the governing equation (1.1), we obtain

$$\begin{aligned} c(\mathbf{x})\sum_{n=1}^{N} u_n(\mathbf{x})\Psi_n'(t) &= \sum_{n=1}^{N} \Delta u_n(\mathbf{x})\Psi_n(t) \\ &\quad + F\left(\mathbf{x}, t, \sum_{n=1}^{N} u_n(\mathbf{x})\Psi_n(t), \sum_{n=1}^{N} \nabla u_n(\mathbf{x})\Psi_n(t)\right) \end{aligned} \tag{3.5}$$

for all $(\mathbf{x},t) \in Q_T$.

REMARK 3.2. *The cut-off number N is chosen numerically such that u^N well approximates the function u, see Section* 5.2 *for more details. Due to the nonlinearity and the ill-posedness of this inverse source problem, studying the convergence of* (3.5) *as $N \to \infty$ is extremely challenging and out of scope of this paper. We only solve Problem* 1.1 *in the approximation context.*

For each $m \in \{1,\ldots,N\}$, multiplying $\Psi_m(t)$ to both sides of (3.5) and then integrating the obtained equation with respect to t give

$$\begin{aligned} c(\mathbf{x})\sum_{n=1}^{N} u_n(\mathbf{x}) \int_0^T \Psi_n'(t)\Psi_m(t)dt &= \sum_{n=1}^{N} \Delta u_n(\mathbf{x}) \int_0^T \Psi_n(t)\Psi_m(t)dt \\ &\quad + \int_0^T F\left(\mathbf{x}, t, \sum_{n=1}^{N} u_n(\mathbf{x})\Psi_n(t), \sum_{n=1}^{N} \nabla u_n(\mathbf{x})\Psi_n(t)\right)\Psi_m(t)dt. \end{aligned} \tag{3.6}$$

Since $\{\Psi_n\}_{n\geq 1}$ is an orthonormal basis, system (3.6) can be rewritten as

$$c(\mathbf{x})\sum_{n=1}^{N} s_{mn}u_n(\mathbf{x}) = \Delta u_m(\mathbf{x}) + F_m\left(\mathbf{x}, U(\mathbf{x}), \nabla U(\mathbf{x})\right) \tag{3.7}$$

for all $\mathbf{x} \in \Omega$ and $m = 1, \dots, N$ where $U = (u_1, u_2, \dots, u_N)^{\mathrm{T}}$,

$$s_{mn} = \int_0^T \Psi_n'(t)\Psi_m(t)dt$$

and

$$F_m(\mathbf{x}, U(\mathbf{x}), \nabla U(\mathbf{x})) = \int_0^T F\left(\mathbf{x}, t, \sum_{n=1}^N u_n(\mathbf{x})\Psi_n(t), \sum_{n=1}^N \nabla u_n(\mathbf{x})\Psi_n(t)\right)\Psi_m(t)dt.$$

Denote by S the matrix $(s_{mn})_{m,n=1}^N$ and $\mathbf{F} = (F_1, F_2, \dots, F_N)^{\mathrm{T}}$. We can rewrite (3.7) as

$$\Delta U(\mathbf{x}) - c(\mathbf{x})SU(\mathbf{x}) + \mathbf{F}(\mathbf{x}, U(\mathbf{x}), \nabla U(\mathbf{x})) = 0 \quad \text{for all } \mathbf{x} \in \Omega. \tag{3.8}$$

We next compute the Cauchy boundary conditions for U_N Due to (3.2),

$$\begin{cases} U(\mathbf{x}) = G(\mathbf{x}) = \left(\int_0^T g(\mathbf{x}, t)\Psi_m(t)dt\right)_{m=1}^N, \\ \partial_\nu U(\mathbf{x}) = Q(\mathbf{x}) = \left(\int_0^T q(\mathbf{x}, t)\Psi_m(t)dt\right)_{m=1}^N \end{cases} \quad m = 1, 2, \dots, N \tag{3.9}$$

for all $\mathbf{x} \in \partial\Omega$, $m = 1, \dots, N$, where $g(\mathbf{x}, t)$ and $q(\mathbf{x}, t)$ are the given data in Problem 1.1.

Combining (3.8) and (3.9), we obtain the system of elliptic equations for $u_m(\mathbf{x})$ for each $m = 1, \dots, N$

$$\begin{cases} \Delta U(\mathbf{x}) - c(\mathbf{x})SU(\mathbf{x}) + \mathbf{F}(\mathbf{x}, U(\mathbf{x}), \nabla U(\mathbf{x})) = 0 & \mathbf{x} \in \Omega, \\ U(\mathbf{x}) = G(\mathbf{x}) & \mathbf{x} \in \partial\Omega, \\ \partial_\nu U(\mathbf{x}) = Q(\mathbf{x}) & \mathbf{x} \in \partial\Omega. \end{cases} \tag{3.10}$$

REMARK 3.3. *Due to the truncation in* (3.3)*, Problem* (3.10) *is not exact. It is an approximation model. Proving the convergence of this approximation model as $N \to \infty$ is extremely challenging. Establishing this result is out of the scope of this paper. The accuracy of* (3.3) *can be verified numerically, see Figure* 1.

Problem 1.1 is reduced to the problem of finding $U = (u_1, u_2, \dots, u_N)^{\mathrm{T}}$ satisfying system (3.10). In fact, if this vector is known, the function $u(\mathbf{x}, t)$ for all $(\mathbf{x}, t) \in Q_T$ can be approximated via (3.3). Then, the solution to Problem 1.1 is given by $p(\mathbf{x}) = u(\mathbf{x}, 0)$ for all $\mathbf{x} \in \Omega$.

Since both Dirichlet and Neumann conditions are imposed in Problem (3.10), it is over-determined. However, since system (3.10) is derived from the Forward Problem, the solution to (3.10) exists whenever the Forward Problem has a solution. In practice, if the measured data contains some noise, the problem might have no solution. Therefore, we compute here the "best-fit" solution to the Problem (3.10). In the next section, we introduce an iterative procedure to find the "best-fit" solution to system (3.10).

3.2. An iterative procedure to solve system (3.10)**.** We introduce an iterative scheme to solve system (3.10). The convergence of this scheme to the true solution of (3.10) will be discussed later in Section 4.

Let $U^{(0)}$ be an arbitrary vector-valued function. Assume by induction that we know $U^{(k-1)}$ for $k \geq 1$. We then find $U^{(k)}$ by solving the following system

$$\left\{\begin{array}{ll} \Delta U^{(k)}(\mathbf{x}) - c(\mathbf{x})SU^{(k)}(\mathbf{x}) = -\mathbf{F}(\mathbf{x}, U^{(k-1)}(\mathbf{x}), \nabla U^{(k-1)}(\mathbf{x})) & \mathbf{x} \in \Omega, \\ U^{(k)}(\mathbf{x}) = G(\mathbf{x}) & \mathbf{x} \in \partial\Omega, \\ \partial_\nu U^{(k)}(\mathbf{x}) = Q(\mathbf{x}) & \mathbf{x} \in \partial\Omega. \end{array}\right. \tag{3.11}$$

Problem (3.11) might not have a solution because it is over-determined. We only compute the "Carleman best-fit" solution $U^{(k)}$ to (3.11) by combining the quasi-reversibility method and a Carleman weight function as follows. The convergence of the sequence of "Carleman best-fit" solutions is one of the important strengths of this paper. Define the set of admissible solutions

$$H = \left\{V \in H^2(\Omega)^N : V|_{\partial\Omega} = G \text{ and } \partial_\nu V|_{\partial\Omega} = Q\right\}. \tag{3.12}$$

REMARK 3.4. *Throughout this paper, we assume that H is nonempty. An example of a set of conditions imposed on G, Q such that H is nonempty is $(G, Q) \in H^{3/2}(\Omega)^N \times H^{1/2}(\Omega)^N$. It corresponds to $(g, q) \in H^{3/2}(\Omega) \times H^{1/2}(\Omega)$.*

Recall β_0 and λ_0 as in Lemma 2.1. Fix $\beta \geq \beta_0$. For each $\lambda > \lambda_0$, we define the following Carleman weighted least squares functional

$$J^{(k)}(V) = \int_\Omega e^{2\lambda b^{-\beta} r^\beta(\mathbf{x})} \left|\Delta V - c(\mathbf{x})SV + \mathbf{F}(\mathbf{x}, U^{(k-1)}, \nabla U^{(k-1)})\right|^2 d\mathbf{x} \tag{3.13}$$

for $V \in H$. Then, we set

$$U^{(k)} = \mathrm{argmin}_{V \in H} J^{(k)}(V). \tag{3.14}$$

See Theorem 4.1 for the existence and uniqueness of the minimizer of $J^{(k)}$.

Recall that the function $U^{(k)}$ is called the "Carleman best fit" solution to (3.11) obtained by the Carleman quasi-reversibility method. The original quasi-reversibility method was introduced by Lattès and Lions in 1969, see [**22**]. We refer readers to [**27, 29, 35**] for using the quasi-reversibility method to solve a linear system of PDEs and [**8**] for a survey of the quasi-reversibility method. In the original quasi-reversibility method, the best fit solutions to (3.11) can be found by minimizing the least-squares functional. Here, we improve this method by imposing a Carleman weight function on the original least-squares functional. By "improve", we mean that the presence of the Carleman weight function plays a crucial role in the convergence analysis in the next section. More precisely, we will show that the sequence $\{U^{(k)}\}_{k\geq 0}$ converges to the true solution to (3.10) as k goes to ∞. The choice of the initial term does not matter.

We summary the procedure to solve Problem 1.1 in the Algorithm.

4. The main theorems

We establish two theorems in this section. The first one proves that the sequence $\{U_N^{(k)}\}_{k\geq 0}$, constructed in Section 3.2 and (3.14), is well-defined. The second one guarantees the convergence of $\{U_N^{(k)}\}_{k\geq 0}$.

THEOREM 4.1. *Fix $\beta \geq \beta_0$ and choose $\lambda > \lambda_0$, where β_0 and λ_0 are in Lemma 2.1 and Corollary 2.1, such that (2.2) holds true. For each $k \geq 1$, the functional $J^{(k)}$, defined in (3.13), has a unique minimizer in H.*

Algorithm 1 The procedure to solve Problem 1.1

1: Choose an orthonormal basis $(\Psi_n)_{n\geq 1}$ and a cut-off number N.
2: Compute the Cauchy data G and Q on $\partial\Omega$ as in (3.9).
3: Choose an arbitrary vector valued function $U^{(0)} \in H^2(\Omega)^N$.
4: By induction, we assume that we knew $U^{(k-1)}$ $k \geq 1$. Solve the system (3.11) by Carleman quasi-reversibility method for the vector valued function $U^{(k)} = \left(u_1^{(k)}, \ldots, u_m^{(k)}\right)^{\mathrm{T}}$.
5: Set the computed source function at step k is

$$p^{(k)}(\mathbf{x}) = u^{(k)}(\mathbf{x}, 0) = \sum_{n=1}^{N} u_n^{(k)}(\mathbf{x})\Psi_n(0)$$

6: Set the computed source function is $p_{\text{comp}} = p^{(k)}$ for $k = \overline{k}$ large enough.

PROOF. The proof of Theorem 4.1 is based on the Riesz Representation Theorem. Since the set of admissible solutions H is nonempty, we can find a vector valued function $\Phi = (\phi_1, \ldots, \phi_N)^{\mathrm{T}}$ in H. Define

$$H_0 = \{U - \Phi \in H^2(\Omega)^N : U \in H\} \tag{4.1}$$

It is obvious that for all $V = (v_1, \ldots, v_N)^{\mathrm{T}} \in H_0$, $V|_{\partial\Omega} = 0$ and $\partial_\nu V|_{\partial\Omega} = 0$. Using the trace theory, we can see that H_0 is a closed subspace of $H^2(\Omega)^N$.

Define the bounded bilinear form $a : H_0 \times H_0 \to \mathbb{R}$ and the bounded linear map $L : H_0 \to \mathbb{R}$ as

$$a(P, Q) = \int_\Omega e^{2\lambda b^{-\beta} r^\beta(\mathbf{x})}\big(\Delta P - c(\mathbf{x})SP\big) \cdot \big(\Delta Q - c(\mathbf{x})SQ\big)d\mathbf{x}$$

and

$$\begin{aligned} LQ = -\int_\Omega e^{2\lambda b^{-\beta} r^\beta(\mathbf{x})}\big(\Delta\Phi(\mathbf{x}) - c(\mathbf{x})S\Phi \\ + \mathbf{F}(\mathbf{x}, U^{(k-1)}, \nabla U^{(k-1)})\big) \cdot \big(\Delta Q(\mathbf{x}) - c(\mathbf{x})SQ\big)d\mathbf{x} \end{aligned} \tag{4.2}$$

for all $P, Q \in H_0$.

We will show that $a(P, Q)$ is an inner product on H_0 and equivalent to the standard norm in $H^2(\Omega)^N$. In fact, for all $P \in H_0$,

$$a(P, P) = \int_\Omega e^{2\lambda b^{-\beta} r^\beta(\mathbf{x})}|\Delta P - c(\mathbf{x})SP|^2 d\mathbf{x} \leq C\|P\|^2_{H^2(\Omega)^N}. \tag{4.3}$$

On the other hand, for all $P \in H_0$, using the inequality $(x - y)^2 \geq \frac{1}{2}x^2 - y^2$, we have

$$\begin{aligned} a(P, P) &= \int_\Omega e^{2\lambda b^{-\beta} r^\beta(\mathbf{x})}|\Delta P - c(\mathbf{x})SP|^2 d\mathbf{x} \\ &\geq \frac{1}{2}\int_\Omega e^{2\lambda b^{-\beta} r^\beta(\mathbf{x})}|\Delta P|^2 d\mathbf{x} - \int_\Omega e^{2\lambda b^{-\beta} r^\beta(\mathbf{x})}|c(\mathbf{x})SP|^2 d\mathbf{x}. \end{aligned} \tag{4.4}$$

Applying the Carleman estimate (2.2) for vector $P \in H_0$, we have

$$
\begin{aligned}
(4.5)\quad a(P,P) \geq C\lambda^{-1} \int_\Omega e^{2\lambda b^{-\beta} r^\beta(\mathbf{x})} |D^2 P|^2 d\mathbf{x} \\
+ C\lambda^3 \int_\Omega e^{2\lambda b^{-\beta} r^\beta} |P|^2 d\mathbf{x} + C\lambda \int_\Omega e^{2\lambda b^{-\beta} r^\beta(\mathbf{x})} |\nabla P|^2 d\mathbf{x} \\
- \int_\Omega e^{2\lambda b^{-\beta} r^\beta(\mathbf{x})} |c(\mathbf{x}) SP|^2 d\mathbf{x}.
\end{aligned}
$$

This estimate leads to

$$
a(P,P) \geq C\|P\|^2_{H^2(\Omega)^N}. \tag{4.6}
$$

Hence, due to (4.3) and (4.6), the bilinear form a defines an inner product on H_0, denoted by $a(\cdot,\cdot)$. This new inner product is equivalent to the traditional one of $H^2(\Omega)^N$.

Recall that L is a bounded linear map. Using the Riesz Representation Theorem for H_0 with the inner product $a(\cdot,\cdot)$, there exists a unique vector W_0 such that

$$
a(W_0, Q) = LQ \text{ for all } Q \in H_0 \tag{4.7}
$$

This means

$$
\begin{aligned}
(4.8)\quad &\int_\Omega e^{2\lambda b^{-\beta} r^\beta(\mathbf{x})} \big(\Delta W_0 - c(\mathbf{x}) S W_0\big)\big(\Delta Q - c(\mathbf{x}) SQ\big) d\mathbf{x} \\
&= -\int_\Omega e^{2\lambda b^{-\beta} r^\beta(\mathbf{x})} \big(\Delta \Phi - c(\mathbf{x}) S\Phi + \mathbf{F}(\mathbf{x}, U^{(k-1)}, \nabla U^{(k-1)})\big) \cdot \big(\Delta Q - c(\mathbf{x}) SQ\big) d\mathbf{x}
\end{aligned}
$$

for all $Q \in H_0$. It implies that

$$
\begin{aligned}
(4.9)\quad \int_\Omega e^{2\lambda b^{-\beta} r^\beta(\mathbf{x})} \big(\Delta W_0 - c(\mathbf{x}) S W_0 + \Delta\Phi - c(\mathbf{x}) S\Phi \\
+ \mathbf{F}(\mathbf{x}, U^{(k-1)}, \nabla U^{(k-1)})\big) \cdot \big(\Delta Q - c(\mathbf{x}) SQ\big) d\mathbf{x} = 0
\end{aligned}
$$

for all $Q \in H_0$.

Let $U_0 = W_0 + \Phi \in H$. It follows from (4.9) that

$$
\begin{aligned}
(4.10)\quad \int_\Omega e^{2\lambda b^{-\beta} r^\beta(\mathbf{x})} \big(\Delta U_0 - c(\mathbf{x}) S U_0 + \\
+ \mathbf{F}(\mathbf{x}, U^{(k-1)}, \nabla U^{(k-1)})\big) \cdot \big(\Delta Q - c(\mathbf{x}) SQ\big) d\mathbf{x} = 0
\end{aligned}
$$

for all $Q \in H_0$. We next claim that U_0 is a minimizer of the functional $J^{(k)}$ defined in (3.13). In fact, for all $U \in H$, define $h = U - U_0 \in H_0$, we have

$$
\begin{aligned}
J^{(k)}(U) - J^{(k)}(U_0) &= J^{(k)}(U_0 + h) - J^{(k)}(U_0) \\
&= \int_\Omega e^{2\lambda b^{-\beta} r^\beta(\mathbf{x})} \Big|\Delta(U_0 + h) - c(\mathbf{x}) S(U_0 + h) + \mathbf{F}\left(\mathbf{x}, U^{(k-1)}, \nabla U^{(k-1)}\right)\Big|^2 d\mathbf{x} \\
&\quad - \int_\Omega e^{2\lambda b^{-\beta} r^\beta(\mathbf{x})} \Big|\Delta U_0 - c(\mathbf{x}) S U_0 + \mathbf{F}\left(\mathbf{x}, U^{(k-1)}, \nabla U^{(k-1)}\right)\Big|^2 d\mathbf{x}
\end{aligned}
$$

Applying $(a+b)^2 = a^2 + 2ab + b^2$, we have

$$J^{(k)}(U) - J^{(k)}(U_0) = \int_\Omega e^{2\lambda b^{-\beta} r^\beta(\mathbf{x})} |\Delta h - c(\mathbf{x})Sh|^2 d\mathbf{x} \tag{4.11}$$

$$+ 2\int_\Omega e^{2\lambda b^{-\beta} r^\beta(\mathbf{x})} \Big(\Delta U_0 - c(\mathbf{x})SU_0 + \mathbf{F}(\mathbf{x}, U^{(k-1)}, \nabla U^{(k-1)})\Big)(\Delta h - c(\mathbf{x})Sh)d\mathbf{x}$$

It follows from (4.10) and (4.11) that

$$J^{(k)}(U) - J^{(k)}(U_0) = \int_\Omega e^{2\lambda b^{-\beta} r^\beta(\mathbf{x})} |\Delta h - c(\mathbf{x})Sh|^2 d\mathbf{x} \geq 0$$

for all $U \in H$. As a result, U_0 is a minimizer of $J^{(k)}$ in H. The uniqueness of the minimizer is due to the strict convexity of $J^{(k)}$ in H. □

We next rigorously prove that the sequence of vectors, $\{U^{(k)}\}_{k\geq 1}$, defined in (3.14) in Section 3.2, converges to the true solution to 3.10. We first consider the simple case when $\|F\|_{C^1(\mathbb{R},\mathbb{R}^d)} < \infty$. The case when $\|F\|_{C^1(\mathbb{R},\mathbb{R}^d)} = \infty$ will follow by using a truncation technique, see Remark 4.3.

THEOREM 4.2. *Let β_0 be as in Lemma 2.1. Fix $\beta \geq \beta_0$ and let λ_0 be the number as in Lemma 2.1 depending only on $\mathbf{x}_0$, Ω, d and β. For all $\lambda \geq \lambda_0$, define the sequence $\{U^{(k)}\}_{k\geq 0}$ as in (3.14) in Section 3.2 where $U^{(0)}$ is an arbitrary function in $H^2(\Omega)^N$. Assume that $\|F\|_{C^1(\mathbb{R},\mathbb{R}^d)} < \infty$. Assume further that the system (3.10) has a unique solution U^*. Then, we have*

(4.12)

$$\int_\Omega e^{2\lambda b^{-\beta} r^\beta(\mathbf{x})} \Big(\lambda^{-2}|D^2(U^{(k)} - U^*)|^2 + |\nabla(U^{(k)} - U^*)|^2 + |U^{(k)} - U^*|^2\Big) d\mathbf{x}$$
$$\leq \Big(\frac{C}{\lambda}\Big)^k \int_\Omega e^{2\lambda b^{-\beta} r^\beta(\mathbf{x})} \Big(\lambda^{-2}|D^2(U^{(0)} - U^*)|^2 + \big|\nabla(U^{(0)} - U^*)\big|^2 + \big|U^{(0)} - U^*\big|^2\Big),$$

for $k = 1, 2, \ldots$ where C is a constant depending only on $\mathbf{x}_0$, Ω, d, and $\|F\|_{C^1(\mathbb{R},\mathbb{R}^d)}$.

PROOF. In this proof, C denotes different positive constants that might change from estimate to estimate. Define

$$H_0 = \{V \in H^2(\Omega)^N : V|_{\partial\Omega} = 0 \text{ and } \partial_\nu V|_{\partial\Omega} = 0\}. \tag{4.13}$$

It is obvious H_0 a closed subspace of $H^2(\Omega)^N$. Since $U^{(k)}$ is the minimizer of $J^{(k)}$ in H, by the variational principle, the Fréchet derivative $DJ^{(k)}(U^{(k)})(h)$ vanishes for all $h \in H_0$. Hence, the following identity holds

$$\Big\langle e^{\lambda b^{-\beta} r^\beta(\mathbf{x})} \Big[\Delta U^{(k)} - c(\mathbf{x})SU^{(k)} + \mathbf{F}\Big(\mathbf{x}, U^{(k-1)}, \nabla U^{(k-1)}\Big)\Big], \tag{4.14}$$
$$e^{\lambda b^{-\beta} r^\beta(\mathbf{x})} \Big[\Delta h(\mathbf{x}) - c(\mathbf{x})Sh\Big]\Big\rangle_{L^2(\Omega)^N} = 0$$

for all $h \in H_0$. On the other hand, since U^* is the solution to system (3.10),

$$\Big\langle e^{\lambda b^{-\beta} r^\beta(\mathbf{x})} \Big[\Delta U^* - c(\mathbf{x})SU^* + \mathbf{F}(\mathbf{x}, U^*, \nabla U^*)\Big], \tag{4.15}$$
$$e^{\lambda b^{-\beta} r^\beta(\mathbf{x})} \Big[\Delta h - c(\mathbf{x})Sh\Big]\Big\rangle_{L^2(\Omega)^N} = 0$$

for all $h \in H_0$. Subtracting (4.15) from (4.14), we obtain

$$\begin{aligned}(4.16)\quad \Big\langle e^{\lambda b^{-\beta} r^\beta(\mathbf{x})}\big[\Delta(U^{(k)} - U^*) - c(\mathbf{x})S(U^{(k)} - U^*) + \mathbf{F}(\mathbf{x}, U^{(k-1)}, \nabla U^{(k-1)})\\ - \mathbf{F}(\mathbf{x}, U^*, \nabla U^*)\big], e^{\lambda b^{-\beta} r^\beta(\mathbf{x})}\big[\Delta h - c(\mathbf{x})Sh\big]\Big\rangle_{L^2(\Omega)^N} = 0\end{aligned}$$

for all $h \in H_0$. Using the test function $h = U^{(k)} - U^* \in H_0$ and Hölder's inequality, we obtain from identity (4.16) that

$$\begin{aligned}(4.17)\quad &\int_\Omega e^{2\lambda b^{-\beta} r^\beta(\mathbf{x})}|\Delta h - c(\mathbf{x})Sh|^2 d\mathbf{x}\\ &\leq \Big[\int_\Omega e^{2\lambda b^{-\beta} r^\beta(\mathbf{x})}|\mathbf{F}(\mathbf{x}, U^{(k-1)}, \nabla U^{(k-1)}) - \mathbf{F}(\mathbf{x}, U^*, \nabla U^*)|^2 d\mathbf{x}\Big]^{1/2}\\ &\qquad\times \Big[\int_\Omega e^{2\lambda b^{-\beta} r^\beta(\mathbf{x})}|\Delta h - c(\mathbf{x})Sh|^2 d\mathbf{x}\Big]^{1/2},\end{aligned}$$

which implies

$$\begin{aligned}(4.18)\quad &\int_\Omega e^{2\lambda b^{-\beta} r^\beta(\mathbf{x})}|\Delta h - c(\mathbf{x})Sh|^2 d\mathbf{x}\\ &\leq \int_\Omega e^{2\lambda b^{-\beta} r^\beta(\mathbf{x})}|\mathbf{F}(\mathbf{x}, U^{(k-1)}, \nabla U^{(k-1)}) - \mathbf{F}(\mathbf{x}, U^*, \nabla U^*)|^2 d\mathbf{x}.\end{aligned}$$

Since F has a finite C^1 norm, so does $\mathbf{F}$. Using the inequality $(a+b)^2 \leq 2(a^2+b^2)$, we obtain from (4.18) that

$$\begin{aligned}(4.19)\quad &\int_\Omega e^{2\lambda b^{-\beta} r^\beta(\mathbf{x})}|\Delta h - c(\mathbf{x})Sh|^2 d\mathbf{x}\\ &\leq C\int_\Omega e^{2\lambda b^{-\beta} r^\beta(\mathbf{x})}[|U^{(k-1)} - U^*|^2 + |\nabla(U^{(k-1)} - U^*)|^2] d\mathbf{x}.\end{aligned}$$

On the other hand, applying the inequality $(a-b)^2 \geq \frac{1}{2}a^2 - b^2$, we get

$$\begin{aligned}(4.20)\quad \int_\Omega e^{2\lambda b^{-\beta} r^\beta(\mathbf{x})}|\Delta h - c(\mathbf{x})Sh|^2 d\mathbf{x} \geq \frac{1}{2}\int_\Omega e^{2\lambda b^{-\beta} r^\beta(\mathbf{x})}\left|\Delta h\right|^2 d\mathbf{x}\\ - \int_\Omega e^{2\lambda b^{-\beta} r^\beta(\mathbf{x})}\left|c(\mathbf{x})Sh\right|^2 d\mathbf{x}.\end{aligned}$$

Using the Carleman estimate (2.2) for the function $h \in H_0$, we have

$$\begin{aligned}(4.21)\quad \int_\Omega e^{2\lambda b^{-\beta} r^\beta(\mathbf{x})}|\Delta h|^2 d\mathbf{x} \geq C\lambda^{-1}\int_\Omega e^{2\lambda b^{-\beta} r^\beta(\mathbf{x})}|D^2 h|^2 d\mathbf{x}\\ + C\lambda^3\int_\Omega e^{2\lambda b^{-\beta} r^\beta}|h|^2 d\mathbf{x} + C\lambda\int_\Omega e^{2\lambda b^{-\beta} r^\beta(\mathbf{x})}|\nabla h|^2 d\mathbf{x}.\end{aligned}$$

Combining (4.19), (4.20) and (4.21), we obtain

$$\begin{aligned}(4.22)\quad \lambda^{-1}\int_\Omega e^{2\lambda b^{-\beta} r^\beta(\mathbf{x})}|D^2 h|^2 d\mathbf{x} + \lambda\int_\Omega e^{2\lambda b^{-\beta} r^\beta}\Big(|h|^2 + |\nabla h|^2\Big) d\mathbf{x}\\ \leq C\int_\Omega e^{2\lambda b^{-\beta} r^\beta(\mathbf{x})}\Big(\left|U^{(k-1)} - U^*\right|^2 + \left|\nabla\left(U^{(k-1)} - U^*\right)\right|^2\Big).\end{aligned}$$

Multiply both sides of (4.22) with $\frac{1}{\lambda}$ and recall $h = U^{(k)} - U^*$. We have

(4.23)
$$\int_\Omega e^{2\lambda b^{-\beta} r^\beta(\mathbf{x})}\Big(\lambda^{-2}|D^2(U^{(k)} - U^*)|^2 + |\nabla(U^{(k)} - U^*)|^2 + |U^{(k)} - U^*|^2\Big) d\mathbf{x}$$
$$\leq \frac{C}{\lambda}\int_\Omega e^{2\lambda b^{-\beta} r^\beta(\mathbf{x})}\Big(\big|\nabla(U^{(k-1)} - U^*)\big|^2 + \big|U^{(k-1)} - U^*\big|^2\Big),$$

which implies

(4.24)
$$\int_\Omega e^{2\lambda b^{-\beta} r^\beta(\mathbf{x})}\Big(\lambda^{-2}|D^2(U^{(k)} - U^*)|^2 + |\nabla(U^{(k)} - U^*)|^2 + |U^{(k)} - U^*|^2\Big) d\mathbf{x}$$
$$\leq \frac{C}{\lambda}\int_\Omega e^{2\lambda b^{-\beta} r^\beta(\mathbf{x})}\Big(\lambda^{-2}|D^2(U^{(k-1)} - U^*)|^2 + \big|\nabla(U^{(k-1)} - U^*)\big|^2 + \big|U^{(k-1)} - U^*\big|^2\Big)$$

By induction, from (4.24), we have (4.12). □

COROLLARY 4.1. *Fix λ large enough such that $\theta = C/\lambda \in (0, 1)$ where C is the constant in (4.12). Then,*

(4.25)
$$\min_{\mathbf{x}\in\overline{\Omega}}\{e^{2\lambda b^{-\beta} r^\beta(\mathbf{x})}\}\int_\Omega\Big(\lambda^{-2}|D^2(U^{(k)} - U^*)|^2 + |\nabla(U^{(k)} - U^*)|^2 + |U^{(k)} - U^*|^2\Big) d\mathbf{x}$$
$$\leq \theta^k \max_{\mathbf{x}\in\overline{\Omega}}\{e^{2\lambda b^{-\beta} r^\beta(\mathbf{x})}\}\int_\Omega\Big(\lambda^{-2}|D^2(U^{(0)} - U^*)|^2 + \big|\nabla(U^{(0)} - U^*)\big|^2 + \big|U^{(0)} - U^*\big|^2\Big)$$

Inequality (4.25) rigorously guarantees that the sequence $\{U^{(k)}\}_{k\geq 1}$ converges to U^* in $H^2(\Omega)^N$. As a result, the sequence $\{p^{(k)}(\mathbf{x})\}_{k\geq 1}$ obtained in Step 5 of Algorithm 1 converges to the true source function $p^*(\mathbf{x})$ in $H^2(\Omega)^N$ where

$$p^*(\mathbf{x}) = \sum_{n=1}^N u_n^*(\mathbf{x})\Psi_n(0) \quad \textit{for all } \mathbf{x} \in \Omega.$$

REMARK 4.1. *One of our contributions in this paper is that we rigorously prove the convergence of the iterative scheme to the true solution of the nonlinear PDEs in $H^2(\Omega)^N$. It is an important improvement in comparison with the main theorem in* [**30, 33**] *which prove the convergence in H^1 only.*

REMARK 4.2. *With the term $\theta \in (0, 1)$, estimate (4.25) is similar to the one in the contraction mapping principle. This explains the title of the paper.*

REMARK 4.3. *In Theorem 4.2, we assume that the function F has a finite C^1 norm. However, in reality, $\|F\|_{C^1}$ might be infinity. We can extend Theorem 4.2 for the case when $\|F\|_{C^1} = \infty$. Recall from the statement of Problem 1.1 that $\|u^*\|_{C^1(\overline{Q_T})} < M$ for some known large number M. Combining with (3.2) and noting that $\|\Psi_n\|_{L^2(0,T)} = 1$, we have $\|u_m^*(\mathbf{x})\|_{C^1(\overline{\Omega})} \leq M\sqrt{T}$. Define the smooth function $\chi \in C^\infty(\mathbb{R} \times \mathbb{R}^d)$ as follows*

$$\chi(s, \mathbf{r}) = \begin{cases} 1 & \textit{if } |s| + |\mathbf{r}| < M\sqrt{T}, \\ \in (0, 1) & \textit{if } M\sqrt{T} \leq |s| + |\mathbf{r}| \leq 2M\sqrt{T}, \\ 0 & \textit{if } |s| + |\mathbf{r}| > 2M\sqrt{T}. \end{cases} \tag{4.26}$$

We then set $\widetilde{\mathbf{F}} = \chi\mathbf{F}$. *Since* $|u_m^*| + |\nabla u_m^*| < M$, *the vector* U^* *solves the following problem*

$$
\begin{cases} \Delta U(\mathbf{x}) - c(\mathbf{x})SU(\mathbf{x}) + \widetilde{\mathbf{F}}(\mathbf{x}, U(\mathbf{x}), \nabla U(\mathbf{x})) = 0 & \mathbf{x} \in \Omega, \\ U(\mathbf{x}) = G(\mathbf{x}) & \mathbf{x} \in \partial\Omega, \\ \partial_\nu U(\mathbf{x}) = Q(\mathbf{x}) & \mathbf{x} \in \partial\Omega. \end{cases} \tag{4.27}
$$

Thus, we can apply Algorithm 1 *for* (4.27) *to compute* $U^*(\mathbf{x})$ *and the source function* $p(\mathbf{x})$.

5. Numerical study

In this section, we perform some numerical results obtained by Algorithm 1. For simplicity purpose, we test our method in the 2-D case, i.e. d=2.

5.1. The forward problem. In order to generated computationally simulated data (1.2), we need to solve numerically the forward problem. Let $\widetilde{R} > R > 0$ be two positive numbers. We define the domains

$$\widetilde{\Omega} = (-\widetilde{R}, \widetilde{R})^2 \quad \text{and} \quad \Omega = (-R, R)^2.$$

We first solving the following problem defined in the bigger domain $\widetilde{\Omega} \times (0, T)$
(5.1)

$$
\begin{cases} c(\mathbf{x})u_t(\mathbf{x}, t) &= \Delta u(\mathbf{x}, t) + F\left(\mathbf{x}, t, u(\mathbf{x}, t), \nabla u(\mathbf{x}, t)\right) & \mathbf{x} \in \widetilde{\Omega}, t \in (0, T) \\ u(\mathbf{x}, 0) &= p(\mathbf{x}) & \mathbf{x} \in \widetilde{\Omega}, \\ u(\mathbf{x}, t) &= 0 & \mathbf{x} \in \partial\widetilde{\Omega}, t \in [\dashv 0, T]. \end{cases}
$$

Let $\mathbf{x} = (x, y) \in \widetilde{\Omega}$. In our numerical tests, the known coefficient function $c(x, y)$ is set as

$$
c(x, y) = 1 + \frac{1}{50}\Big(3(1 - x)^2 e^{-x^2 - (y+1)^2} - 10\left(\frac{x}{5} - x^3 - y^5\right) e^{-x^2 - y^2} - \frac{1}{3} e^{-(x+1)^2 - y^2}\Big). \tag{5.2}
$$

This function is a scale of the "peaks" function in Matlab. The values of coefficient function $c(x, y)$ vary in the range of $[0.8693, 1.1618]$. We solve (5.1) by the finite difference method using the explicit scheme. Afterward, we extract the data on the boundary of the domain Ω for the simulated data:

$$g(\mathbf{x}, t) = u(\mathbf{x}, t) \text{ and } q(\mathbf{x}, t) = \partial_\nu u(\mathbf{x}, t) \text{ for } \mathbf{x} \in \partial\Omega, t \in [0, T]$$

5.2. Implementation. On the domain $\overline{\Omega}$, we arrange an $N_{\mathbf{x}} \times N_{\mathbf{x}}$ uniform grid

$$\mathcal{G} = \left\{(x_i, y_j) : x_i = -R + (i - 1)h, y_j = -R + (j - 1)h, 1 \le i, j \le N_{\mathbf{x}}\right\}$$

where $N_{\mathbf{x}}$ is the number of grid points, $h = \dfrac{2R}{N_{\mathbf{x}} - 1}$ is the mesh spacing.

REMARK 5.1. *In our computations, we set* $R_1 = 6$, $R = 1$, $T = 1.5, N_{\mathbf{x}} = 80$.

In the following, we present the implementation of Algorithm 1 to solve Problem 1.1 .

Step 1. We employ the orthonormal basis $\{\Psi_n\}_{n \ge 1}$ as in (2.2). The cut-off number N in our approximation of $u(\mathbf{x}, t)$ in (3.3) is chosen as follows. We first choose a test (Test 1 in Section 5.3) as a reference test. Then, for each $N \geq 1$, we compute the error function

$$e_N(\mathbf{x}) = \left| u^*(\mathbf{x}, 0) - \sum_{n=1}^{N} u_n^*(\mathbf{x})\Psi_n(0) \right| \quad \mathbf{x} \in \Omega \tag{5.3}$$

and choose N such that $\|e_N\|_{L^\infty(\Omega)}$ is small enough. Figure 1 presents the values of the error e_N when $N = 15, 35$ and 40. It is obvious that increasing the value of N reduces the error. With $N = 40$, the error is small enough. We, therefore, choose $N = 40$ for all numerical tests.

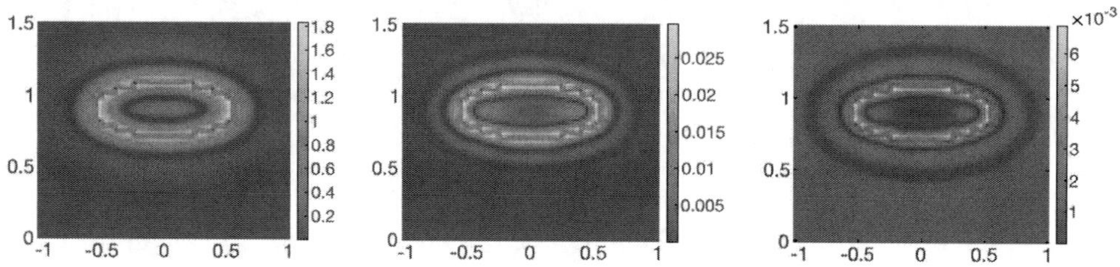

(A) $N = 15$ (B) $N = 35$ (C) $N = 40$

FIGURE 1. The function e_N defined in (5.3) where $u^*(\mathbf{x}, 0) = p(\mathbf{x})$ is the source given in Test 1. It is evident that the larger N, the better approximation in (3.3).

REMARK 5.2. *The basis $\{\Psi_n\}_{n\geq 1}$ in (2.2) was successfully used frequently in our research group. The reason for us to employ this basis rather than the well-known traditional "sine and cosine" basis of the Fourier series is that the first elements of the latter basis is a constant. Thus, $u_1(\mathbf{x})\Psi_1'(t)$ vanishes. Hence, we might lose some information of $u_1(\mathbf{x})$ in when plugging (3.4) into (1.1). This leads to a significant error in computation because the first term in the series $\sum_{n=1}^{\infty} u_n(\mathbf{x})\Psi_n(t)$ is the most important one.*

Step 2. We compute the "indirect data" $G(\mathbf{x})$ and $Q(\mathbf{x})$ on $\partial\Omega$ as in (3.9). Denote $G^*(\mathbf{x})$ and $Q^*(\mathbf{x})$ as the noiseless data. The noisy data is set as follows, for $\mathbf{x} \in \partial\Omega$

$$G^\delta(\mathbf{x}) = G^*(\mathbf{x})\,(1 + \delta\text{rand})\,, \quad Q^\delta(\mathbf{x}) = Q^*(\mathbf{x})\,(1 + \delta\text{rand})$$

for all $\mathbf{x} \in \partial\Omega$, where $\delta > 0$ represents for the noise level and "rand" is the function taking uniformly distributed random numbers in the range $[-1, 1]$. In our numerical tests, a significant noise with noise level $\delta = 20\%$ is involved in the data.

Step 3. For simplicity, we choose the initial solution $U^{(0)} = \overrightarrow{0}$.

Step 4. Given $U^{(k-1)}$ for $k \geq 1$, we minimize the functional $J^{(k)}$, defined in (3.13), by the optimization package of MATLAB with the command "lsqlin". The implementation for the quasi-reversibility method to minimize similar functionals was described in [**25**, §5.3] and in [**27**, §5]. We do not repeat this process here. We then set $U^{(k)} = \text{argmin}_{V\in H} J^{(k)}(V)$.

The scripts for other steps of Algorithm 1 can be written easily.

REMARK 5.3. *We choose λ, β in the Carleman Weight Function defined in* 2.1 *by a trial-error process that is similar to the one in* **[25]**. *Just as in* **[25]**, *we choose a reference numerical test in which we know the true solution (Test 1 in section* 5.3*). We then tried several values of λ, β until obtained a satisfactory numerical result for that reference test. Next, we have used the same values of these parameters for all other tests. In our computation, $\lambda = 40, \beta = 20$.*

5.3. Numerical results. In this section, we present three (3) numerical tests to verify the efficiency of Algorithm 1. The noise level of the data in these tests is 20%.

Test 1. In this test, the source function is defined as follows

$$p_{\text{true}} = \begin{cases} 8 & 0.2x^2 + (y-0.2)^2 < 0.25^2, \\ 0 & \text{otherwise.} \end{cases}$$

The function F has the form of

$$F(\mathbf{x}, t, s, \mathbf{r}) = s + \sqrt{|\mathbf{r}|} \text{ for all } \mathbf{x} \in \mathbb{R}^d, t \in [0, T], s \in \mathbb{R}, \mathbf{r} \in \mathbb{R}^d.$$

The true and the reconstructed source functions, p^* and p_{comp} respectively, are displayed in Figure 2. The true source includes an elliptic placed at point $(x, y) = (0, 0.2)$ with contrast 8, see 2a. It is evident that our numerical method can successfully reconstruct the source function with a fast convergence. In more details, Figure 2b shows the computed source function which clearly indicates the position of the ellipse. The maximal value of the inclusion is 8.5022 (relative error 6.28%). Although the contrast is high, our method provides a good approximation of the true source without requesting a good initial solution. Besides, Figure 2c illustrates the consecutive relative errors which are computed as follows

$$err_k = \frac{\|p^{(k)} - p^{(k-1)}\|_{L^\infty(\Omega)}}{\|p^{(k)}\|_{L^\infty(\Omega)}} \quad \text{for } k = 1, 2, \dots, 8$$

It is noticeable that our numerical method converges rapidly. The consecutive errors go to zero quickly. Actually, after only five (5) iterations, we can obtain a stable solution for the source function. This fact verifies numerically the rate of convergence estimate $O(\theta^n)$ as $n \to \infty$ in (4.12) and (4.25).

Test 2. In this test, the source function is defined as follows

$$p_{\text{true}} = \begin{cases} 6 & (x+0.5)^2 + (y+0.5)^2 < 0.35^2, \\ 8 & (x+0.5)^2 + (y-0.5)^2 < 0.35^2, \\ 10 & (x-0.5)^2 + (y+0.5)^2 < 0.35^2, \\ 0 & \text{otherwise.} \end{cases}$$

The function F has the form of

$$F(\mathbf{x}, t, s, \mathbf{r}) = s(1-s) + \frac{1}{2}\left(|r_1| - |r_2|\right) \text{ for all } \mathbf{x} \in \mathbb{R}^d, t \in [0, T], s \in \mathbb{R}, \mathbf{r} \in \mathbb{R}^d$$

where $\mathbf{r} = \begin{pmatrix} r_1 \\ r_2 \end{pmatrix}$.

The true and the reconstructed source functions, p^* and p_{comp} respectively, are displayed in Figure 3. The true source includes three ball with different contrasts, see 3a. The computed source function in Figure 3b clearly indicates the position of balls and it is a good approximation of the true one. The upper left ball has the true value 8 and the maximal computed one 8.3344 (relative error 4.18%). The lower left ball has the true value 6 and the maximal computed one 6.2892 (relative

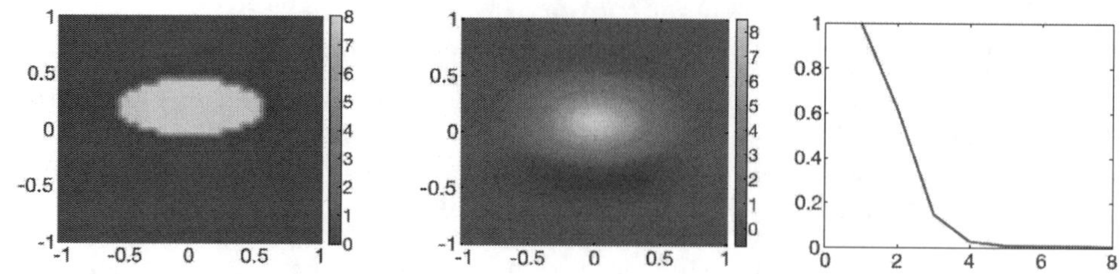

(A) (B) (C)

FIGURE 2. Test 1. The reconstruction of the source function. The initial solution of the source function $p^{(0)}(\mathbf{x}) = 0, \forall \mathbf{x} \in \Omega$. (a) The function p_{true}. (b) The computed source function p_{comp} obtained by Step 6 of Algorithm 1. (c) The consecutive relative error $\dfrac{\|p^{(k)} - p^{(k-1)}\|_{L^\infty(\Omega)}}{\|p^{(k)}\|_{L^\infty(\Omega)}}$ for $k = 1, 2, \ldots, 8$. The noise level of the data in this test is 20%.

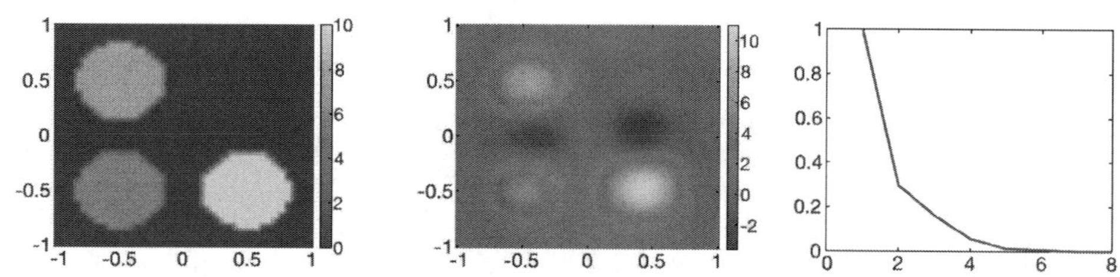

(A) (B) (C)

FIGURE 3. Test 2. The reconstruction of the source function. The initial solution of the source function $p^{(0)}(\mathbf{x}) = 0, \forall \mathbf{x} \in \Omega$. (a) The function p_{true}. (b) The computed source function p_{comp} obtained by Step 6 of Algorithm 1. (c) The consecutive relative error $\dfrac{\|p^{(k)} - p^{(k-1)}\|_{L^\infty(\Omega)}}{\|p^{(k)}\|_{L^\infty(\Omega)}}$ for $k = 1, 2, \ldots, 8$. The noise level of the data in this test is 20%.

error 4.82%). The lower right ball has the true value 10 and the maximal computed one 11.0032 (relative error 10.03%). Besides, the Figure 2c illustrates our numerical method converges rapidly after only five (5) iterations.

Test 3. In this test, the source function is defined as follows

$$p_{\text{true}} = \begin{cases} 1 & 0.4^2 < x^2 + y^2 < 0.8^2, \\ 0 & \text{otherwise.} \end{cases}$$

The function F has the form of

$$F(\mathbf{x}, t, s, \mathbf{r}) = \sqrt{r_1^2 + r_2^2 + 1} \text{ for all } \mathbf{x} \in \mathbb{R}^d, t \in [0, T], s \in \mathbb{R}, \mathbf{r} \in \mathbb{R}^d$$

where $\mathbf{r} = \begin{pmatrix} r_1 \\ r_2 \end{pmatrix}$.

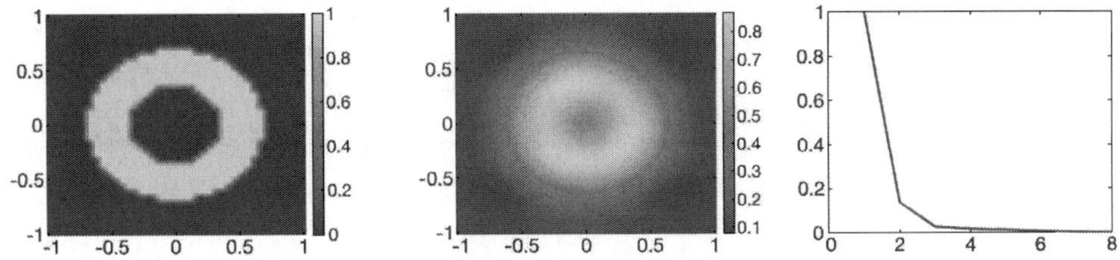

(A) (B) (C)

FIGURE 4. Test 3. The reconstruction of the source function. The initial solution of the source function $p^{(0)}(\mathbf{x}) = 0, \forall \mathbf{x} \in \Omega$. (a) The function p_{true}. (b) The computed source function p_{comp} obtained by Step 6 of Algorithm 1. (c) The consecutive relative error $\dfrac{\|p^{(k)} - p^{(k-1)}\|_{L^\infty(\Omega)}}{\|p^{(k)}\|_{L^\infty(\Omega)}}$ for $k = 1, 2, \ldots, 8$. The noise level of the data in this test is 20%.

The true and the reconstructed source functions, p^* and p_{comp} respectively, are displayed in Figure 4. The true source includes a ring with contrast 1, see 4a. The Figure 4b shows the computed source function which clearly indicates the position of the ring and the void inside. The maximal value of the ring is 0.9226 (relative error 7.74%). Besides, the consecutive relative errors in the Figure 2c illustrates the fast convergence of our numerical method. The convergence is obtained after seven (7) iterations.

6. Concluding remarks

In this paper, we introduce a new iterative method to solve an inverse source problem for a nonlinear parabolic equations. The first step of our numerical method is to derive a system of nonlinear elliptic equations whose solutions yields directly the solution for the inverse source problem. We then propose an iterative scheme to solve that nonlinear system. Our numerical method is global convergent in the sense that:

(i) It provides a good approximation to the true source function.
(ii) It does not require any knowledge of the true source function. This means a good initial guess is not necessary.

In our computation, we start our iterative scheme from the zero vector. The convergence of our scheme was proved rigorously. Numerical results are present to verify our theoretical part.

Acknowledgment. The author sincerely appreciates Dr. Loc H. Nguyen for many fruitful discussions that strongly improve the mathematical results and the presentation of this paper.

References

[1] A. El Badia and T. Ha-Duong, *On an inverse source problem for the heat equation. Application to a pollution detection problem*, J. Inverse Ill-Posed Probl. **10** (2002), no. 6, 585–599, DOI 10.1515/jiip.2002.10.6.585. MR1967440

[2] Anatoly B. Bakushinskii, Michael V. Klibanov, and Nikolaj A. Koshev, *Carleman weight functions for a globally convergent numerical method for ill-posed Cauchy problems for some quasilinear PDEs*, Nonlinear Anal. Real World Appl. **34** (2017), 201–224, DOI 10.1016/j.nonrwa.2016.08.008. MR3567958

[3] L. Beilina and M. V. Klibanov. *Approximate Global Convergence and Adaptivity for Coefficient Inverse Problems.* Springer, New York, 2012.

[4] A. L. Bukhgeĭm and M. V. Klibanov, *Uniqueness in the large of a class of multidimensional inverse problems* (Russian), Dokl. Akad. Nauk SSSR **260** (1981), no. 2, 269–272. MR630135

[5] W. De Tannenberg, *Sur les équations aux dérivées partielles du premier ordre à deux variables indépendantes, qui admettent un groupe continu de transformations* (French), Ann. Fac. Sci. Toulouse Sci. Math. Sci. Phys. **5** (1891), no. 1, B1–B40. MR1508114

[6] Michael V. Klibanov, *Global convexity in a three-dimensional inverse acoustic problem*, SIAM J. Math. Anal. **28** (1997), no. 6, 1371–1388, DOI 10.1137/S0036141096297364. MR1474219

[7] Michael V. Klibanov, *Estimates of initial conditions of parabolic equations and inequalities via lateral Cauchy data*, Inverse Problems **22** (2006), no. 2, 495–514, DOI 10.1088/0266-5611/22/2/007. MR2216411

[8] Michael V. Klibanov, *Carleman estimates for the regularization of ill-posed Cauchy problems*, Appl. Numer. Math. **94** (2015), 46–74, DOI 10.1016/j.apnum.2015.02.003. MR3342583

[9] Michael V. Klibanov, *Convexification of restricted Dirichlet-to-Neumann map*, J. Inverse Ill-Posed Probl. **25** (2017), no. 5, 669–685, DOI 10.1515/jiip-2017-0067. MR3707100

[10] Michael V. Klibanov and Olga V. Ioussoupova, *Uniform strict convexity of a cost functional for three-dimensional inverse scattering problem*, SIAM J. Math. Anal. **26** (1995), no. 1, 147–179, DOI 10.1137/S0036141093244039. MR1311886

[11] Michael V. Klibanov and Aleksandr E. Kolesov, *Convexification of a 3-D coefficient inverse scattering problem*, Comput. Math. Appl. **77** (2019), no. 6, 1681–1702, DOI 10.1016/j.camwa.2018.03.016. MR3926836

[12] Michael V. Klibanov, Aleksandr E. Kolesov, and Dinh-Liem Nguyen, *Convexification method for an inverse scattering problem and its performance for experimental backscatter data for buried targets*, SIAM J. Imaging Sci. **12** (2019), no. 1, 576–603, DOI 10.1137/18M1191658. MR3928344

[13] Michael V. Klibanov, Aleksandr E. Kolesov, Anders Sullivan, and Lam Nguyen, *A new version of the convexification method for a 1D coefficient inverse problem with experimental data*, Inverse Problems **34** (2018), no. 11, 115014, 29, DOI 10.1088/1361-6420/aadbc6. MR3863309

[14] Michael V. Klibanov, Thuy T. Le, and Loc H. Nguyen, *Numerical solution of a linearized travel time tomography problem with incomplete data*, SIAM J. Sci. Comput. **42** (2020), no. 5, B1173–B1192, DOI 10.1137/19M1299487. MR4152678

[15] Michael V. Klibanov, Loc H. Nguyen, Anders Sullivan, and Lam Nguyen, *A globally convergent numerical method for a 1-D inverse medium problem with experimental data*, Inverse Probl. Imaging **10** (2016), no. 4, 1057–1085, DOI 10.3934/ipi.2016032. MR3610751

[16] Michael V. Klibanov and Jingzhi Li, *Inverse problems and Carleman estimates—global uniqueness, global convergence and experimental data*, Inverse and Ill-posed Problems Series, vol. 63, De Gruyter, Berlin, [2021] ©2021, DOI 10.1515/9783110745481. MR4368863

[17] Michael V. Klibanov, Jingzhi Li, and Wenlong Zhang, *Convexification for the inversion of a time dependent wave front in a heterogeneous medium*, SIAM J. Appl. Math. **79** (2019), no. 5, 1722–1747, DOI 10.1137/18M1236034. MR4001767

[18] Michael Klibanov, Jingzhi Li, and Wenlong Zhang, *Convexification of electrical impedance tomography with restricted Dirichlet-to-Neumann map data*, Inverse Problems **35** (2019), no. 3, 035005, 33, DOI 10.1088/1361-6420/aafecd. MR3916465

[19] Michael V. Klibanov, Jingzhi Li, and Wenlong Zhang, *Convexification for an inverse parabolic problem*, Inverse Problems **36** (2020), no. 8, 085008, 32, DOI 10.1088/1361-6420/ab9893. MR4151409

[20] Micheal V. Klibanov and Loc H. Nguyen, *PDE-based numerical method for a limited angle X-ray tomography*, Inverse Problems **35** (2019), no. 4, 045009, 32, DOI 10.1088/1361-6420/ab0133. MR3939496

[21] Olga A. Ladyzhenskaya and Nina N. Ural'tseva, *Linear and quasilinear elliptic equations*, Academic Press, New York-London, 1968. Translated from the Russian by Scripta Technica, Inc; Translation editor: Leon Ehrenpreis. MR0244627

[22] R. Lattès and J.-L. Lions, *The method of quasi-reversibility. Applications to partial differential equations*, Modern Analytic and Computational Methods in Science and Mathematics, No. 18, American Elsevier Publishing Co., Inc., New York, 1969. Translated from the French edition and edited by Richard Bellman. MR0243746

[23] M. M. Lavrent′ev, V. G. Romanov, and S. P. Shishat·skiĭ, *Ill-posed problems of mathematical physics and analysis*, Translations of Mathematical Monographs, vol. 64, American Mathematical Society, Providence, RI, 1986. Translated from the Russian by J. R. Schulenberger; Translation edited by Lev J. Leifman. MR847715

[24] Thuy T. Le, Michael V. Klibanov, Loc H. Nguyen, Anders Sullivan, and Lam Nguyen, *Carleman contraction mapping for a 1D inverse scattering problem with experimental time-dependent data*, Inverse Problems **38** (2022), no. 4, Paper No. 045002, 31, DOI 10.1088/1361-6420/ac50b8. MR4399523

[25] Thuy Thi Thu Le and Loc Hoang Nguyen, *A convergent numerical method to recover the initial condition of nonlinear parabolic equations from lateral Cauchy data*, J. Inverse Ill-Posed Probl. **30** (2022), no. 2, 265–286, DOI 10.1515/jiip-2020-0028. MR4401759

[26] Thuy T. Le and Loc H. Nguyen, *The gradient descent method for the convexification to solve boundary value problems of quasi-linear PDEs and a coefficient inverse problem*, J. Sci. Comput. **91** (2022), no. 3, Paper No. 74, 23, DOI 10.1007/s10915-022-01846-3. MR4414399

[27] Thuy T. Le, Loc H. Nguyen, Thi-Phong Nguyen, and William Powell, *The quasi-reversibility method to numerically solve an inverse source problem for hyperbolic equations*, J. Sci. Comput. **87** (2021), no. 3, Paper No. 90, 23, DOI 10.1007/s10915-021-01501-3. MR4254715

[28] Jingzhi Li, Masahiro Yamamoto, and Jun Zou, *Conditional stability and numerical reconstruction of initial temperature*, Commun. Pure Appl. Anal. **8** (2009), no. 1, 361–382, DOI 10.3934/cpaa.2009.8.361. MR2449114

[29] Qitong Li and Loc Hoang Nguyen, *Recovering the initial condition of parabolic equations from lateral Cauchy data via the quasi-reversibility method*, Inverse Probl. Sci. Eng. **28** (2020), no. 4, 580–598, DOI 10.1080/17415977.2019.1643850. MR4076242

[30] D-L. Nguyen, L. H. Nguyen, and T. Truong. The Carleman-based contraction principle to reconstruct the potential of nonlinear hyperbolic equations. *preprint arXiv:arXiv:2204.06060*, 2022.

[31] Hoai-Minh Nguyen and Loc Hoang Nguyen, *Cloaking using complementary media for the Helmholtz equation and a three spheres inequality for second order elliptic equations*, Trans. Amer. Math. Soc. Ser. B **2** (2015), 93–112, DOI 10.1090/btran/7. MR3418646

[32] Loc Hoang Nguyen, *A new algorithm to determine the creation or depletion term of parabolic equations from boundary measurements*, Comput. Math. Appl. **80** (2020), no. 10, 2135–2149, DOI 10.1016/j.camwa.2020.09.010. MR4154907

[33] L. H. Nguyen. The Carleman-contraction method to solve quasi-linear elliptic equations. *arXiv preprint arXiv:2203.12694*, 2022.

[34] Loc H. Nguyen and Michael V. Klibanov, *Carleman estimates and the contraction principle for an inverse source problem for nonlinear hyperbolic equations*, Inverse Problems **38** (2022), no. 3, Paper No. 035009, 19, DOI 10.1088/1361-6420/ac4d09. MR4385429

[35] Loc H. Nguyen, Qitong Li, and Michael V. Klibanov, *A convergent numerical method for a multi-frequency inverse source problem in inhomogenous media*, Inverse Probl. Imaging **13** (2019), no. 5, 1067–1094, DOI 10.3934/ipi.2019048. MR4027044

[36] M. H. Protter, *Unique continuation for elliptic equations*, Trans. Amer. Math. Soc. **95** (1960), 81–91, DOI 10.2307/1993331. MR113030

[37] Alexey V. Smirnov, Michael V. Klibanov, and Loc H. Nguyen, *On an inverse source problem for the full radiative transfer equation with incomplete data*, SIAM J. Sci. Comput. **41** (2019), no. 5, B929–B952, DOI 10.1137/19M1253605. MR4002730

[38] Vo Anh Khoa, Grant W. Bidney, Michael V. Klibanov, Loc H. Nguyen, Lam H. Nguyen, Anders J. Sullivan, and Vasily N. Astratov, *Convexification and experimental data for a 3D inverse scattering problem with the moving point source*, Inverse Problems **36** (2020), no. 8, 085007, 34, DOI 10.1088/1361-6420/ab95aa. MR4151408

[39] Vo Anh Khoa, Grant W. Bidney, Michael V. Klibanov, Loc H. Nguyen, Lam H. Nguyen, Anders J. Sullivan, and Vasily N. Astratov, *An inverse problem of a simultaneous reconstruction of the dielectric constant and conductivity from experimental backscattering data*, Inverse Probl. Sci. Eng. **29** (2021), no. 5, 712–735, DOI 10.1080/17415977.2020.1802447. MR4258641

[40] Vo Anh Khoa, Michael Victor Klibanov, and Loc Hoang Nguyen, *Convexification for a three-dimensional inverse scattering problem with the moving point source*, SIAM J. Imaging Sci. **13** (2020), no. 2, 871–904, DOI 10.1137/19M1303101. MR4097812

Department of Mathematics and Statistics, University of North Carolina at Charlotte, Charlotte, North Carolina 28223

Email address: tle55@uncc.edu

Contemporary Mathematics
Volume **784**, 2023
https://doi.org/10.1090/conm/784/15752

Regularization of the factorization method with applications to inverse scattering

Isaac Harris

ABSTRACT. Here we discuss a regularized version of the factorization method for positive operators acting on a Hilbert Space. The factorization method is a *qualitative* reconstruction method that has been used to solve many inverse shape problems. In general, qualitative methods seek to reconstruct the shape of an unknown object using little to no a priori information. The regularized factorization method presented here seeks to avoid numerical instabilities in the inversion algorithm. This allows one to recover unknown structures in a computationally simple and analytically rigorous way. We will discuss the theory and application of the regularized factorization method to examples coming from acoustic inverse scattering. Numerical examples will also be presented using synthetic data to show the applicability of the method.

1. Introduction

In this paper, we will discuss a regularized version of the factorization method as well as its applications scattering theory. We will briefly review the theoretical framework that was developed in [**14**]. The factorization method (see for e.g. [**6, 11, 12, 18, 19, 21, 23, 25**]) is a method used to solve inverse shape problems and fall under the category of *qualitative methods*. Qualitative methods are otherwise referred to as non-iterative or direct methods. In many applications it is optimal to use qualitative methods rather than applying non-linear optimization techniques for two reasons: first is that optimization methods require a priori information (to construct an initial guess) that may not be readily available such as the number of regions to be recovered, second is that these methods can be computationally expensive and highly ill-conditioned. All qualitative methods seek to recover the shape of an unknown region from little a priori information by relating the support of the region to the range of the 'measured' data operator. These methods where first introduced in [**9**] and are frequently used in non-destructive testing where one is given measurements on the surface (or exterior) of an object and one tries to reconstruct interior structures. This has many applications in the area of medical imaging and non-destructive testing in engineering.

The factorization method solves the inverse shape problem by appealing to Picard's criteria for compact operators. To this end, a range test is used to determine

---

2020 *Mathematics Subject Classification.* Primary 35J05, 35Q81, 46C07.

The research of I. Harris is partially supported by the NSF DMS Grant 2107891.

the support of the unknown region denoted D. In general, we have that

$$z \in D \iff \ell_z \in \text{Range}\big(A^{1/2}\big)$$

where ℓ_z is known and the positive compact operator A is given by the measurements. In order to apply Picard's criteria to reconstruct the unknown region a series is computed where one divides by the sequence of eigenvalues (or singular values) of the compact operator A. Since this sequence tends to zero (usually rapidly) this could result in numerical instabilities. Therefore, we will develop a regularization strategy for the factorization method motivated by the previous works in [**1–3, 16, 26**]. In [**1, 2**] the linear sampling method was studied by appealing to the analytical techniques in the factorization method and then applying a suitable regularization strategy. Whereas in [**3**] a new qualitative method known as the generalized linear sampling method was developed and uses a specific cost-functional for the regularization scheme in applying the linear sampling method. What we present here is mainly influenced by [**16, 26**]. Loosely speaking, we have the result that for a positive compact operator $A : X \to X^*$ where X is a Hilbert space and X^* is the corresponding dual-space then

$$\ell \in \text{Range}\big(A^{1/2}\big) \iff \liminf_{\alpha \to 0} \langle x_\alpha \, , A x_\alpha \rangle_{X \times X^*} < \infty$$

where x_α (defined below) is the regularized solution to $Ax = \ell$. The regularization scheme that is used to compute x_α can be taken to be any of the standard techniques i.e. Tikhonov regularization, Spectral cutoff and Landweber iteration. Here $\langle \cdot \, , \cdot \rangle_{X \times X^*}$ is the sesquilinear dual-pairing between X and X^*. The main analytical tool one needs to prove this result is the spectral decomposition for the given positive compact operator $A : X \to X^*$.

The preceding sections are organized as follows. First, we will briefly discuss the theory behind the regularized version of the factorization method for a positive compact operator $A : X \to X^*$ where X is a Hilbert space. The analysis presented here was initially studied in [**26**]. Then, we consider two inverse shape problems coming from inverse scattering. First, we will consider the problem of recovering an isotropic scatterer from far-field measurements. Lastly, we will consider the problem of recovering a sound soft scatterer from near-field measurements.

2. Regularized Factorization Method

In this section, we will discuss the theoretical framework that was developed in [**14**] for the regularized factorization method. The analysis here generalizes the main result in [**16**]. To begin, we assume that we have a given data operator denoted by $A : X \to X^*$ acting on the Hilbert space X that is positive and compact. Again, we note that here we take the notation that X^* denotes the dual space of X as well as $\langle \cdot \, , \cdot \rangle_{X \times X^*}$ denoting the sesquilinear dual-product between X and X^*. Furthermore, assume that there is a separable Hilbert pivoting space H with dense inclusions $X \subseteq H \subseteq X^*$ i.e. a Gelfand triple of Hilbert spaces.

Now in [**14**] it is proven that the operator A has a spectral decomposition provided that either X is a complex Hilbert space or the bilinear form

$$(x, y) \longmapsto \langle y \, , Ax \rangle_{X \times X^*} \quad \text{for any} \quad x, y \in X$$

is symmetric. Under these assumptions we have that

$$Ax = \sum \lambda_n (x, x_n)_X \, \ell_n \quad \text{or} \quad Ax = \sum \lambda_n \langle x, \ell_n \rangle_{X \times X^*} \, \ell_n \tag{2.1}$$

for any $x \in X$. Here the decreasing sequence $\lambda_n \in \mathbb{R}_{>0}$ converges to zero whereas $\{x_n\}_{n\in\mathbb{N}}$ is an orthonormal basis of X and $\{\ell_n\}_{n\in\mathbb{N}}$ is an orthonormal basis of X^*. Moreover, $\{\ell_n\}_{n\in\mathbb{N}}$ is the corresponding dual-basis for $\{x_n\}_{n\in\mathbb{N}}$ such that

$$\langle x_m \, , \ell_n \rangle_{X\times X^*} = \delta_{mn} \quad \text{for any} \quad n, m \in \mathbb{N}.$$

This gives that $\{\lambda_n; x_n; \ell_n\} \in \mathbb{R}_{>0} \times X \times X^*$ is the singular value decomposition of the compact operator A. Now just as in [**24**] we can define the regularized solution of $Ax = \ell$ to be x_α which is given by

$$x_\alpha = \sum \frac{\phi(\lambda_n;\alpha)}{\lambda_n} \overline{\langle x_n, \ell \rangle}_{X\times X^*} \, x_n. \tag{2.2}$$

The real-valued function $\phi(t;\alpha)$ denotes the filter associated with a given regularization technique. Here we will assume that $\phi(t;\alpha) : \big(0, \lambda_1\big] \to \mathbb{R}_{\geq 0}$ satisfies that for all $0 < t \leq \lambda_1$

$$\lim_{\alpha\to 0} \phi(t;\alpha) = 1 \quad \text{and} \quad \phi(t;\alpha) \leq C_{\text{reg}} \quad \text{for all } \alpha > 0.$$

Now, due to the fact that A is positive and compact we have that there is a bounded linear 'square root' operator denoted $Q : X \to H$ such that $A = Q^*Q$ where the adjoint Q^* is defined by

$$(Qx, h)_H = \langle x, Q^*h \rangle_{X\times X^*} \quad \text{for all} \quad h \in H \text{ and } x \in X$$

see Theorem 2.2 of [**14**] for details. We also obtain that

$$\ell \in \text{Range}(Q^*) \iff \sum \frac{1}{\lambda_n} \left|\langle x_n, \ell \rangle_{X\times X^*}\right|^2 < \infty$$

from Theorem 2.2 of [**14**] which is proven in a similar as was Picard's Criteria (see for e.g. Theorem 1.28 of [**5**]). Then by appealing to the properties of the filter function $\phi(t;\alpha)$ it can be shown that

$$\ell \in \text{Range}(Q^*) \iff \liminf_{\alpha\to 0} \langle x_\alpha \, , Ax_\alpha \rangle_{X\times X^*} < \infty.$$

This can be done by using the fact that

$$\langle x_\alpha \, , Ax_\alpha \rangle_{X\times X^*} = \sum \frac{\phi^2(\lambda_n;\alpha)}{\lambda_n} |\langle x_n, \ell \rangle_{X\times X^*}|^2 \tag{2.3}$$

along with some simple estimates of the above quantity using the properties of the filter function. Some common filter functions are given by

$$\phi(t;\alpha) = \frac{t^2}{t^2+\alpha}, \quad \phi(t;\alpha) = 1 - \big(1-\beta t^2\big)^{1/\alpha} \quad \text{and} \quad \phi(t;\alpha) = \begin{cases} 1, & t^2 \geq \alpha, \\ 0, & t^2 < \alpha \end{cases} \tag{2.4}$$

which corresponds to Tikhonov regularization, Landweber iteration (with $\alpha = 1/m$ for some $m \in \mathbb{N}$ and constant $\beta < 1/\lambda_1^2$) and the Spectral cutoff respectively. It is clear that these filter functions satisfy the above constraints (see for e.g. [**20**]).

Note, that the classical factorization method (i.e. without regularization) is given by using (2.3) with $\alpha = 0$. Formally, this would imply that $\phi(t;0) = 1$ and therefore one would be dividing by the singular values λ_n. This is not numerically stable since $\lambda_n \to 0$ as $n \to \infty$. This will be seen in one of our numerical examples provided in a later section.

Now, assume that the operator $A : X \to X^*$ also has the following factorization

$$A = S^*TS \quad \text{where} \quad S : X \to V \quad \text{and} \quad T : V \to V^*$$

with V also being a Hilbert space. Here the adjoint operator $S^* : V^* \to X^*$ is given by

$$\langle Sx, v\rangle_{V\times V^*} = \langle x, S^*v\rangle_{X\times X^*} \quad \text{for all} \quad v \in V \text{ and } x \in X. \tag{2.5}$$

Furthermore, we assume T is bounded and strictly coercive on Range(S) i.e.

$$\beta\|Sx\|^2_V \leq \langle Sx, TSx\rangle_{V\times V^*} \quad \text{for all } x \in X.$$

If we assume that S is a compact and injective then we have that $A : X \to X^*$ is positive and compact. From the previous discussion, this implies that $A = Q^*Q$ where Q is the 'square root' of the operator. Notice, that we have the estimate

$$\beta\|Sx\|^2_V \leq \|Qx\|^2_H = \langle x\, , Ax\rangle_{X\times X^*} = \langle Sx, TSx\rangle_{V\times V^*} \leq \|T\|_{V\to V^*}\|Sx\|^2_V$$

for all $x \in X$ by appealing to the boundedness and coercivity of the operator T. We can conclude, by Theorem 1 of [**10**] that Range$\big(Q^*\big)$ = Range(S^*). Putting everything together we have the following result.

THEOREM 2.1. *Let $A : X \to X^*$ have the factorization $A = S^*TS$ such that $S : X \to V$ and $T : V \to V^*$ are bounded linear operators where X and V are Hilbert spaces. Assume that S is compact and injective as well as T being strictly coercive on Range(S). Then we have that*

$$\ell \in \textit{Range}(S^*) \iff \liminf_{\alpha\to 0}\langle x_\alpha\, , Ax_\alpha\rangle_{X\times X^*} < \infty$$

where x_α is the regularized solution given by (2.2) to $Ax = \ell$.

PROOF. For details of the proof see [**14**]. □

Notice, that the result in Theorem 2.1 can be reformulated using the spectral decomposition of A such that

$$\ell \in \text{Range}(S^*) \iff \liminf_{\alpha\to 0}\sum \frac{\phi^2(\lambda_n;\alpha)}{\lambda_n}|(\ell_n,\ell)_{X^*}|^2 < \infty \tag{2.6}$$

where we have used (2.3). Here $\phi(t;\alpha)$ again denotes the filter function used to find the regularized solution to $Ax = \ell$. From this we have related the range of S^* to the spectral decomposition of A just as in the traditional factorization method but we have a regularization step. This allows one to have a rigorous range characterization without having unstable numerical reconstructions due to the fact that λ_n tend to zero rapidly. In the preceding section we will see how to apply Theorem 2.1 to inverse shape problems coming from inverse scattering. We note that this method has been used (without proof) for numerical examples in [**7, 28**] for recovering scatterers with the classical factorization method.

3. Applications to Inverse Scattering

In this section, we will see how the theory developed in section 2 can be applied to solving inverse shape problems. Here we are interested in problems coming from the area of inverse scattering. This comes up in many areas of engineering and medical imaging. The goal is to recover the shape of an object using the measured scattering data with little to no a priori information about the object. The scatterer will be illuminated by an incident wave and we will show how to recover the scatterer from the measured scattering data using the regularized factorization method.

3.1. An example with far-field measurements. We will now consider the inverse shape problem of reconstructing an isotropic scatterer using far-field measurements. This problem has been studied by many researchers with many interesting reconstruction methods for e.g. [**3**, **4**, **22**]. Here we will apply Theorem 2.1 to solve the inverse shape problem as well as provide some numerical examples. The classical factorization method was studied for this problem in [**22**].

To begin, let $D \subset \mathbb{R}^m$ (for $m = 2$ or 3) denote the unknown inhomogeneous isotropic scattering region with Lipschitz boundary. Here we assume that the incident plane wave given by $u^i(x) = \mathrm{e}^{\mathrm{i}kx\cdot d}$ is used to illuminate the scatterer where $d \in \mathbb{S} =$ unit sphere/circle. The parameter $k > 0$ denotes that wave number. The incident plane wave's interaction with the scatterer D results in the radiating scattered field $u^s(x, d)$ that satisfies

$$\Delta u^s + k^2(1+q)u^s = -k^2 q u^i \quad \text{in } \mathbb{R}^m \tag{3.1}$$

$$\partial_r u^s - \mathrm{i}k u^s = \mathcal{O}\big(r^{-(m+1)/2}\big) \quad \text{as } r \to \infty. \tag{3.2}$$

Here (3.2) is the Sommerfeld radiation condition and is assumed to hold uniformly with respect to the angular direction(s) with $r = |x|$. The contrast $q \in L^\infty(\mathbb{R}^m)$ defines the deviation in the refractive index from the background. Therefore, we let $n(x) = 1 + q(x)$ denote the refractive index which is the material parameter with the presence of the scatterer such that $\mathrm{supp}(q) = D$.

Assuming that there is a constant $q_{\min}$ where

$$\Re(q) \geq q_{\min} > 0 \quad \text{and} \quad \Im(q) \geq 0 \quad \text{for a.e. } x \in D$$

then the analysis in chapter 8 of [**8**] implies that (3.1)–(3.2) has a unique solution $u^s \in H^1_{loc}(\mathbb{R}^m)$. Since u^s is a radiating solution to Helmholtz equation in $\mathbb{R}^m \setminus \overline{D}$ we have the expansion

$$u^s(x, d) = \gamma \frac{\mathrm{e}^{\mathrm{i}kr}}{r^{(m-1)/2}} \left\{ u^\infty(\hat{x}, d) + \mathcal{O}\left(\frac{1}{r}\right) \right\} \quad \text{as } r \to \infty$$

where $u^\infty(\hat{x}, d)$ denotes the corresponding *far-field pattern* for (3.1)–(3.2). This quantity depending on the incident direction d and the measurement direction $\hat{x} = x/r$. The constant

$$\gamma = \frac{\mathrm{e}^{\mathrm{i}\pi/4}}{\sqrt{8\pi k}} \text{ in } \mathbb{R}^2 \quad \text{and} \quad \gamma = \frac{1}{4\pi} \text{ in } \mathbb{R}^3.$$

For this model we will assume that the far-field pattern is measured from the scattered field far away from the scatterer D. This implies that we have access to the measured far-field operator

$$F : L^2(\mathbb{S}) \longrightarrow L^2(\mathbb{S}) \quad \text{such that} \quad (Fg)(\hat{x}) = \int_{\mathbb{S}} u^\infty(\hat{x}, d) g(d) \, \mathrm{d}s(d). \tag{3.3}$$

In order to solve the inverse shape problem of recovering D from the knowledge of F we will appeal to Theorem 2.1 along with the factorization analysis in [**22**].

We now derive and use the factorization the far-field operator F to solve the inverse problem. For this, motivated by (3.1)–(3.2) we consider the problem

$$\Delta w + k^2(1+q)w = -k^2 q f \quad \text{in} \quad \mathbb{R}^m$$

along with (3.2) for any $f \in L^2(D)$. Therefore, we note that $w \in H^1_{loc}(\mathbb{R}^m)$ satisfies (see for e.g. [**8**])

$$w(x) = k^2 \int_D q(y)\Phi(x,y)\big[w(y) + f(y)\big]\,\mathrm{d}y \quad \text{for any} \quad x \in \mathbb{R}^m. \tag{3.4}$$

Here Φ denotes the fundamental solution for Helmholtz equation given by

$$\Phi(\cdot\,,y) = \begin{cases} \frac{\mathrm{i}}{4} H_0^{(1)}(k|\cdot - y|) & \text{for} \quad m = 2, \\ \dfrac{\mathrm{e}^{\mathrm{i}k|\cdot - y|}}{4\pi|\cdot - y|} & \text{for} \quad m = 3 \end{cases} \tag{3.5}$$

where $H_0^{(1)}$ is the first kind Hankel function of order zero. Using the fact that

$$\Phi(x,y) = \gamma \frac{\mathrm{e}^{\mathrm{i}k|x|}}{|x|^{(m-1)/2}} \left\{ \mathrm{e}^{-\mathrm{i}k\hat{x}\cdot y} + \mathcal{O}\left(\frac{1}{|x|}\right)\right\} \quad \text{as} \quad |x| \to \infty$$

from (3.4) we can conclude that the far-field pattern for w is given by

$$w^\infty(\hat{x}) = k^2 \int_D q(y)\mathrm{e}^{-\mathrm{i}k\hat{x}\cdot y}\big[w(y) + f(y)\big]\,\mathrm{d}y. \tag{3.6}$$

From this, we define the operator

$$H : L^2(\mathbb{S}) \longrightarrow L^2(D) \quad \text{such that} \quad Hg = \int_{\mathbb{S}} \mathrm{e}^{\mathrm{i}ky\cdot d} g(d)\mathrm{d}s(d)\Big|_D$$

and its adjoint

$$H^* : L^2(D) \longrightarrow L^2(\mathbb{S}) \quad \text{such that} \quad H^*\varphi = \int_D \mathrm{e}^{-\mathrm{i}k\hat{x}\cdot y}\varphi(y)\mathrm{d}y$$

for any $g \in L^2(\mathbb{S})$ and $\varphi \in L^2(D)$. Lastly, we define the bounded linear operator $T : L^2(D) \to L^2(D)$ such that

$$Tf = k^2 q[w + f]\big|_D \quad \text{for any} \quad f \in L^2(D).$$

It is well-known that F corresponds to the far-field pattern when u^i is replaced by Hg. The representation (3.6) implies that

$$Fg = k^2 \int_D q(y)\mathrm{e}^{-\mathrm{i}k\hat{x}\cdot y}\big[w_g(y) + (Hg)(y)\big]\,\mathrm{d}y$$

where w_g is the scattered field for $f = Hg$. We now have the factorization

$$F = H^*TH$$

where the operators T and H are as defined above.

Notice, that we have a symmetric factorization of the far-field operator that is needed to apply the theory in section 2. We now need that the operators used in the factorization do indeed satisfy the assumptions of Theorem 2.1. To this end, it is well known that H is is compact and injective as well as

$$\ell_z = \mathrm{e}^{-\mathrm{i}k\hat{x}\cdot z} \quad \text{satisfies} \quad \ell_z \in \text{Range}(H^*) \iff z \in D.$$

The last piece of the puzzle is the coercivity of the middle operator. As it stands, the middle operator T is not strictly coercive on the range of H. In order to solve this problem we consider the operator

$$F_\sharp = \big|\Re(F)\big| + \big|\Im(F)\big|.$$

where

$$\Re(F) = \frac{1}{2}(F + F^*) \quad \text{and} \quad \Im(F) = \frac{1}{2\mathrm{i}}(F - F^*).$$

Note, that $\Re(F)$ and $\Im(F)$ are self-adjoint compact operators by definition which implies that the absolute value can be compute via the spectral decomposition i.e. the Hilbert-Schmidt Theorem. From the analysis of the operator T (see for e.g. chapter 4 [**25**] for details) we have that

$$F_\sharp = H^* T_\sharp H$$

where the new operator $T_\sharp$ is strictly coercive on $L^2(D)$. Therefore, by appealing to Theorem 2.1 we have that

$$z \in D \iff \liminf_{\alpha \to 0} (g_z^\alpha, F_\sharp g_z^\alpha)_{L^2(\mathbb{S})} < \infty \tag{3.7}$$

provided that g_z^α is the regularized solution to $F_\sharp g = \ell_z$.

Numerical examples: We now give some numerical reconstructions using (3.7) in two dimensions. To this end, we assume that the scatterer has small area. Then, we can exploit the Born approximation for the scattered field to simplify the calculations of the synthetic data. Therefore, we will compute the synthetic far-field data using the approximation

$$u^\infty(\hat{x}, d) \approx k^2 \int_D q(y) \mathrm{e}^{-\mathrm{i}ky\cdot(\hat{x}-d)}\, \mathrm{d}y.$$

In all of the preceding examples we will take a constant contrast in the scatterer as well as a fixed wave number given by $q = 1 + \mathrm{i}$ and $k = 4$, respectively. We let the boundary of the scatterer to be given by

$$\partial D = r(\theta)\, (\cos(\theta), \sin(\theta)) \quad \text{for} \quad 0 \leq \theta \leq 2\pi.$$

Here the radial function $r(\theta)$ is given by either

$$r(\theta) = 0.5\left(|\sin(\theta)|^{10} + 0.1|\cos(\theta)|^{10}\right)^{-1/10} \quad \text{or} \quad r(\theta) = 0.5\big(1 - 0.25\sin(4\theta)\big)$$

for a rounded square shaped scatterer or star shaped scatterer, respectively.

In Figure 1–4, we plot the discretized version of the reciprocal to (3.7) in order to recover the scatterer. For this, we will take a fixed regularization parameter $\alpha = 10^{-6}$ in all our examples. Now, we need to define the discretized far-field operator with random noise added as

$$\mathbf{F}_\delta = \left[u^\infty(\hat{x}_i, d_j)\,(1 + \delta E_{i,j})\right]_{i,j=1}^{64}$$

with random complex-valued matrix $\mathbf{E}$ satisfying $\|\mathbf{E}\|_2 = 1$. Here, we take $\hat{x}_i, d_j$ to be equally spaced points on the unit circle given by

$$\hat{x}_i = d_i = (\cos\theta_i, \sin\theta_i) \quad \text{with} \quad \theta_i = 2\pi(i-1)/64.$$

Therefore, following [**14**] we have that the imaging functional that discretizes the reciprocal to (3.7) is given by (see also (2.3))

$$W(z) = \left[\sum_{j=1}^{64} \frac{\phi^2(\sigma_j;\alpha)}{\sigma_j} \big|(\mathbf{u}_j, \boldsymbol{\ell}_z)\big|^2\right]^{-1} \quad \text{with} \quad \boldsymbol{\ell}_z = [\mathrm{e}^{-\mathrm{i}k\hat{x}_i\cdot z}]_{i=1}^{64}.$$

Here σ_j are the singular values and $\mathbf{u}_j$ are the left singular vectors of

$$\mathbf{F}_{\delta,\sharp} = \big|\Re(\mathbf{F}_\delta)\big| + \big|\Im(\mathbf{F}_\delta)\big|$$

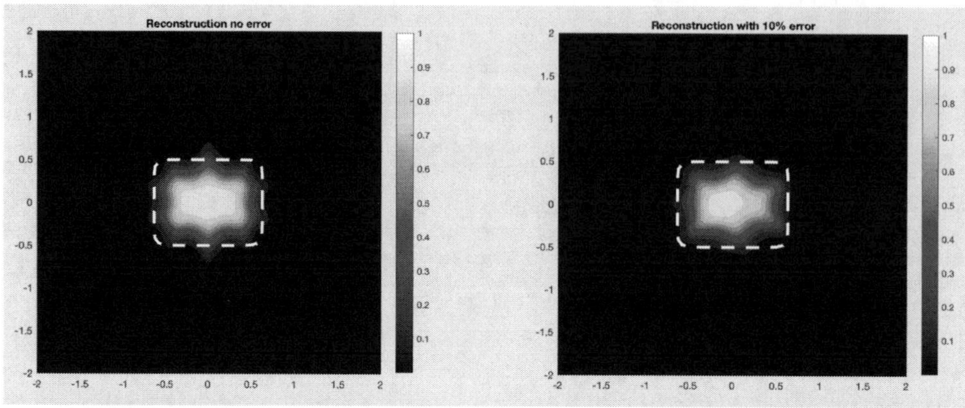

FIGURE 1. Reconstruction of the rounded square shaped scatterer with the Tikhonov filter given in (2.4). Left: reconstruction with no added noise and Right: reconstruction with 10% added noise.

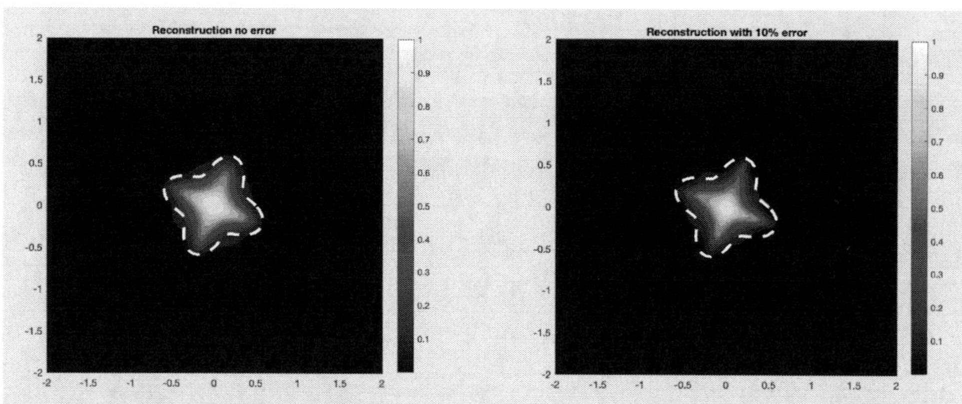

FIGURE 2. Reconstruction of the star shaped scatterer with the Tikhonov filter given in (2.4). Left: reconstruction with no added noise and Right: reconstruction with 10% added noise.

and the filter function $\phi(t;\alpha)$ is given by (2.4). To reiterate, the absolute value of a self-adjoint matrix is given by its eigenvalue decomposition.

By Theorem 2.1 and (3.7) we expect that $W(z) > 0$ for $z \in D$ and $W(z) \approx 0$ for $z \notin D$. In the following examples for this section, we plot the imaging function $W(z)$ along with the true shape of the scatterer given by the dotted lines.

We will also check the influence of the filter function on the numerical reconstruction. The numerical examples in **[14]** seem to suggest that the reconstruction does not depend heavily on the regularization scheme used. We test that here where we present the reconstructed star shaped scatterer with multiple filter functions.

In Figure 2–4, we reconstruct the star shaped scatterer with the three filter functions given in (2.4). As we can see, the choice of filter function seems to cause little to no differences in the numerical reconstruction.

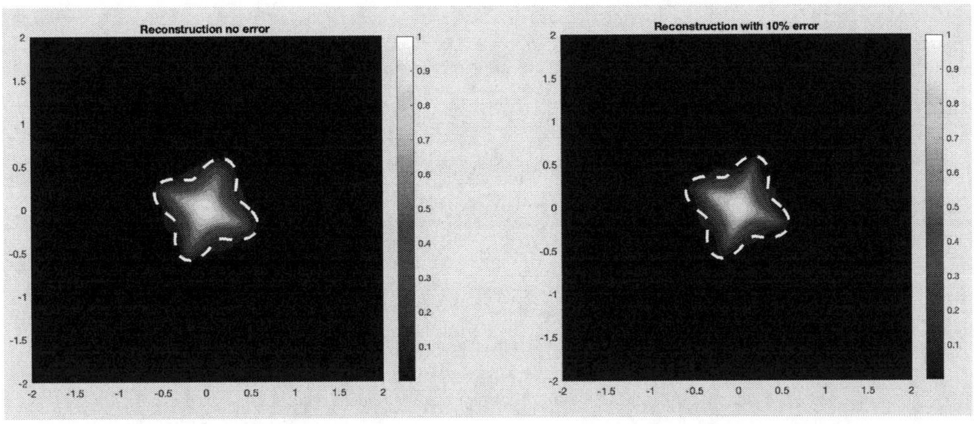

FIGURE 3. Reconstruction of the star shaped scatterer with the Spectral cutoff filter given in (2.4). Left: reconstruction with no added noise and Right: reconstruction with 10% added noise.

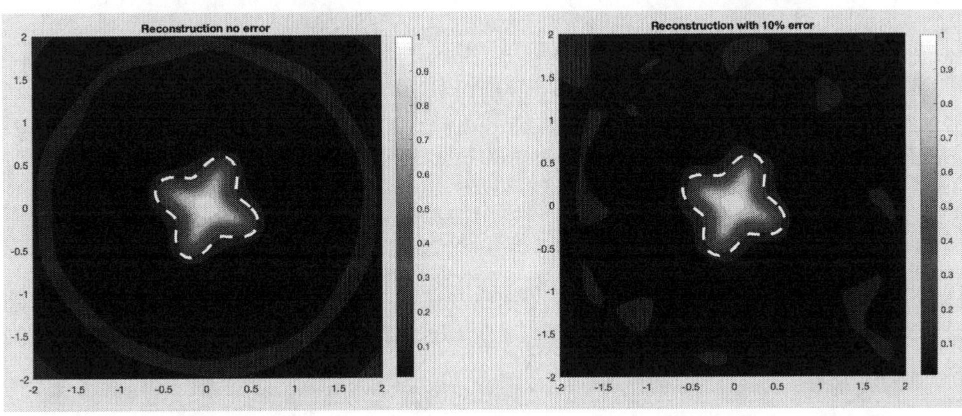

FIGURE 4. Reconstruction of the star shaped scatterer with the Landweber filter given in (2.4). Left: reconstruction with no added noise and Right: reconstruction with 10% added noise.

Lastly, we wish to show that the regularization step is needed to insure stable reconstructions. In Figure 5–6, we again recover the the star shaped scatterer. Here we plot the imaging functional $W(z)$ without regularization (i.e. $\alpha = 0$ corresponding to the classical factorization method) and with regularization (i.e. $\alpha = 10^{-3}$). We give the reconstruction when no error is added to the data and 10% random noise is added to the data. As we can see, the case without regularization fails to recover the scatterer when noise is added to the far-field data.

3.2. An example with near-field measurements. We will now consider the inverse shape problem of reconstructing a sound soft scatterer using near-field measurements. One of the main difficulties when using a factorization method with near-field measurements is the fact that the near-field operator does not have a symmetric factorization as is needed to apply Theorem 2.1. Due to this, researchers have developed analytical tools for post-processing the near-field measurements to

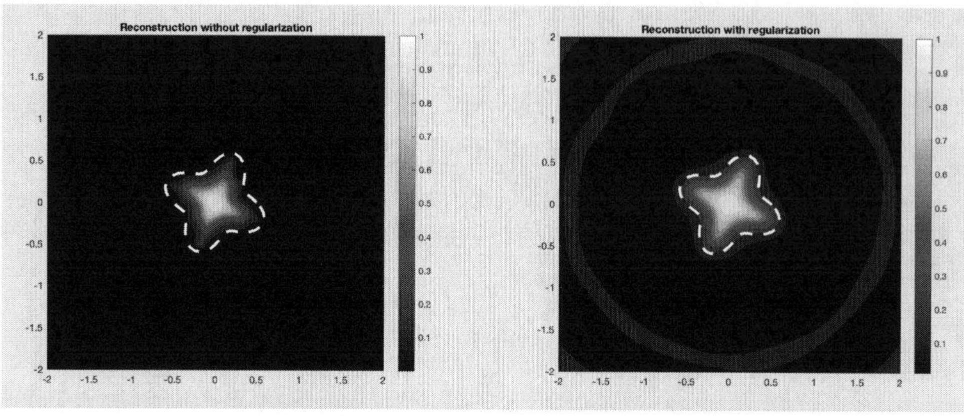

FIGURE 5. Reconstruction of the star shaped scatterer with the Tikhonov filter given in (2.4). Left: reconstruction without regularization and Right: reconstruction with regularization. Here no error is added to the far-field data.

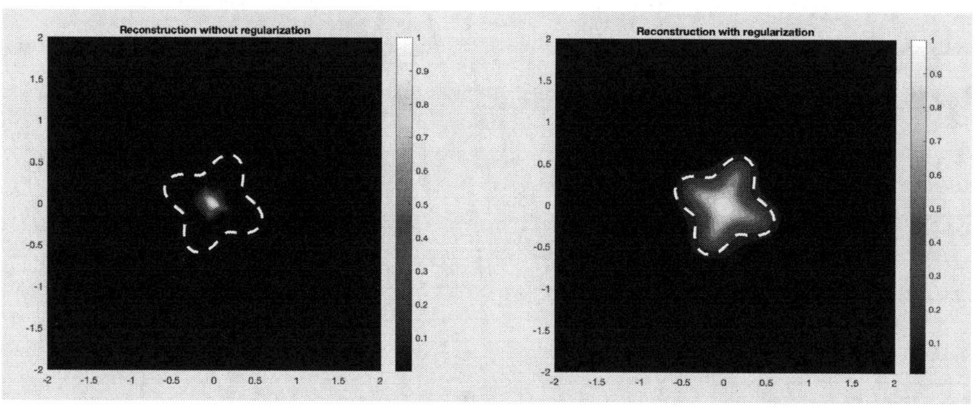

FIGURE 6. Reconstruction of the star shaped scatterer with the Tikhonov filter given in (2.4). Left: reconstruction without regularization and Right: reconstruction with regularization. Here 10% error is added to the far-field data.

give the corresponding operator a symmetric factorization. One way to achieve this is by using the Outgoing-to-Incoming operator. This was done in [**17**] for the classical factorization method using near-field measurements. In [**16**] non-physical sources where used to insure that the near-field operator admits a symmetric factorization. Recently, in [**13**] a Dirichlet-to-Far-Field operator was used to convert the near-field measurements into far-field measurements for a direct sampling inversion method. This is advantageous due to the fact that the far-field operator usually admits a symmetric factorization as we have seen in the previous section.

Now, we will formulate the inverse scattering problem under consideration and then apply the Dirichlet-to-Far-Field operator to the measurements in order to apply Theorem 2.1. To this end, we will assume that the unknown scatterer is denoted by $D \subset \mathbb{R}^m$ where $\partial D \in \mathcal{C}^2$ is a closed curve/surface such that $\mathbb{R}^m \setminus \overline{D}$ is connected. Again, we let $k > 0$ denote the associated wave number. The scatterer is illuminated by a point source incident field $u^i(\cdot\,,y) = \Phi(\cdot\,,y)$ given by (3.5) where y is the location of the point source on the curves/surface Γ. We will assume that the scatterer D is contained in the region inclosed by the Γ such that $\text{dist}(\Gamma, D) > 0$. Therefore, the radiating scattered field $u^s(\cdot\,,y) \in H^1_{\text{loc}}(\mathbb{R}^m \setminus \overline{D})$ satisfies

$$\Delta u^s + k^2 u^s = 0 \ \text{ in } \mathbb{R}^m \setminus \overline{D} \quad \text{and} \quad u^s(\cdot\,,y) = -u^i(\cdot\,,y) \ \text{ on } \partial D \tag{3.8}$$

along with the radiation condition (3.2). It is well known that for every $y \in \Gamma$ there is a unique scattered field. So we may assume that the scattered field $u^s(x,y)$ is known/measured for all $x, y \in \Gamma$. Therefore, we now define the so-called near-field operator

$$N : L^2(\Gamma) \longrightarrow L^2(\Gamma) \quad \text{given by} \quad (Ng)(x) = \int_\Gamma u^s(x,y) g(y)\, \mathrm{d}s(y).$$

Here in inverse shape problem is to recover D from the knowledge of the near-field operator N.

To this end, just as in the previous example we begin by deriving a factorization for the near-field operator. In order to continue, we make the assumption that k^2 is not a Dirichlet eigenvalue of the negative Laplacian in D. For this, motivated by (3.8) we consider the problem

$$\Delta w + k^2 w = 0 \ \text{ in } \mathbb{R}^m \setminus \overline{D} \quad \text{and} \quad w = -f \ \text{ on } \partial D$$

along with (3.2) for any $f \in H^{1/2}(\partial D)$. From the analysis done in [**13**] we have that the solution $w \in H^1_{\text{loc}}(\mathbb{R}^m \setminus \overline{D})$ has the integral representation

$$w(x) = -\int_{\partial D} \Phi(x,\omega)\big[S^{-1}f\big](\omega)\, \mathrm{d}s(\omega) \quad \text{for any} \quad x \in \mathbb{R}^m \setminus \overline{D} \tag{3.9}$$

where

$$S : H^{-1/2}(\partial D) \longrightarrow H^{1/2}(\partial D) \quad \text{such that} \quad S\varphi = \int_{\partial D} \Phi(\cdot\,,\omega)\varphi(\omega)\, \mathrm{d}s(\omega)\Big|_{\partial D}$$

for any $\varphi \in H^{-1/2}(\partial D)$. By the assumption on k it is known that S has a bounded inverse (chapter 1 in [**25**]) which implies that (3.9) is well defined. Now, we define the bounded linear operator

$$M : L^2(\Gamma) \longrightarrow L^2(\partial D) \quad \text{given by} \quad Mg = \int_\Gamma \Phi(\cdot\,,y) g(y)\, \mathrm{d}s(y)\Big|_{\partial D} \tag{3.10}$$

and the dual-operator

$$M^\top : L^2(\partial D) \longrightarrow L^2(\Gamma) \quad \text{given by} \quad M^\top\varphi = \int_{\partial D} \Phi(\omega,\cdot)\varphi(\omega)\, \mathrm{d}s(\omega)\Big|_\Gamma. \tag{3.11}$$

Notice, that the dual-operator is with respect to the *bilinear* L^2 dual-product $\langle\cdot\,,\cdot\rangle_{L^2}$ such that

$$\langle \varphi, Mg\rangle_{L^2(\partial D)} = \langle M^\top\varphi, g\rangle_{L^2(\Gamma)} \quad \text{for all} \quad g \in L^2(\Gamma) \text{ and } \varphi \in L^2(\partial D).$$

For the mapping properties of the operators M and $M^\top$ see chapter 6 in [**27**]. It is well-known that the near-field operator is the trace on Γ for the solution to (3.8) provided that the incident field $u^i(\cdot\,,y)$ is replaced by Mg. Therefore, by equation (3.9) we have that

$$Ng = -\int_{\partial D} \Phi(x,\omega)\big[S^{-1}Mg\big](\omega)\, \mathrm{d}s(\omega)\Big|_\Gamma \quad \text{for all} \quad g \in L^2(\Gamma).$$

By the definition of the operator M and its dual-operator we can conclude that

$$N = -M^\top S^{-1} M. \tag{3.12}$$

Notice, that (3.12) is not a symmetric factorization as in Theorem 2.1 due to the transpose rather than the adjoint. This implies that we must continue our analysis of the near-field operator in order to continue.

We could employ the so-called Outgoing-to-Incoming operator as in [**17**]. From this the classical factorization method was studied in in [**17**] for three types of scatterers using near-field measurements. More recently, in [**13**] it has been shown that the near-field data can be transformed into the far-field data for the corresponding problem. We will now, use the analysis in [**13**] to augment the measurements to use the regularized factorization method for this problem.

We now define the Dirichlet-to-Far-Field operator which is a main component of the analysis. Now, let $v \in H^1_{\text{loc}}(\mathbb{R}^m \setminus \overline{\text{Int}(\Gamma)})$ be the unique solution to

$$\Delta v + k^2 v = 0 \quad \text{in} \quad \mathbb{R}^m \setminus \overline{\text{Int}(\Gamma)} \quad \text{with} \quad v|_\Gamma = f \tag{3.13}$$

along with the radiation condition (3.2) for any $f \in H^{1/2}(\Gamma)$. Then, just as in [**15**] we can define Dirichlet-to-Far-Field operator given by

$$\mathcal{Q} : H^{1/2}(\Gamma) \longrightarrow L^2(\mathbb{S}) \quad \text{such that} \quad (\mathcal{Q}f)(\hat{x}) = v^\infty(\hat{x}), \quad \forall\, \hat{x} \in \mathbb{S}. \tag{3.14}$$

Now, we have that

$$(\mathcal{Q} M^\top \varphi)(\hat{x}) = \int_{\partial D} \mathrm{e}^{-\mathrm{i}k\hat{x}\cdot\omega} \varphi(\omega) \mathrm{d}s(\omega) \quad \text{for any} \quad \varphi \in L^2(\partial D) \tag{3.15}$$

by the asymptotic relations for the fundamental solution as $|x| \to \infty$. Therefore, by (3.15) we define the bounded linear operator

$$H : L^2(\mathbb{S}) \longrightarrow L^2(\partial D) \quad \text{given by} \quad (Hg)(\omega) = \int_{\mathbb{S}} \mathrm{e}^{\mathrm{i}k\omega\cdot\hat{x}} g(\hat{x}) \mathrm{d}s(\hat{x}) \Big|_{\partial D}$$

for any $g \in L^2(\mathbb{S})$. This corresponds to the trace of Herglotz wave function on the boundary of the scatterer. From this, we see that $\mathcal{Q} M^\top = H^*$. Now take the transpose the expression to obtain that $(H^*)^\top = M\mathcal{Q}^\top$. We can now show that

$$\begin{aligned} \left((H^*)^\top g\right)(\omega) &= \int_{\mathbb{S}} \mathrm{e}^{-\mathrm{i}k\omega\cdot\hat{x}} g(\hat{x}) \mathrm{d}s(\hat{x}) \Big|_{\partial D} \\ &= \int_{\mathbb{S}} \mathrm{e}^{\mathrm{i}k\omega\cdot\hat{x}} g(-\hat{x}) \mathrm{d}s(\hat{x}) \Big|_{\partial D} \\ &= (H\mathcal{R}g)(\omega) \end{aligned}$$

where the operator

$$\mathcal{R} : L^2(\mathbb{S}) \longrightarrow L^2(\mathbb{S}) \quad \text{is given by} \quad (\mathcal{R}g)(\hat{x}) = g(-\hat{x}).$$

Clearly, $\mathcal{R}$ is a bounded linear operator with $\mathcal{R} = \mathcal{R}^{-1}$. From this we can conclude that $H = M\mathcal{Q}^\top\mathcal{R}$. By the definition of $\mathcal{Q}$ and $\mathcal{R}$ we obtain that

$$\mathcal{Q}N\mathcal{Q}^\top\mathcal{R} = -H^* S^{-1} H \quad \text{where} \quad \mathcal{Q}N\mathcal{Q}^\top\mathcal{R} : L^2(\mathbb{S}) \longrightarrow L^2(\mathbb{S}) \tag{3.16}$$

by appealing to the factorization in (3.12).

We can now relate the transformed operator $\mathcal{Q}N\mathcal{Q}^\top\mathcal{R}$ to the far-field operator for the scattering problem (3.8) where the incident field is given by a plane wave. Indeed, by (3.16) and equation (1.55) in [**25**] we have that $\mathcal{Q}N\mathcal{Q}^\top\mathcal{R} = F$ where F is the corresponding far-field operator. This is important for our analysis here since F does have a symmetric factorization. From Theorem 1.15 in [**25**] we have that

$$F = GS^*G^* \quad \text{which implies that} \quad \mathcal{Q}N\mathcal{Q}^\top\mathcal{R} = GS^*G^*$$

where S^* is the adjoint of S defined above. The operator G maps the trace on ∂D to the far-field pattern for w where

$$G : H^{1/2}(\partial D) \longrightarrow L^2(\mathbb{S}) \quad \text{is given by} \quad Gw\big|_{\partial D} = w^\infty.$$

Recall, $w \in H^1_{\text{loc}}(\mathbb{R}^m \setminus \overline{D})$ can be written using (3.9) and satisfies (3.2). Just as in the previous section we consider

$$\big(\mathcal{Q}N\mathcal{Q}^\top\mathcal{R}\big)_\sharp = \big|\Re\big(\mathcal{Q}N\mathcal{Q}^\top\mathcal{R}\big)\big| + \big|\Im\big(\mathcal{Q}N\mathcal{Q}^\top\mathcal{R}\big)\big|.$$

By appealing to Lemma 1.14 of [**25**] we again can conclude that we have the factorization

$$\big(\mathcal{Q}N\mathcal{Q}^\top\mathcal{R}\big)_\sharp = GS^*_\sharp G^*$$

where $S^*_\sharp$ is a strictly coercive operator. The last piece we need to complete the puzzle is the fact that G^* is compact and injective (see Theorem 1.15 of [**25**]) along with

$$\ell_z = \mathrm{e}^{-\mathrm{i}k\hat{x}\cdot z} \quad \text{satisfies} \quad \ell_z \in \text{Range}(G) \iff z \in D.$$

Therefore, by appealing to Theorem 2.1 we have that

$$z \in D \iff \liminf_{\alpha\to 0} \Big(g_z^\alpha, \big(\mathcal{Q}N\mathcal{Q}^\top\mathcal{R}\big)_\sharp g_z^\alpha\Big)_{L^2(\mathbb{S})} < \infty \tag{3.17}$$

provided that g_z^α is the regularized solution to $\big(\mathcal{Q}N\mathcal{Q}^\top\mathcal{R}\big)_\sharp g = \ell_z$.

In order to apply (3.17) we need to compute the operators $\mathcal{Q}$ and $\mathcal{R}$. To due so, assume that $\Gamma = \partial B(0;\rho)$ for fixed $\rho > 0$ in two dimensions then by appealing to separation of variables and the asymptotic expansions for Hankel functions to obtain a formula for $\mathcal{Q}$. Indeed, we can use the fact that

$$v(r,\theta) = \sum_{|n|=0}^{\infty} \frac{f_n}{H_n^{(1)}(k\rho)} H_n^{(1)}(kr)\mathrm{e}^{\mathrm{i}n\theta} \quad \text{for all} \quad n \in \mathbb{Z}$$

where f_n are the Fourier coefficients for f along with the asymptotic formula

$$H_n^{(1)}(kr) = \sqrt{\frac{2}{\pi kr}}\mathrm{e}^{\mathrm{i}kr - \mathrm{i}n\pi/2 - \mathrm{i}\pi/4} + \mathcal{O}(r^{-3/2}) \quad \text{as} \quad r \to \infty$$

to derive a computable formula for $\mathcal{Q}$. From [**15**] we have that the explicit formula

$$(\mathcal{Q}f)(\theta) = \int_0^{2\pi} Q(\theta,\phi)f(\phi)\mathrm{d}\phi \quad \text{where} \quad Q(\theta,\phi) = \frac{(1-\mathrm{i})}{2\pi\sqrt{\pi k}} \sum_{|n|=0}^{\infty} \frac{\mathrm{e}^{\mathrm{i}n(\theta-\phi-\pi/2)}}{H_n^{(1)}(k\rho)}$$

with $f(\phi) = f\big(\rho(\cos\phi\,,\,\sin\phi)\big)$. Here, the constant radius ρ is assumed to be large enough such that $D \subset B(0;\rho)$. When $\Gamma \neq \partial B(0;\rho)$ we can define Dirichlet-to-Far-Field operator $\mathcal{Q}$ by using boundary integral equations. See Section 2 of [**25**] for a detailed construction. Now we need an explicit formula for the operator $\mathcal{R}$. To this end, we can use the expression given in [**13**] to write $\mathcal{R}$ as an integral operator with explicit kernel function. This expression uses the fact that $\hat{x} = (\cos\theta\,,\,\sin\theta)$ for $\theta \in [0, 2\pi)$ in two dimensions. Now, by appealing to sum of angles formula we obtain

$$-\cos(\theta) = \cos(\theta+\pi) \quad \text{and} \quad -\sin(\theta) = \sin(\theta+\pi)$$

which implies that

$$(\mathcal{R}g)(\theta) = g(\theta + \pi).$$

This formula uses the notation that $g(\theta) = g\big((\cos\theta\,,\,\sin\theta)\big)$. Therefore, by using the Fourier series representation for g we can conclude that

$$(\mathcal{R}g)(\theta) = \int_0^{2\pi} R(\theta,\phi)g(\phi)\mathrm{d}\phi \quad \text{where} \quad R(\theta,\phi) = \frac{1}{2\pi}\sum_{|n|=0}^{\infty} \mathrm{e}^{\mathrm{i}n(\theta-\phi+\pi)}.$$

For either $\mathcal{Q}$ or $\mathcal{R}$ we have that the series for the kernel function can be truncated to approximate the operators. It is shown in [**13, 15**] that the truncated series is a valid approximation for both operators.

Numerical examples: Just as in the previous section, we will provide some numerical examples of (3.17) for recovering a sound soft scatterer. To this end, we again assume that the boundary of the scatterer is given by

$$\partial D = r(\theta)\,(\cos(\theta), \sin(\theta)) \quad \text{for} \quad 0 \leq \theta \leq 2\pi.$$

Here we take $r(\theta)$ to be given by either

$$r(\theta) = 0.25(2 + 0.5\cos(3\theta)) \quad \text{or} \quad r(\theta) = 0.75\sqrt{0.75\cos^2(\theta) + 0.07\sin^2(\theta)}$$

for an acorn shaped scatterer or peanut shaped scatterer, respectively. In all our examples we take $\Gamma = \partial B(0; 5)$ (i.e. the disk with radius=5) and the wave number $k = 4$. The location of the sources and receivers will be given by

$$x_i = y_i = 5(\cos\theta_i, \sin\theta_i) \quad \text{with} \quad \theta_i = 2\pi(i-1)/64.$$

This corresponds to 64 equally spaced points on the disk.

Now, we need to compute the scattering data $u^s(x_i, y_j)$ by solving (3.8). Therefore, we use the fact that the scattered field is given by the series expansion

$$u^s(x, y_j) = \sum_{|n|=0}^{\infty} c_n(y_j) H_n^{(1)}(k|x|)\mathrm{e}^{\mathrm{i}n\theta_x} \quad \text{for each} \quad j = 1, \cdots, 64.$$

The above representation is given by using separation of variables in $\mathbb{R}^2 \setminus \overline{D}$ for the Helmholtz equation. Notice, that the radiation condition (3.2) is satisfied by the asymptotic formula for the Hankel functions $H_n^{(1)}(k|x|)$ as $|x| \to \infty$. Just as in [**13**] we will truncate the above series representation for $|n| = 0, \cdots, 15$ and solve for the series coefficients $c_n(y_j)$ such that

$$u^s(\widetilde{x}, y_j) = -\Phi(\widetilde{x}, y_j) \quad \text{for all} \quad \tilde{x} \in \partial D$$

and for each $j = 1, \cdots, 64$. This is done by insuring that the above equality holds for each $\widetilde{x}_i = r(\theta_i)\,(\cos(\theta_i), \sin(\theta_i))$ for each $i = 1, \cdots, 64$. So we solve the resulting 64×31 linear system of equations for each series coefficient $c_n(y_j)$. Once we have solved for the coefficients we have that the approximate scattering data on the measurement curve Γ is given by

$$u^s(x_i, y_j) \approx \sum_{|n|=0}^{15} c_n(y_j)\mathrm{H}_n^{(1)}(5k)\mathrm{e}^{\mathrm{i}n\theta_i}.$$

The discretized near-field operator with random noise added is given by

$$\mathbf{N}_\delta = [u^s(x_i, y_j)\,(1 + \delta E_{i,j})]_{i,j=1}^{64}$$

with random complex-valued matrix $\mathbf{E}$ satisfying $\|\mathbf{E}\|_2 = 1$.

Another piece we need in order to apply (3.17) is the discretization of the operators $\mathcal{Q}$ and $\mathcal{R}$. From the definition given earlier in this section we have that these operators can be written as integral operators with an explicit kernel given by an infinite series. To approximate the operators, we must first truncate the series representation for the kernel functions. Therefore, we now let

$$\widetilde{Q}(\theta,\phi) = \frac{(1-\mathrm{i})}{2\pi\sqrt{\pi k}} \sum_{|n|=0}^{10} \frac{\mathrm{e}^{\mathrm{i}n(\theta-\phi-\pi/2)}}{H_n^{(1)}(5k)} \quad \text{and} \quad \widetilde{R}(\theta,\phi) = \frac{1}{2\pi} \sum_{|n|=0}^{10} \mathrm{e}^{\mathrm{i}n(\theta-\phi+\pi)}.$$

This corresponds to the truncated series for the kernel functions. We note that in [**13, 15**] the approximate property of the truncated series approximates was established. Using that

$$(\mathcal{Q}f)(\theta) \approx \int_0^{2\pi} \widetilde{Q}(\theta,\phi) f(\phi)\mathrm{d}\phi \quad \text{and} \quad (\mathcal{R}g)(\theta) \approx \int_0^{2\pi} \widetilde{R}(\theta,\phi) g(\phi)\mathrm{d}\phi$$

we can employ a standard 64 point Riemann sum collocation approximation for the integrals. From this we obtain a 64×64 discretization of the operators given by

$$\mathbf{Q} = \left[\widetilde{Q}(\theta_i,\theta_j)\right]_{i,j=1}^{64} \quad \text{and} \quad \mathbf{R} = \left[\widetilde{R}(\theta_i,\theta_j)\right]_{i,j=1}^{64}.$$

Again, we have taken $\theta_i = 2\pi(i-1)/64$ for $i = 1, \cdots, 64$.

Now that we have the discretized operators we define the discretized far-field transform of the near-field operator as given by $\mathbf{Q}\mathbf{N}_\delta\mathbf{Q}^\top\mathbf{R}$. Therefore, we again let σ_j be the singular values and $\mathbf{u}_j$ be the left singular vectors of the matrix

$$\left(\mathbf{Q}\mathbf{N}_\delta\mathbf{Q}^\top\mathbf{R}\right)_\sharp = \left|\Re(\mathbf{Q}\mathbf{N}_\delta\mathbf{Q}^\top\mathbf{R})\right| + \left|\Im(\mathbf{Q}\mathbf{N}_\delta\mathbf{Q}^\top\mathbf{R})\right|.$$

By (3.17) we have that the imaging functional

$$W(z) = \left[\sum_{j=1}^{64} \frac{\phi^2(\sigma_j;\alpha)}{\sigma_j} \left|(\mathbf{u}_j, \boldsymbol{\ell}_z)\right|^2\right]^{-1} \quad \text{with } \boldsymbol{\ell}_z = [\mathrm{e}^{-\mathrm{i}k\hat{x}_i\cdot z}]_{i=1}^{64}$$

with filter function $\phi(t;\alpha)$ given by (2.4) can be used to approximate the scatterer D. Just as in the previous section we have that $W(z) > 0$ for $z \in D$ and $W(z) \approx 0$ for $z \notin D$ by appealing to Theorem 2.1. In all our examples, we take the fixed regularization parameter $\alpha = 10^{-6}$ and plot the imaging functional.

As we see, in Figures 7–10 the filter function used in the reconstruction gives little to no difference computational results. Just as in the previous section, this implies that the reconstruction seems to not be sensitive with respect to the regularization scheme used in the imaging functional.

4. Conclusions

In this paper, we have discussed a regularized version of the factorization method as well as its application to inverse scattering. Note, that this method was originally used in an application to diffuse optical tomography in [**26**]. We

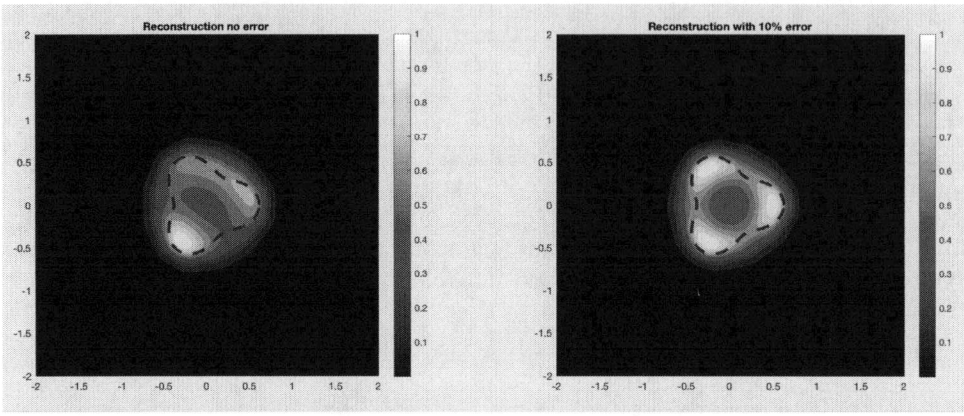

FIGURE 7. Reconstruction of the ball shaped scatterer with the Tikhonov filter given in (2.4). Left: reconstruction with no added noise and Right: reconstruction with 10% added noise.

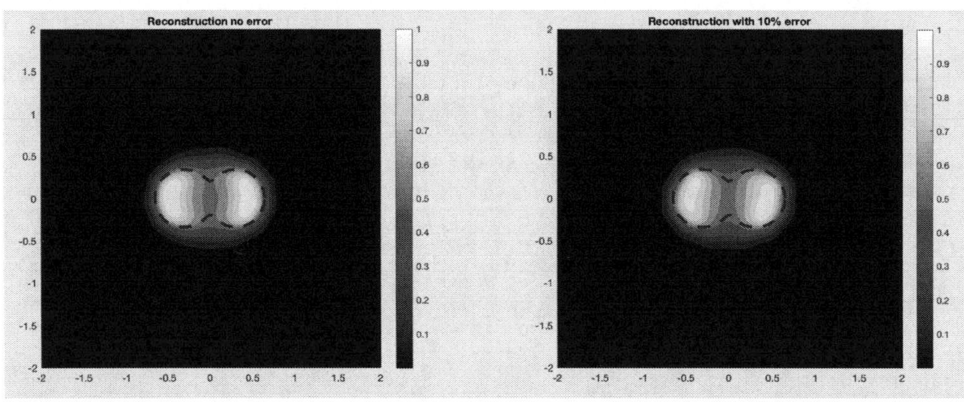

FIGURE 8. Reconstruction of the peanut shaped scatterer with the Tikhonov filter given in (2.4). Left: reconstruction with no added noise and Right: reconstruction with 10% added noise.

have seen that this method gives a new theoretically valid and analytically rigorous method for solving inverse shape problems. Moreover, we have applied this method to both near and far-field data sets. From our numerical investigation we see that the choice of regularization scheme seems to have little effect on the reconstructions. A future direction of this research can be to provide theoretical justification of the regularized factorization method for a perturbed data operator. Also, the question of how to pick the regularization parameter is still open. In all our examples, we take the regularization parameter ad-hoc but one should determine a discrepancy principle to optimize the resolution of the imaging functional. From this, the main novelty of this paper is two fold. First, we have given a more extensive numerical study of this new qualitative reconstruction method. We have also, given another analytical and computation method for applying a factorization method for the near-field operator that lacks the symmetric factorization needed in the typical

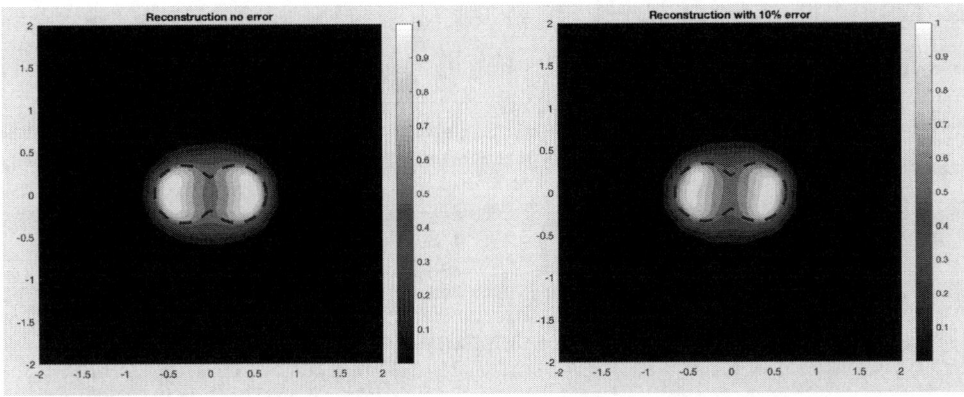

FIGURE 9. Reconstruction of the peanut shaped scatterer with the Spectral cutoff filter given in (2.4). Left: reconstruction with no added noise and Right: reconstruction with 10% added noise.

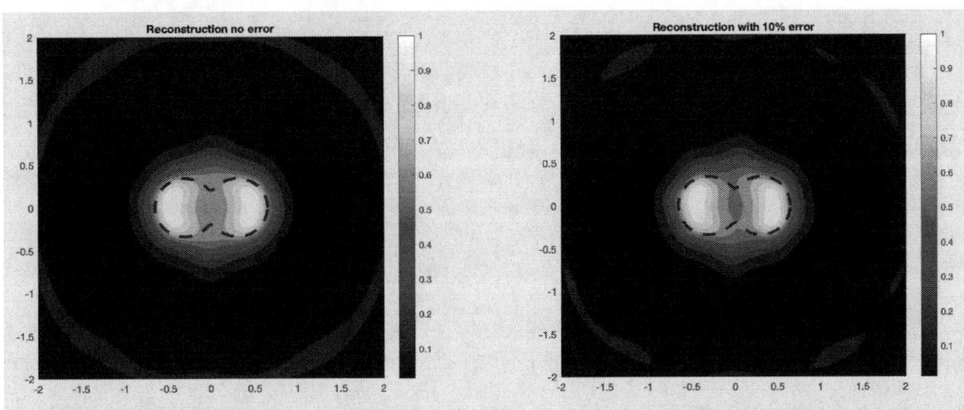

FIGURE 10. Reconstruction of the peanut shaped scatterer with the Landweber filter given in (2.4). Left: reconstruction with no added noise and Right: reconstruction with 10% added noise.

analysis. Also, we can apply the regularized factorization method to other imaging modalities such as electrical impedance tomography.

Acknowledgments: The research of I. Harris is partially supported by the NSF DMS Grant 2107891.

References

[1] Tilo Arens, *Why linear sampling works*, Inverse Problems **20** (2004), no. 1, 163–173, DOI 10.1088/0266-5611/20/1/010. MR2044612

[2] T. Arens and A. Lechleiter, *Indicator functions for shape reconstruction related to the linear sampling method*, SIAM J. Imaging Sci. **8** (2015), no. 1, 513–535, DOI 10.1137/140965740. MR3317780

[3] Lorenzo Audibert and Houssem Haddar, *A generalized formulation of the linear sampling method with exact characterization of targets in terms of farfield measurements*, Inverse Problems **30** (2014), no. 3, 035011, 20, DOI 10.1088/0266-5611/30/3/035011. MR3168273

[4] Gang Bao and Peijun Li, *Inverse medium scattering for the Helmholtz equation at fixed frequency*, Inverse Problems **21** (2005), no. 5, 1621–1641, DOI 10.1088/0266-5611/21/5/007. MR2173413

[5] Fioralba Cakoni, David Colton, and Houssem Haddar, *Inverse scattering theory and transmission eigenvalues*, CBMS-NSF Regional Conference Series in Applied Mathematics, vol. 88, Society for Industrial and Applied Mathematics (SIAM), Philadelphia, PA, 2016, DOI 10.1137/1.9781611974461.ch1. MR3601119

[6] Fioralba Cakoni, Houssem Haddar, and Armin Lechleiter, *On the factorization method for a far field inverse scattering problem in the time domain*, SIAM J. Math. Anal. **51** (2019), no. 2, 854–872, DOI 10.1137/18M1214809. MR3932624

[7] Mathieu Chamaillard, Nicolas Chaulet, and Houssem Haddar, *Analysis of the factorization method for a general class of boundary conditions*, J. Inverse Ill-Posed Probl. **22** (2014), no. 5, 643–670, DOI 10.1515/jip-2013-0013. MR3276233

[8] David Colton and Rainer Kress, *Inverse acoustic and electromagnetic scattering theory*, 3rd ed., Applied Mathematical Sciences, vol. 93, Springer, New York, 2013, DOI 10.1007/978-1-4614-4942-3. MR2986407

[9] David Colton and Andreas Kirsch, *A simple method for solving inverse scattering problems in the resonance region*, Inverse Problems **12** (1996), no. 4, 383–393, DOI 10.1088/0266-5611/12/4/003. MR1402098

[10] Mary R. Embry, *Factorization of operators on Banach space*, Proc. Amer. Math. Soc. **38** (1973), 587–590, DOI 10.2307/2038955. MR312287

[11] Bastian Gebauer, *The factorization method for real elliptic problems*, Z. Anal. Anwend. **25** (2006), no. 1, 81–102, DOI 10.4171/ZAA/1279. MR2216883

[12] J. Guo, G. Nakamura, and H. Wang, The factorization method for recovering cavities in a heat conductor, preprint (2019) arXiv:1912.11590

[13] I. Harris, Direct methods for recovering sound soft scatterers from point source measurements. *Computation* **9(11)** 120 (2021).

[14] Isaac Harris, *Regularization of the factorization method applied to diffuse optical tomography*, Inverse Problems **37** (2021), no. 12, Paper No. 125010, 19, DOI 10.1088/1361-6420/ac37f9. MR4344730

[15] Isaac Harris, Dinh-Liem Nguyen, and Thi-Phong Nguyen, *Direct sampling methods for isotropic and anisotropic scatterers with point source measurements*, Inverse Probl. Imaging **16** (2022), no. 5, 1137–, DOI 10.3934/ipi.2022015. MR4475846

[16] I. Harris and S. Rome, *Near field imaging of small isotropic and extended anisotropic scatterers*, Appl. Anal. **96** (2017), no. 10, 1713–1736, DOI 10.1080/00036811.2017.1284312. MR3659663

[17] Guanghui Hu, Jiaqing Yang, Bo Zhang, and Haiwen Zhang, *Near-field imaging of scattering obstacles with the factorization method*, Inverse Problems **30** (2014), no. 9, 095005, 25, DOI 10.1088/0266-5611/30/9/095005. MR3257998

[18] N. Hyvönen, Application of a weaker formulation of the factorization method to the characterization of absorbing inclusions in optical tomography, *Inverse Problems*, **21** 1331 (2005).

[19] N. Hyvönen, Characterizing inclusions in optical tomography, *Inverse Problems*, **21** 737–751 (2004).

[20] Andreas Kirsch, *An introduction to the mathematical theory of inverse problems*, 2nd ed., Applied Mathematical Sciences, vol. 120, Springer, New York, 2011, DOI 10.1007/978-1-4419-8474-6. MR3025302

[21] Andreas Kirsch, *Characterization of the shape of a scattering obstacle using the spectral data of the far field operator*, Inverse Problems **14** (1998), no. 6, 1489–1512, DOI 10.1088/0266-5611/14/6/009. MR1662460

[22] Andreas Kirsch, *The MUSIC algorithm and the factorization method in inverse scattering theory for inhomogeneous media*, Inverse Problems **18** (2002), no. 4, 1025–1040, DOI 10.1088/0266-5611/18/4/306. MR1929280

[23] Andreas Kirsch, *The factorization method for a class of inverse elliptic problems*, Math. Nachr. **278** (2005), no. 3, 258–277, DOI 10.1002/mana.200310239. MR2110531

[24] Andreas Kirsch, *An introduction to the mathematical theory of inverse problems*, 2nd ed., Applied Mathematical Sciences, vol. 120, Springer, New York, 2011, DOI 10.1007/978-1-4419-8474-6. MR3025302

[25] Andreas Kirsch and Natalia Grinberg, *The factorization method for inverse problems*, Oxford Lecture Series in Mathematics and its Applications, vol. 36, Oxford University Press, Oxford, 2008. MR2378253

[26] Armin Lechleiter, *A regularization technique for the factorization method*, Inverse Problems **22** (2006), no. 5, 1605–1625, DOI 10.1088/0266-5611/22/5/006. MR2261257

[27] William McLean, *Strongly elliptic systems and boundary integral equations*, Cambridge University Press, Cambridge, 2000. MR1742312

[28] Dinh-Liem Nguyen, *Shape identification of anisotropic diffraction gratings for TM-polarized electromagnetic waves*, Appl. Anal. **93** (2014), no. 7, 1458–1476, DOI 10.1080/00036811.2013.835041. MR3207767

Department of Mathematics, Purdue University, West Lafayette, IN 47907
Email address: harri814@purdue.edu

Contemporary Mathematics
Volume **784**, 2023
https://doi.org/10.1090/conm/784/15753

Sampling type method combined with deep learning for inverse scattering with one incident wave

Thu Le, Dinh-Liem Nguyen, Vu Nguyen, and Trung Truong

ABSTRACT. We consider the inverse problem of determining the geometry of penetrable objects from scattering data generated by one incident wave at a fixed frequency. We first study an orthogonality sampling type method which is fast, simple to implement, and robust against noise in the data. This sampling method has a new imaging functional that is applicable to data measured in near field or far field regions. The resolution analysis of the imaging functional is analyzed where the explicit decay rate of the functional is established. A connection with the orthogonality sampling method by Potthast is also studied. The sampling method is then combined with a deep neural network to solve the inverse scattering problem. This combined method can be understood as a network using the image computed by the sampling method for the first layer and followed by the U-net architecture for the rest of the layers. The fast computation and the knowledge from the results of the sampling method help speed up the training of the network. The combination leads to a significant improvement in the reconstruction results initially obtained by the sampling method. The combined method is also able to invert some limited aperture experimental data without any additional transfer training.

1. Introduction

We consider an inhomogeneous medium that fills up a bounded Lipschitz domain $D \subset \mathbb{R}^n$ ($n = 2$ or 3). Suppose that the medium is characterized by bounded function $\eta(y)$ satisfying $\eta = 0$ in $\mathbb{R}^n \setminus \overline{D}$. Consider the incident plane wave

$$u_{\text{in}}(x) = e^{ikx\cdot d}, \quad x \in \mathbb{R}^n, \quad d \in \mathbb{S}^{n-1} := \{x \in \mathbb{R}^n : |x| = 1\},$$

where $k > 0$ is the wave number and d is the direction vector of propagation. The scattering of u_{in} by the inhomogeneous medium is described by the following model problem

$$\Delta u + k^2(1 + \eta(x))u = 0, \quad x \in \mathbb{R}^n, \tag{1.1}$$

$$u = u_{\text{sc}} + u_{\text{in}}, \tag{1.2}$$

$$\lim_{r\to\infty} r^{\frac{n-1}{2}} \left(\frac{\partial u_{\text{sc}}}{\partial r} - iku_{\text{sc}}\right) = 0, \quad r = |x|, \tag{1.3}$$

2020 *Mathematics Subject Classification.* Primary 35R30, 78A46; Secondary 65C20.

Key words and phrases. Sampling method, deep learning, U-net, inverse scattering, one incident wave, Cauchy data.

This work was supported in part by NSF Grant DMS-1812693.

where u is the total field, u_{sc} is the scattered field, and the Sommerfeld radiation condition (1.3) holds uniformly for all directions $x/|x| \in \mathbb{S}^{n-1}$. If $\mathbb{R}^n \backslash \overline{D}$ is connected and $\mathrm{Re}\,(\eta) \geq 0$, this scattering problem is known to have a unique weak solution $u_{\mathrm{sc}} \in H^1_{\mathrm{loc}}(\mathbb{R}^n)$, see [**11**].

Inverse problem. Let Ω be a bounded Lipschitz domain such that $\overline{D} \subset \Omega$ and denote by $\nu(x)$ the outward normal unit vector to $\partial\Omega$ at x. Given u_{sc} and $\partial u_{\mathrm{sc}}/\partial\nu$ on $\partial\Omega$ determine D.

In this paper we develop a sampling type method and combine this method with deep learning to solve the inverse problem above. There has been a large body of literature on both theoretical and numerical studies on this inverse problem and its variants, see [**10, 11**] and references therein. To numerically solve the inverse problem sampling methods are known to be fast, non-iterative and do not require advanced a priori information about the unknown target. More details about sampling methods can be found in [**6, 23, 28**] and references therein.

The sampling method studied in this paper is inspired by the orthogonality sampling method (OSM) by Potthast [**29**]. The OSM is also studied in other works under the name of direct sampling method. The OSM is attractive and promising thanks to its computational efficiency. For instance, the OSM is a regularization free method that is very robust against noise in the data, and its implementation only involves an evaluation of vector products. However, the theoretical analysis of the OSM is far less developed compared with that of the linear sampling method and the factorization method. Also most of the published results deal with the case of far-field measurements, see e.g., [**1, 13, 15, 17, 18, 24, 27, 29**]. There have been only a few results on the OSM concerning the case of near-field measurements. The near-field OSM studied in [**3**] is applicable to the 2D case with circular measurement boundaries. The 3D case was studied in [**20**] under the small volume hypothesis of well-separated inhomogeneities. The sampling method proposed in this paper can be applied to near-field data (or far-field data) and it is not limited to those conditions. However, it requires Cauchy data instead of only scattered field data. Using the Lippmann-Schwinger equation and the Helmholtz integral representation we show that the imaging functional is equal to a functional that involves Bessel functions $J_0(k|y-z|)$ or $j_0(k|y-z|)$ where z is a sampling point and y is inside the unknown scatterer. These Bessel functions are crucial for justifying the behavior of the imaging functional.

It is known that in the case of scattering data generated by one incident wave the OSM can only provide reasonable reconstructions for small scattering objects [**29**]. It is also the case of other reconstruction methods. This motivates us to combine our orthogonality sampling type method with deep learning to obtain better reconstruction results. The orthogonality sampling type method is combined with deep learning in the following way. This combined method can be understood as a network that uses the image computed by the sampling method for the first layer, followed by the U-net structure for the remaining layers. The combination of deep learning to physics-based reconstruction methods to solve inverse problems has been studied, see, e.g., [**4, 7, 14, 19, 21, 26, 32**]. In particular, we refer to the review paper [**7**] and references therein for recent advances in using deep learning for solving inverse scattering problems. To our knowledge (see also [**7**]) most of the studies on deep learning for inverse scattering consider data generated by multiple incident fields while the the data for inverse scattering problem considered in this paper is associated with only one incident field at a fixed wave number. Our inverse problem has a minimal amount of data. In addition, the imaging functional for the proposed sampling method is a new functional and it has an advantage over the imaging functional of the OSM that it can be applied to near field data or far field data. Furthermore, to our knowledge the combination of this sampling method (or orthogonality sampling methods) and deep learning for solving inverse scattering problems has not been studied before.

In this paper we exploit a deep neural network of U-Net-Xception style [**8**]. We train this network using simulated data sets of scattering objects with elliptical geometry. The fast computation and the information from the results of the sampling method help speed up the training of the network. Our numerical study shows that the network is able to predict some scattering objects with geometries different from those of the training and testing data sets. The combination leads to a significant improvement in the reconstruction results initially obtained by the sampling method. We also demonstrate that the combined method is also able to invert some experimental data from the Fresnel Institute ([**5**]) without any additional transfer learning.

The paper is organized as follows. The orthogonality sampling type method and its analysis are presented in Section 2. Section 3 is dedicated to a description of the deep neural network used to combine with the sampling method. The numerical study for simulated data and real data is presented in Sections 4 and 5, respectively.

2. An orthogonality sampling type method

In this section we develop the an orthogonality sampling type method for the inverse scattering problem. For $x, y \in \mathbb{R}^3$ and $x \neq y$, we denote by $\Phi(x,y)$ the free-space Green's function of the scattering problem (1.1)–(1.3) which is given by

$$\Phi(x,y) = \begin{cases} \frac{i}{4} H_0^{(1)}(k|x-y|), & n = 2, \\ \frac{\exp(ik|x-y|)}{4\pi|x-y|}, & n = 3. \end{cases} \tag{2.1}$$

It is well known that problem (1.1)–(1.3) is equivalent to the Lippmann-Schwinger equation (see, e.g. [**11**])

$$u_{\mathrm{sc}}(x) = k^2 \int_D \Phi(x,y)\eta(y)u(y)\mathrm{d}y, \quad x \in \mathbb{R}^3.$$

The following Helmholtz integral representation is important to the analysis of the sampling method.

LEMMA 2.1. *Assume that W is bounded Lipschitz domain in $\mathbb{R}^n$ such that $\mathbb{R}^n \setminus \overline{W}$ is connected. Let ν be a unit outward normal vector on ∂W. For any function $w \in H^2(W)$ we have*

$$w(x) = \int_{\partial W} \left(\frac{\partial w(y)}{\partial \nu} \Phi(x,y) - w(y) \frac{\Phi(x,y)}{\partial \nu(y)} \right) \mathrm{d}s(y) - \int_W (\Delta w + k^2 w)\Phi(x,y)\mathrm{d}y. \tag{2.2}$$

PROOF. We refer to [**11**, Chapter 2] for a proof of the lemma. □

We also need to the Funk-Hecke formula for the analysis in this section.

LEMMA 2.2. *Let Y_m be the spherical harmonics of order m and j_m be the spherical Bessel functions of first kind and order m. We have*

$$\int_{\mathbb{S}^{n-1}} e^{-ikx\cdot\hat{z}} Y_n(\hat{z}) ds(\hat{z}) = \frac{4\pi}{i^m} j_m(k|x|) Y_m\left(\frac{x}{|x|}\right). \tag{2.3}$$

PROOF. A proof of this lemma can be found in [**11**, Chapter 2]. □

Now we define the imaging functional as

$$\mathcal{I}(z) := \left|\int_{\partial\Omega} \left(\frac{\partial \mathrm{Im}\,\Phi(x,z)}{\partial\nu(x)} u_{\mathrm{sc}}(x) - \mathrm{Im}\,\Phi(x,z)\frac{\partial u_{\mathrm{sc}}(x)}{\partial\nu(x)}\right) \mathrm{d}s(x)\right|^{\rho}, \tag{2.4}$$

where $\rho = 1$ or $\rho = 2$.

This functional $\mathcal{I}(z)$ aims to determine D in Ω and in the numerical implementation we evaluate $\mathcal{I}(z)$ for a finite set of sampling points z. We expect that $\mathcal{I}(z)$ is relatively large as $z \in D$ and that it is small as z is outside D. The behavior of $\mathcal{I}(z)$ is analyzed in the following theorem.

THEOREM 2.3. *The imaging functional $\mathcal{I}$ satisfies*

$$\mathcal{I}(z) = \left|k^2 \int_D \mathrm{Im}\,\Phi(y,z)\eta(y)u(y)\mathrm{d}y\right|^{\rho}.$$

PROOF. From the Lippmann-Schwinger equation we obtain that

$$\frac{\partial u_{\mathrm{sc}}(x)}{\partial\nu(x)} = k^2 \int_D \frac{\partial\Phi(x,y)}{\partial\nu(x)}\eta(y)u(y)\mathrm{d}y.$$

Therefore, a substitution leads to

$$\begin{aligned}
&\int_{\partial\Omega} \left(\frac{\partial \mathrm{Im}\,\Phi(x,z)}{\partial\nu(x)} u_{\mathrm{sc}}(x) - \mathrm{Im}\,\Phi(x,z)\frac{\partial u_{\mathrm{sc}}(x)}{\partial\nu(x)}\right) \mathrm{d}s(x)\\
&= \int_{\partial\Omega} \left(\frac{\partial \mathrm{Im}\,\Phi(x,z)}{\partial\nu(x)} k^2 \int_D \Phi(x,y)\eta(y)u(y)\mathrm{d}y - \mathrm{Im}\,\Phi(x,z)k^2 \int_\Omega \frac{\partial\Phi(x,y)}{\partial\nu(x)}\eta(y)u(y)\mathrm{d}y\right) \mathrm{d}s(x)\\
&= k^2 \int_D \int_{\partial\Omega} \left(\frac{\partial \mathrm{Im}\,\Phi(x,z)}{\partial\nu(x)} \Phi(y,x) - \mathrm{Im}\,\Phi(x,z)\frac{\partial\Phi(y,x)}{\partial\nu(x)}\right) \mathrm{d}s(x)\,\eta(y)u(y)\mathrm{d}y.
\end{aligned} \tag{2.5}$$

Since $\Delta\mathrm{Im}\,\Phi(z,y) + k^2\mathrm{Im}\,\Phi(z,y) = 0$ for all $z, y \in \mathbb{R}^n$ and $\mathrm{Im}\,\Phi(z,y)$ is regular function, the Helmholtz integral representation (2.2) implies that

$$\int_{\partial\Omega} \left(\frac{\partial \mathrm{Im}\,\Phi(x,z)}{\partial\nu(x)} \Phi(y,x) - \mathrm{Im}\,\Phi(x,z)\frac{\partial\Phi(y,x)}{\partial\nu(x)}\right) \mathrm{d}s(x) = \mathrm{Im}\,\Phi(y,z).$$

Therefore, substituting this identity in (2.5) implies

$$\int_{\partial\Omega} \left(\frac{\partial \mathrm{Im}\,\Phi(x,z)}{\partial\nu(x)} u_{\mathrm{sc}}(x) - \mathrm{Im}\,\Phi(x,z)\frac{\partial u_{\mathrm{sc}}(x)}{\partial\nu(x)}\right) \mathrm{d}s(x) = k^2 \int_D \mathrm{Im}\,\Phi(y,z)\eta(y)u(y)\mathrm{d}y.$$

Now the proof can be completed by substituting this equation in (2.4). □

We know that

$$\mathrm{Im}\,\Phi(y,z) = \begin{cases} J_0(k|z-y|), & n = 2,\\ \frac{k}{4\pi} j_0(k|z-y|), & n = 3. \end{cases} \tag{2.6}$$

Since $J_0(k|y-z|)$ and $j_0(k|y-z|)$ peak when $k|y-z| = 0$, we expect from Theorem 1 that $\mathcal{I}(z)$ takes larger values for $z \in D$ and much smaller values outside D. Further, from the asymptotic behavior of $J_0(k|y-z|)$ and $j_0(k|y-z|)$ as $|y-z| \to \infty$ we can easily estimate, for $z \notin D$, that

$$\mathcal{I}(z) = O\left(\frac{1}{\operatorname{dist}(z,D)^{\rho(n-1)/2}}\right) \quad \text{as } \operatorname{dist}(z,D) \to \infty.$$

This incomplete analysis is common for imaging functionals of orthogonality sampling methods (see, e.g., [**1**,**3**,**17**,**18**,**27**,**29**]). To our knowledge a complete analysis for orthogonality sampling methods is still an open problem.

We now show a relation between $\mathcal{I}(z)$ and the imaging functional suggested by Potthast in [**29**]. To this end, we recall that the scattered field u_{sc} has the asymptotic behavior

$$u_{\mathrm{sc}}(x) = \frac{e^{ik|x|}}{|x|^{(n-1)/2}}(u^\infty(\hat{x}) + O(1/|x|)), \quad |x| \to \infty,$$

for all $\hat{x} \in \mathbb{S}^{n-1}$. The function $u^\infty(\hat{x})$ is called the scattering amplitude or the far-field pattern of $u_{\mathrm{sc}}(x)$. The original imaging functional of the orthogonality sampling method suggested in [**29**] is given by

$$\mathcal{I}_{\mathrm{OSM}}(z) = \left|\int_{\mathbb{S}^{n-1}} e^{ikz\cdot\hat{x}} u^\infty(\hat{x})\mathrm{d}s(\hat{x})\right|.$$

THEOREM 2.4. *The two imaging functionals are related by*

$$\mathcal{I}(z) = \gamma\, |\mathcal{I}_{\mathrm{OSM}}(z)|^\rho$$

where

$$\gamma = \begin{cases} \left(\frac{\sqrt{\pi}e^{i\pi/4}}{\sqrt{2k}}\right)^\rho, & n=2,\\ \left(\frac{4\pi}{k}\right)^\rho, & n=3.\end{cases}$$

PROOF. From the Lippmann-Schwinger equation for u_{sc} it is known that its scattering amplitude is given by

$$u^\infty(\hat{x}) = \alpha k^2 \int_D e^{-iky\cdot\hat{x}}\eta(y)u(y)\mathrm{d}y$$

with the constant

$$\alpha = \begin{cases} \frac{e^{i\pi/4}}{\sqrt{8\pi k}}, & n=2,\\ \frac{1}{4\pi}, & n=3.\end{cases}$$

Substituting $u^\infty(\hat{x})$ in $\mathcal{I}_{\mathrm{OSM}}(z)$ implies

$$\mathcal{I}_{\mathrm{OSM}}(z) = \left|\alpha k^2 \int_D \int_{\mathbb{S}^{n-1}} e^{ik\hat{x}\cdot(z-y)}\mathrm{d}s(\hat{x})\eta(y)u(y)\mathrm{d}y\right|. \tag{2.7}$$

Now using (2.6) and the Funk-Hecke formula (2.3) we obtain that

$$\int_{\mathbb{S}^{n-1}} e^{ik\hat{x}\cdot(z-y)}\mathrm{d}s(\hat{x}) = \begin{cases} 2\pi \operatorname{Im}\Phi(z,y), & n=2,\\ \frac{(4\pi)^2}{k}\operatorname{Im}\Phi(z,y), & n=3.\end{cases} \tag{2.8}$$

Substituting (2.8) in (2.7) and comparing with Theorem 1 we complete the proof. □

Theorem 2 also implies that $\mathcal{I}(z)$ is stable against noise in the data following the stability of $\mathcal{I}_{\mathrm{OSM}}(z)$ [**29**].

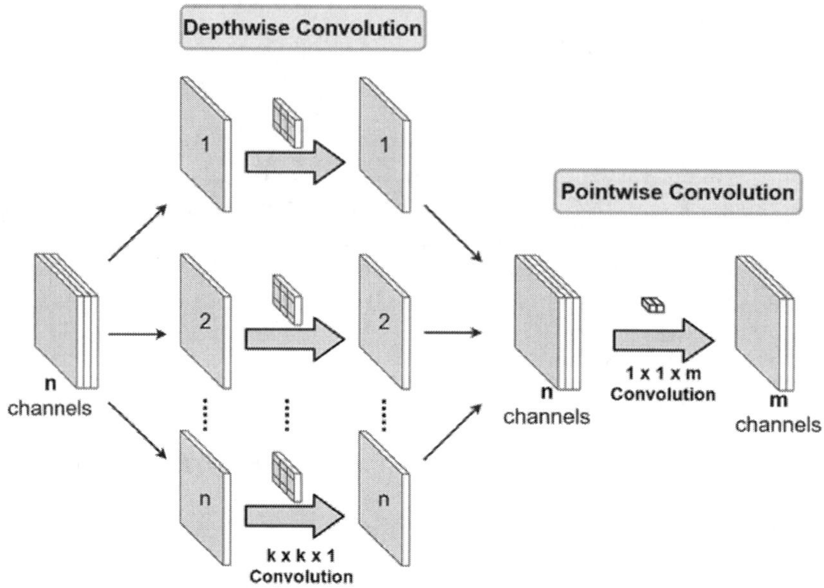

FIGURE 1. Architecture of Depthwise Separable Convolution.

REMARK 2.5. If $\partial\Omega$ is the boundary of some ball with large radius, we can approximate $\partial u_{\mathrm{sc}}/\partial\nu \approx iku_{\mathrm{sc}}$ using the radiation condition. Then $\mathcal{I}(z)$ can be modified to handle the far-field data $u_{\mathrm{sc}}(x)$ as follows

$$\mathcal{I}(z) = \left| \int_{\partial\Omega} \left(\frac{\partial \mathrm{Im}\, \Phi(x,z)}{\partial\nu(x)} u_{\mathrm{sc}}(x) - ik\mathrm{Im}\, \Phi(x,z) u_{\mathrm{sc}}(x) \right) \mathrm{d}s(x) \right|^{\rho}. \tag{2.9}$$

3. The deep neural network

In this section, we use the sampling method along with a deep neural network of U-Net-Xception style to form a combined method to solve the inverse problem. Numerical simulation was done for both simulated and experimental scattering data.

3.1. U-Net Xception style. Xception is an efficient architecture which relies on two main points: Depthwise Separable Convolution and Shortcuts between Convolution blocks. Xception architecture is an alternative to classical convolution layers, which performs more efficiently in terms of computation time and accuracy. See details of Xception architecture in [**8**].

Firstly, Depthwise Separable Convolution includes two subnetworks: Depthwise Convolution and Pointwise Convolution. Compared to conventional convolution layers, Depthwise Separable Convolution does not need to compute convolution operations across all channels, which will eliminate the number of parameters in the model and makes it simpler. The overall architecture of Depthwise Separable Convolution is described in Figure 1.

Secondly, there are residual (shortcut/skip) connections in Xception networks. Many state-of-the-art Deep neural networks (Resnet, YOLOv3, MobileNetV2, Cascading Residual Network) have shown that accuracy is much higher when residual

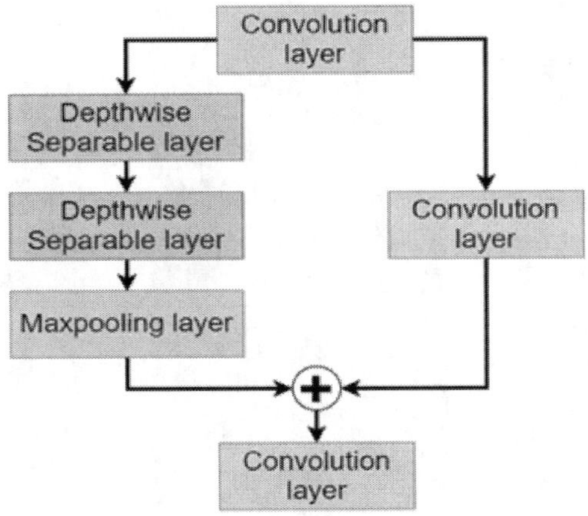

FIGURE 2. Residual connections within Xception networks.

connections are used, see [**2, 16, 30, 31**]. The pattern of residual connection in U-Net Xception architecture is shown in Figure 2.

3.2. Training the network. Given some scattering object and an incident wave we use the spectral method studied in [**25**] to solve the direct problem and generate the associated scattering data. The image of the scattering object is called a true image. Applying the sampling method to the scattering data we obtain a computed image of the scattering object. We call this image a preliminary image.

The training data set includes $60,000$ pairs of images, each pair consists of a true image and a preliminary image. The true images are those of either one or two ellipses whose center and radii are randomly generated within the square domain $(-2,2)^2$. More precisely, if the object consists of only one ellipse, the center is randomly generated within the square $[-0.8,0.8]^2$ and the radii are randomly generated within the interval $[0.1,1]$; if the object consists of two ellipses, the first one is generated in the same manner as before, while the second one has its center randomly generated within the square $[-1,1]^2$ and radii randomly generated within the interval $[0.1,0.5]$. Note that the two ellipses are allowed to overlap each other. The wave number associated with the data set is $k = 6$, which means the wavelength is about 1. As for the incident wave, we use $u^{in}(x) = e^{ik(x_1 \cos\theta + x_2 \sin\theta)}$ where $\theta = 90^o$ for 30,000 images and $\theta = 45^o$ for the other 30,000 images. The Cauchy scattering data are computed on the circle of radius 100. We did not add any artificial noise to the scattering data for the training and testing of the network. See Figure 3 for some examples of training data pairs.

All images are uniformly partitioned into 160×160 pixels. Let $I = \{1,2,\dots,160^2\}$. The neural network is initialized with random parameters. Using a preliminary image as input, the network produces the corresponding output $\hat{y} = \{\hat{y}_i\}_{i\in I}$ that will then be used along with the true image $y = \{y_i\}_{i\in I}$ to calculate the loss function

$$L = -\sum_{i\in I}(y_i \log \hat{y}_i + (1-y_i)\log(1-\hat{y}_i)).$$

After that, the parameters within the convolution layers are tuned using Adam method (see [**22**]) in order to minimize the loss function. This minimization process is done for one batch of size 32 at a time, with the loss function for each batch being

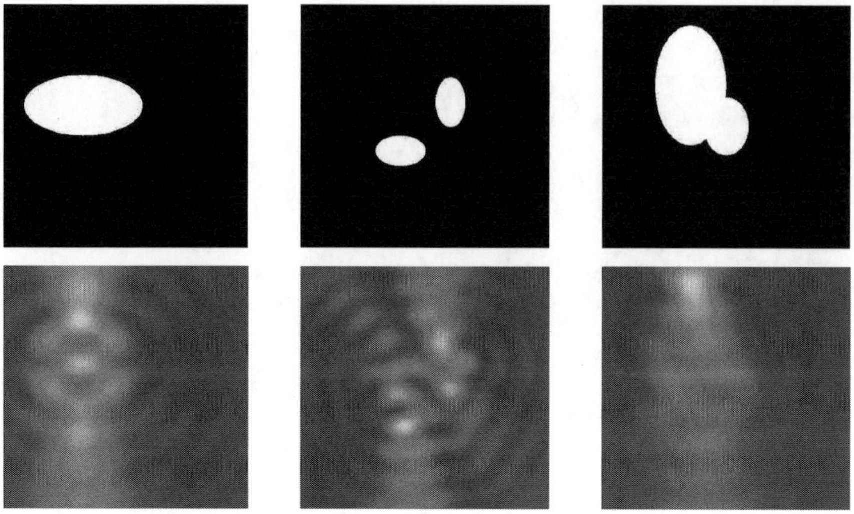

FIGURE 3. Examples of three training data pairs: true images (top), preliminary images (bottom).

the sum of those of individual training pairs in the batch. The initial learning rate is 10^{-2}, and will be divided by 10 if the validation loss does not decrease or validation accuracy does not increase after 5 epochs until it reaches 10^{-6}. The neural network's architecture can be briefly described by the diagram in Figure 4.

Out of 60,000 pairs of images, 50,000 were used to train the network, 5,000 were used for validation and 5,000 are used for testing. In Figure 5, we show the result of training loss versus validation loss and training accuracy versus validation accuracy. The loss value at each epoch is the average value of L over all paired images, whereas the accuracy value is the ratio of the number of pixels in the predicted image that match their counterparts in the true image to the total number of pixels. Those values for validation were computed over $5,000$ images in the validation set, and those values for training were computed over $50,000$ images in the training set.

In order to enrich the training data set, we used *data augmentation*. More specifically, we produced five copies of each training pair by applying five types of transformations: horizontal flip, vertical flip, 90° rotation, -90° rotation and zoom. Zoom transformations were done by first upscaling the image by 5%, then using a random frame of the same size as the original image to crop it. See Figure 6 for an example of such augmentations.

Our choice for some hyperparameters was based on direct comparison to other choices in terms of validation loss and accuracy. For example, we trained the model with $10,000$ pairs of images through 5 epochs with different input image sizes, then we chose $160 \times 160 \times 3$ because it gave the best values for loss and accuracy. We did a similar experiment for two different optimization methods, which are Adam method and stochastic gradient descent with momentum (SGD), and Adam method performed better for this problem. Detailed results for the comparisons are given in Table 1 and 2. As for the other hyperparameters such as the number of hidden layers and units, we followed the same choice as in [**9**]. With this choice,

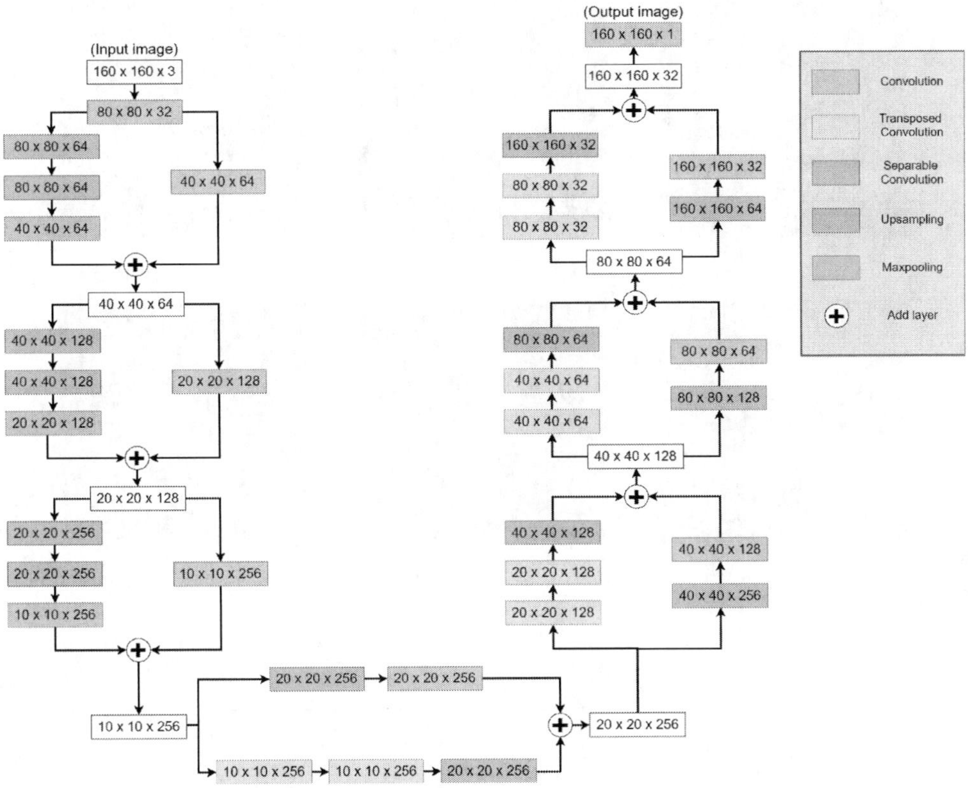

FIGURE 4. Architecture of the neural network.

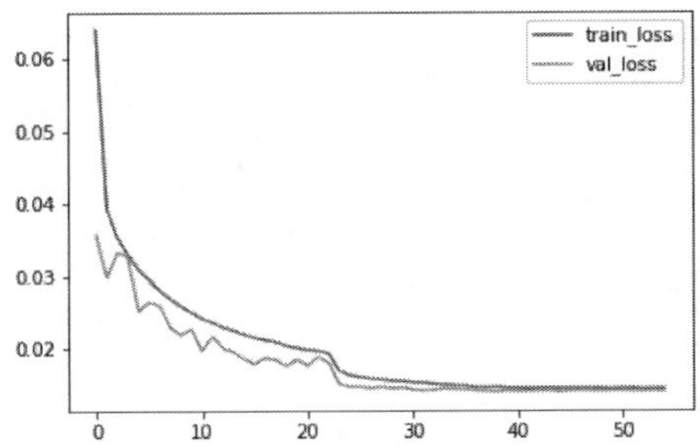

Loss

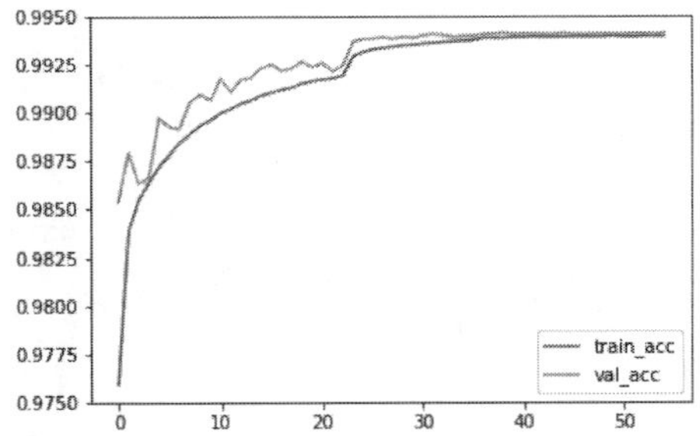

Accuracy

FIGURE 5. Training and validation results through 50 epochs.

we were able to get very good loss and accuracy values, see Figure 5. Therefore, no adjustment was needed.

We also compared the performance of the U-Net Xception style with a modified U-Net in [**12**]. The result in Table 3 shows that U-Net Xception style works better for our problem.

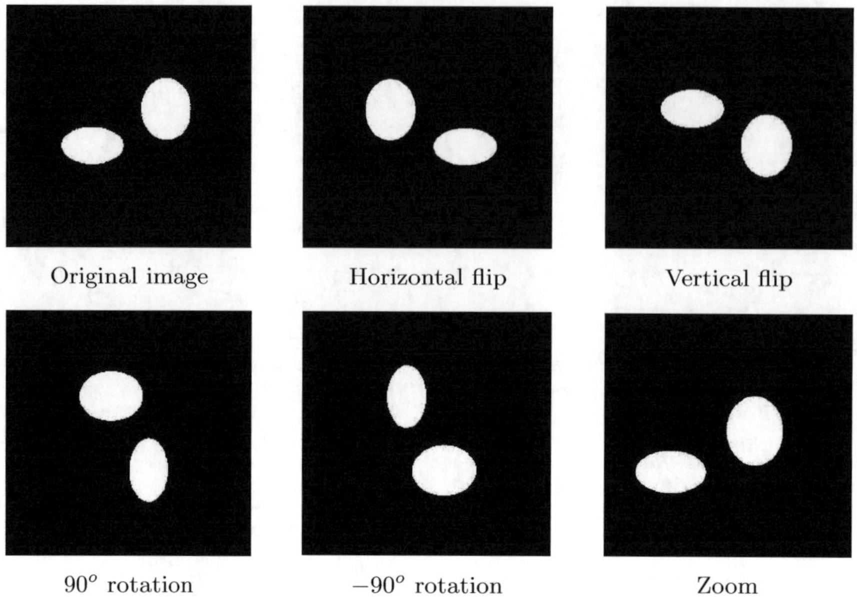

FIGURE 6. An example of data augmentation

Input image size	Validation accuracy	Validation loss
$32 \times 32 \times 3$	0.9718	0.0677
$64 \times 64 \times 3$	0.9762	0.0579
$128 \times 128 \times 3$	0.9728	0.0713
$160 \times 160 \times 3$	**0.9822**	**0.0452**
$192 \times 192 \times 3$	0.9749	0.0708

TABLE 1. Input image size comparison

Optimization method	Validation accuracy	Validation loss
SGD	0.9751	0.0622
Adam	**0.9822**	**0.0452**

TABLE 2. Optimization method comparison

Model	Validation accuracy	Validation loss
Modified U-Net	0.9876	0.0299
U-Net Xception style	**0.9932**	**0.0175**

TABLE 3. Model comparison

4. Numerical study for simulated data

In this section we use the combined method to reconstruct objects from simulated scattering data. We consider the types of objects varying from those similar to the training data set to those entirely different. We also tested different wave numbers and incident wave directions.

4.1. Implementation of the sampling method. In our numerical study the sampling domain is chosen as $[-2,2]^2$. The sampling domain contains the unknown scatterer D and is uniformly discretized by 64 sampling points in each direction. The imaging functional $\mathcal{I}(z)$ in (2.4) is evaluated at each sampling point z of the sampling domain. As mentioned in the section of training the neural network the data $(u_{\text{sc}}, \partial u_{\text{sc}}/\partial \nu)$ are given at 32 uniformly distributed points on the circle of radius 100. With the data given, the evaluation of $\mathcal{I}(z)$ is simple since after a change of variables using polar coordinates we numerically evaluate a single integral using the rectangular rule. In the numerical simulation the imaging functional is normalized by dividing by its (numerical) maximal value. Note that we make sure that the sampling point z and the data point x are away from each other so that the integrands are all regular functions. Recall that the scattering data are measured on $\partial\Omega$ meaning that we should look for the unknown scatterer D in a sampling domain that is strictly contained in Ω.

4.2. Reconstruction of objects in the testing data set. In this part, we present some reconstruction results of the combined method for objects in the testing data set. These objects were generated in a way that is similar to that of the training data sets, meaning the objects consist of one or two ellipses. The wave number k and incident direction θ were those the neural network was trained with. Moreover, the scattering data generated by solving the direct problem was not perturbed by noise.

In Figures 7–8, four test objects are presented: one large disk, two ellipses of similar sizes, two ellipses of different sizes and two overlapping ellipses. We can see that the method was able to reconstruct very well the geometry of these test objects. As it was done for the validation and training data sets in Figure 5, the accuracy for 5000 pairs of images in the testing data set is also computed as 99.4%. This once again confirms that the training process was effective as indicated by the training and validation accuracy in Figure 5.

4.3. Reconstruction of objects with no elliptical shapes. In this part, we test the method for other types of objects with noisy scattering data. We added $5\%, 7\%, 10\%$ and 15% of artificial noise to the scattering data obtained from solving the direct problem with wave number $k = 6$. Recall that the inverse problem with

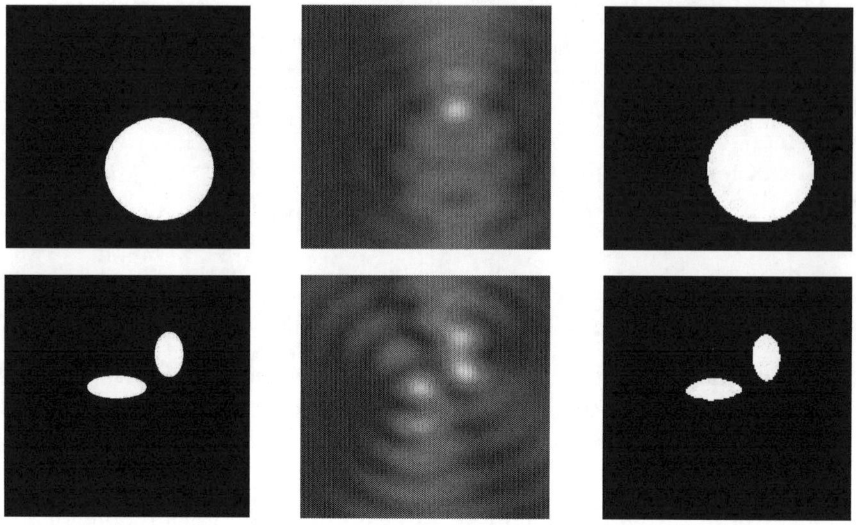

True objects Sampling method Combined method

FIGURE 7. Reconstruction of objects in the testing data set with $k = 6$ and $\theta = 90^o$: one disk (top) and two ellipses of similar sizes (bottom).

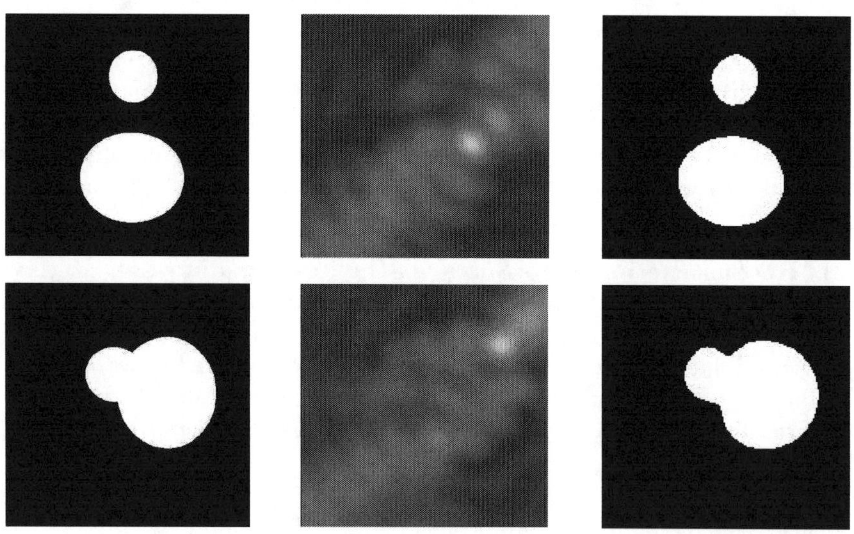

True objects Sampling method Combined method

FIGURE 8. Reconstruction of objects in the testing data set with $k = 6$ and $\theta = 45^o$: two ellipses of different sizes (top) and two overlapping ellipses (bottom).

one incident wave is extremely ill-posed, and there were no artificial noise added to the scattering data in the training process. The tests with noisy data for non-elliptical objects are to show that the network can generalize its training to more

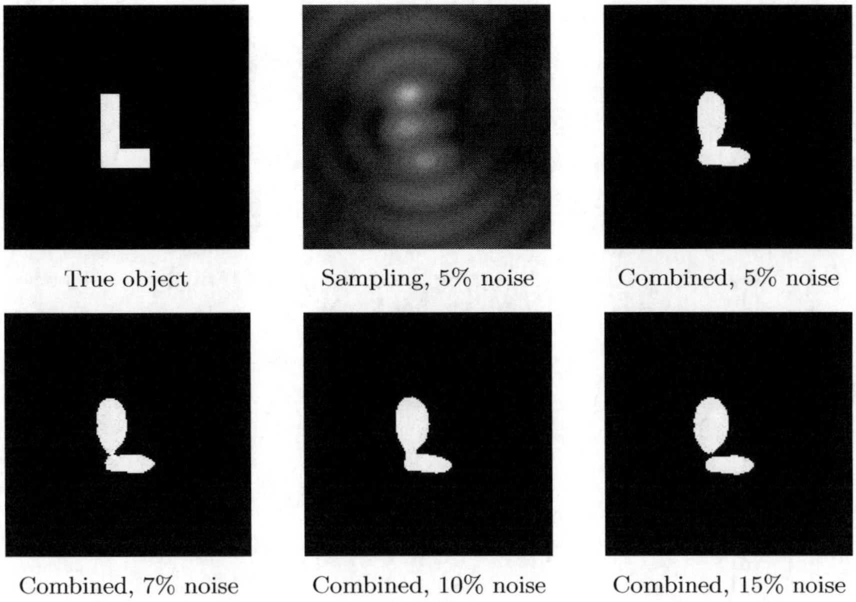

FIGURE 9. Reconstruction of an L-shaped object with $k = 6$, $\theta = 90^o$, and different noise levels added to the scattering data: 5%, 7%, 10% and 15%.

realistic situations that are unseen from the training process. We first used the same incident direction as in training, $\theta = 90^o$, to reconstruct an L-shaped object, see Figure 9. The combined method was able to provide reasonable reconstructions for the L-shaped object with different levels of noise.

Next, in Figures 10–12, we test the method with different non-elliptical shapes and noisy scattering data associated with an incident direction different from which was used in the training process. This is to further assert its flexibility. Recall that we trained the neural network with $\theta = 90^o$ and $\theta = 45^o$. We reconstructed a peanut-shaped object, a T-shaped object and an object consisting of a disk and a rectangle using incident direction $\theta = 220^o$, see Figure 10–12. Again, we were able to get satisfactory results regardless of the circumstances.

For scattering data from a single incident direction and a fixed wave number, it is almost impossible not only for sampling methods but also for many other inversion methods to reconstruct the shape and location of extended objects. Therefore, these results show that the deep neural network has significantly improved the reconstruction capability of the imaging functional under lack of data. Moreover, they also show that the combined method is flexible, meaning it can work well in some situations that are unseen from the training process.

5. Numerical study for experimental data

The results from simulated scattering data show that the proposed deep learning approach for OSM is capable of reconstructing different types of geometries in different setups. Therefore in this section, we demonstrate the effectiveness of our method when it comes to experimental data.

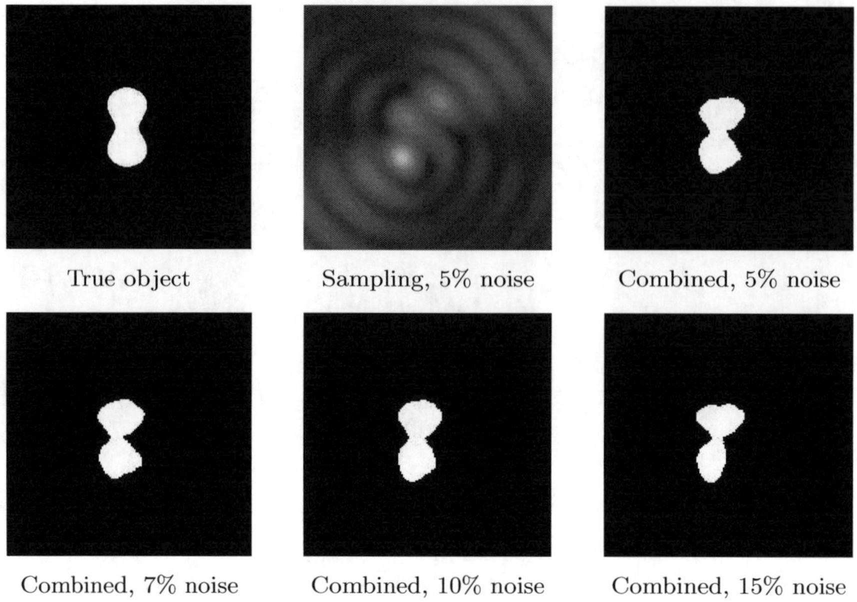

FIGURE 10. Reconstruction of a peanut-shaped object with $k = 6$, $\theta = 220^o$, and different noise levels added to the scattering data: 5%, 7%, 10% and 15%.

We used the data sets for two-dimensional homogeneous objects provided by Fresnel Institute [**5**]. Unlike the simulated data that are computed on an entire circle of radius 100 these experimental data are measured on the arc of a circular sector of angle 240° (from 60° to 300° with step size 5°). The radius of the circle is 19 (about 760 mm). Two data sets are investigated: the first one named *dielTM_dec4f.exp* deals with a de-centered circular cross section of radius 15 mm, and the second one is *rectTM_cent.exp* concerning a centered rectangular cross section of dimensions 25.4×12.7 mm^2. We refer to [**5**] for detailed descriptions of the experimental setup.

We rescale 40 millimeters to be 1 unit of length in our MATLAB simulations. No other adaptation to the experimental system was necessary. The two scatterers were both tested at wave frequency 8 GHz (wave number k is about 6.7) and with incident direction $\theta = 90^o$. See reconstructions using our combined method in Figure 13 and Figure 14.

We see that the proposed method can handle limited-aperture experimental data very well without any additional transfer learning. For the first object, we are able to achieve a reasonable reconstruction of both shape and location of the target. The second object is outside the training process, but the inverted result still can capture highly accurate information for its location and provide a pretty good estimate for its shape.

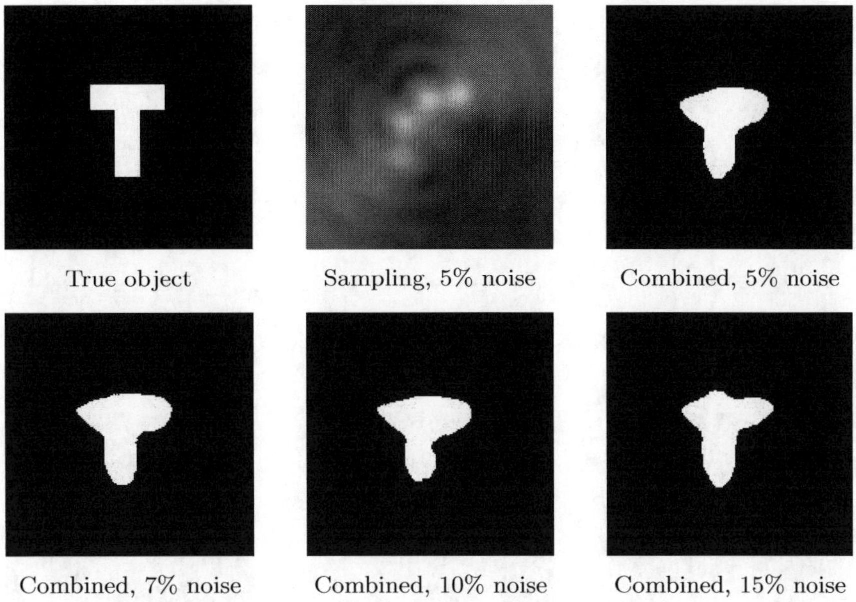

FIGURE 11. Reconstruction of a T-shaped object with $k = 6$, $\theta = 220^o$, and different noise levels added to the scattering data: 5%, 7%, 10% and 15%.

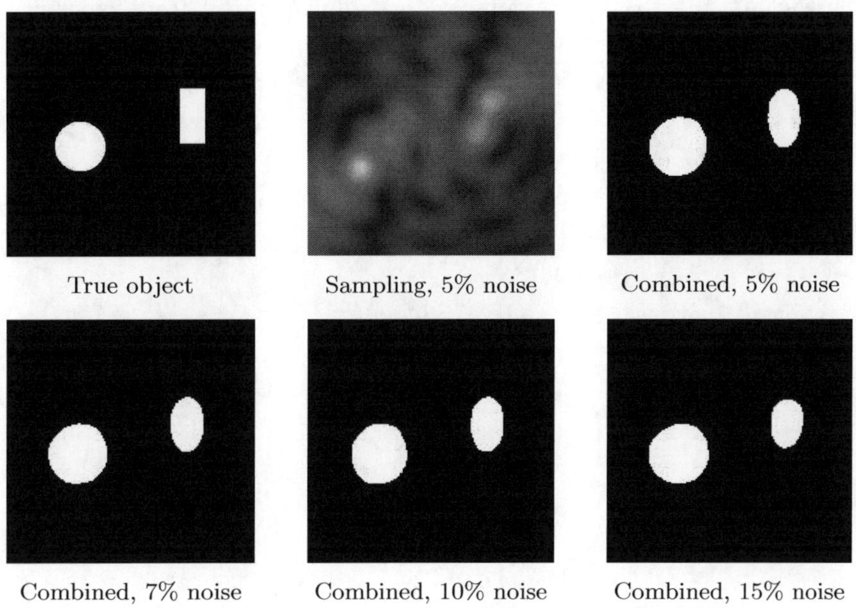

FIGURE 12. Reconstruction of a rectangle and a disk with $k = 6$, $\theta = 220^o$, and different noise levels added to the scattering data: 5%, 7%, 10% and 15%.

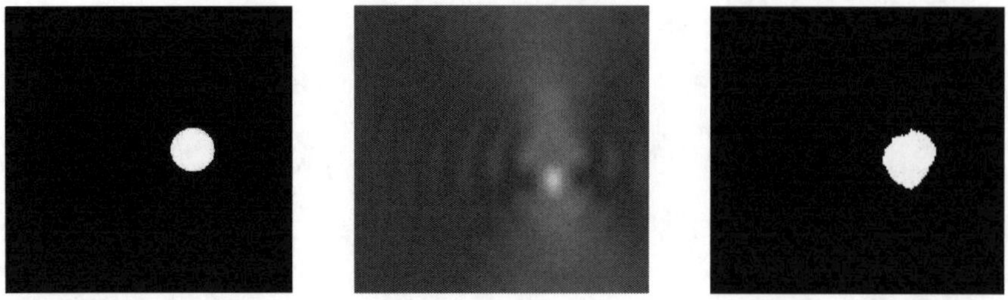

True object Sampling method Combined method

FIGURE 13. Reconstruction of a circular object from experimental data at 8 GHz, $\theta = 90^o$.

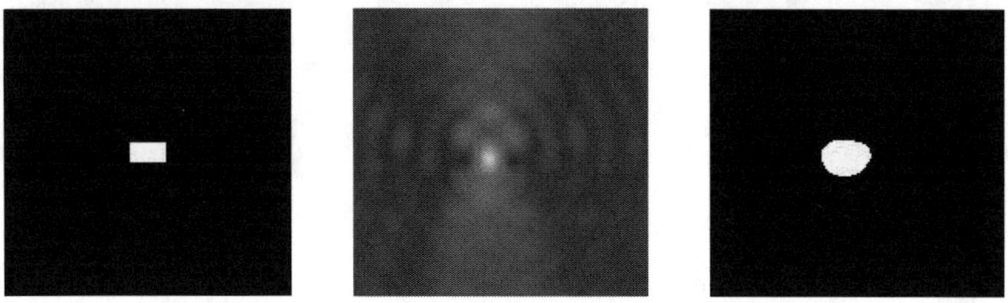

True object Sampling method Combined method

FIGURE 14. Reconstruction of a rectangular object from experimental data at 8 GHz, $\theta = 90^o$.

References

[1] Chi Young Ahn, Seongje Chae, and Won-Kwang Park, *Fast identification of short, sound-soft open arcs by the orthogonality sampling method in the limited-aperture inverse scattering problem*, Appl. Math. Lett. **109** (2020), 106556, 8, DOI 10.1016/j.aml.2020.106556. MR4109343

[2] N. Ahn, B. Kang, and K.-A. Sohn, *Fast, accurate, and lightweight super-resolution with cascading residual network*, `arXiv:1803.08664` (2018).

[3] M. Akinci, M. Cayoren, and I. Akduman, *Near-field orthogonality sampling method for microwave imaging: Theory and experimental verification*, IEEE Trans. Microw. Theory Tech. **64** (2016), 2489.

[4] Stephan Antholzer, Markus Haltmeier, and Johannes Schwab, *Deep learning for photoacoustic tomography from sparse data*, Inverse Probl. Sci. Eng. **27** (2019), no. 7, 987–1005, DOI 10.1080/17415977.2018.1518444. MR3938045

[5] Kamal Belkebir and Marc Saillard, *Guest editors' introduction: Testing inversion algorithms against experimental data: inhomogeneous targets*, Inverse Problems **21** (2005), no. 6, S1–S3, DOI 10.1088/0266-5611/21/6/S01. MR2183641

[6] Fioralba Cakoni, David Colton, and Peter Monk, *The linear sampling method in inverse electromagnetic scattering*, CBMS-NSF Regional Conference Series in Applied Mathematics, vol. 80, Society for Industrial and Applied Mathematics (SIAM), Philadelphia, PA, 2011, DOI 10.1137/1.9780898719406. MR2731610

[7] X. Chen, Z. Wei, M. Li, and P. Rocca, *A review of deep learning approaches for inverse scattering problems*, Progress In Electromagnetics Research **167** (2020), 67–81.

[8] F. Chollet, *Xception: Deep learning with depthwise separable convolutions*, `arXiv:1610.02357` (2016).

[9] F. Chollet et al., *Keras*, `https://github.com/keras-team/keras-io/blob/master/examples/vision/oxford_pets_image_segmentation.py`, 2021.

[10] David Colton, Joe Coyle, and Peter Monk, *Recent developments in inverse acoustic scattering theory*, SIAM Rev. **42** (2000), no. 3, 369–414, DOI 10.1137/S0036144500367337. MR1786932

[11] David Colton and Rainer Kress, *Inverse acoustic and electromagnetic scattering theory*, 3rd ed., Applied Mathematical Sciences, vol. 93, Springer, New York, 2013, DOI 10.1007/978-1-4614-4942-3. MR2986407

[12] M. Daoust et al., *Tensorflow*, `https://github.com/tensorflow/docs/blob/master/site/en/tutorials/images/segmentation.ipynb`, 2021.

[13] Roland Griesmaier, *Multi-frequency orthogonality sampling for inverse obstacle scattering problems*, Inverse Problems **27** (2011), no. 8, 085005, 23, DOI 10.1088/0266-5611/27/8/085005. MR2819947

[14] S. J. Hamilton and A. Hauptmann, *Deep d-bar: Real-time electrical impedance tomography imaging with deep neural networks*, IEEE Trans. Medical Imaging **37** (2018), 2367–2377.

[15] Isaac Harris and Dinh-Liem Nguyen, *Orthogonality sampling method for the electromagnetic inverse scattering problem*, SIAM J. Sci. Comput. **42** (2020), no. 3, B722–B737, DOI 10.1137/19M129783X. MR4101374

[16] Dong Yu and Li Deng, *Automatic speech recognition*, Signals and Communication Technology, Springer, London, 2015. A deep learning approach; With a foreword by Sadaoki Furui. MR3559045

[17] Kazufumi Ito, Bangti Jin, and Jun Zou, *A direct sampling method to an inverse medium scattering problem*, Inverse Problems **28** (2012), no. 2, 025003, 11, DOI 10.1088/0266-5611/28/2/025003. MR2876854

[18] Kazufumi Ito, Bangti Jin, and Jun Zou, *A direct sampling method for inverse electromagnetic medium scattering*, Inverse Problems **29** (2013), no. 9, 095018, 19, DOI 10.1088/0266-5611/29/9/095018. MR3104934

[19] Kyong Hwan Jin, Michael T. McCann, Emmanuel Froustey, and Michael Unser, *Deep convolutional neural network for inverse problems in imaging*, IEEE Trans. Image Process. **26** (2017), no. 9, 4509–4522, DOI 10.1109/TIP.2017.2713099. MR3670561

[20] Sangwoo Kang and Marc Lambert, *Structure analysis of direct sampling method in 3D electromagnetic inverse problem: near- and far-field configuration*, Inverse Problems **37** (2021), no. 7, Paper No. 075002, 27, DOI 10.1088/1361-6420/abfe4e. MR4281488

[21] Yuehaw Khoo and Lexing Ying, *SwitchNet: a neural network model for forward and inverse scattering problems*, SIAM J. Sci. Comput. **41** (2019), no. 5, A3182–A3201, DOI 10.1137/18M1222399. MR4018415

[22] D. Kingma and J. Ba, *Adam: A method for stochastic optimization*, International Conference on Learning Representations (2014).

[23] Andreas Kirsch and Natalia Grinberg, *The factorization method for inverse problems*, Oxford Lecture Series in Mathematics and its Applications, vol. 36, Oxford University Press, Oxford, 2008. MR2378253

[24] Thu Le, Dinh-Liem Nguyen, Hayden Schmidt, and Trung Truong, *Imaging of 3D objects with experimental data using orthogonality sampling methods*, Inverse Problems **38** (2022), no. 2, Paper No. 025007, 20, DOI 10.1088/1361-6420/ac3d85. MR4374010

[25] Armin Lechleiter and Dinh-Liem Nguyen, *A trigonometric Galerkin method for volume integral equations arising in TM grating scattering*, Adv. Comput. Math. **40** (2014), no. 1, 1–25, DOI 10.1007/s10444-013-9295-2. MR3158012

[26] L. Li, L. G. Wang, F. L. Teixeira, C. Liu, A. Nehorai, and T. J. Cui, *Deepnis: Deep neural network for nonlinear electromagnetic inverse scattering*, IEEE Trans. Antennas and Propagation **67** (2019), 1819–1825.

[27] Dinh-Liem Nguyen, *Direct and inverse electromagnetic scattering problems for bi-anisotropic media*, Inverse Problems **35** (2019), no. 12, 124001, 27, DOI 10.1088/1361-6420/ab382d. MR4039925

[28] Roland Potthast, *A survey on sampling and probe methods for inverse problems*, Inverse Problems **22** (2006), no. 2, R1–R47, DOI 10.1088/0266-5611/22/2/R01. MR2216404

[29] Roland Potthast, *A study on orthogonality sampling*, Inverse Problems **26** (2010), 074015.

[30] J. Redmon and A. Farhadi, *Yolov3: An incremental improvement*, arXiv:1804.02767 (2018).

[31] M. Sandler, A. Howard, M. Zhu, A. Zhmoginov, and L.-C. Chen, *Mobilenetv2: Inverted residuals and linear bottlenecks*, arXiv:1801.04381 (2019).

[32] J. Xiao, J. Li, Y. Chen, F. Han, and Q. H. Liu, *Fast electromagnetic inversion of inhomogeneous scatterers embedded in layered media by born approximation and 3-D U-Net*, IEEE Geoscience and Remote Sensing Letters **17** (2020), 1677–1681.

Department of Mathematics, Kansas State University, Manhattan, Kansas 66506
Email address: thule@ksu.edu

Department of Mathematics, Kansas State University, Manhattan, Kansas 66506
Email address: dlnguyen@ksu.edu

INRS, University of Quebec, Montreal, Quebec, Canada
Email address: Thai.Vu.Nguyen@inrs.ca

Department of Mathematics, Kansas State University, Manhattan, Kansas 66506
Email address: trungt@ksu.edu

Contemporary Mathematics
Volume **784**, 2023
https://doi.org/10.1090/conm/784/15754

Fast numerical solutions to direct and inverse scattering for bi-anisotropic periodic Maxwell's equations

Dinh-Liem Nguyen and Trung Truong

ABSTRACT. This paper is concerned with the numerical solution to the direct and inverse electromagnetic scattering problem for bi-anisotropic periodic structures. The direct problem can be reformulated as an integro-differential equation. We study the existence and uniqueness of solution to the latter equation and analyze a spectral Galerkin method to solve it. This spectral method is based on a periodization technique which allows us to avoid the evaluation of the quasiperiodic Green's tensor and to use the fast Fourier transform in the numerical implementation of the method. For the inverse problem, we study the orthogonality sampling method to reconstruct the periodic structures from scattering data generated by only two incident fields. The sampling method is fast, simple to implement, regularization free, and very robust against noise in the data. Numerical examples for both direct and inverse problems are presented to examine the efficiency of the numerical solvers.

1. Introduction

We study in this paper numerical methods for solving both direct and inverse electromagnetic scattering problems for bi-anisotropic periodic media. This work is motivated by applications of electromagnetic scattering from complex periodic media (e.g. bi-anisotropic media, chiral media) in optics and metamaterials; and nondestructive evaluations [**9, 13, 15, 27**].

Results on the well-posedness of the direct problem for both bounded and periodic scattering media can be found in [**2, 5, 7, 8, 26, 29, 30**]. To our knowledge results on numerical methods for direct and inverse scattering problems for electromagnetic complex media are limited. The authors in [**3**] studied numerical analysis of a finite element method for solving the direct problem for periodic chiral media. Spectral Galerkin methods for the integro-differential equation formulation of the direct scattering problem were studied in [**24**] for chiral bounded objects and in [**23**] for bi-anisotropic bounded objects. The advantages of the spectral Galerkin method studied in [**23**] are that this spectral method, which is based on a periodization technique, allows us to avoid the evaluation of the quasiperiodic Green's tensor and to use the fast Fourier transform in the numerical implementation of

2020 *Mathematics Subject Classification.* Primary 65R20; Secondary 35R30, 78M22.

Key words and phrases. Spectral Galerkin method, Maxwell's equations, bi-anisotropic periodic media, inverse scattering, orthogonality sampling method.

This work was supported in part by NSF Grant DMS-1812693.

the method. In this paper we study this spectral method to solve the direct problem for bi-anisotropic periodic media by extending the results in [**21**, **23**]. More precisely, we follow the periodization process of [**21**] for the integro-differential equation. The Garding estimates for the integro-differential equation can be done similarly as [**23**]. The uniqueness of solution, the mapping properties of the periodized integro-differential operators, and the numerical examples for the Galerkin method are the new ingredients of the study for the direct scattering problem in this paper.

For the inverse problem, we refer to [**6**, **17**, **19**, **20**] for some uniqueness results in the case of bounded scattering objects. To numerically solve the inverse problem, the factorization method was studied in [**12**] for the case of chiral bounded objects. This method was also studied in [**22**, **25**] for periodic chiral and bi-anisotropic periodic media. However, the factorization method requires the scattering data associated with multiple incident fields. In this paper we implement the orthogonality sampling method (OSM) to solve the inverse problem with only a few incident fields. The OSM we implement is also very stable against noise in the data. The OSM was introduced by Potthast in [**28**] and has attracted increasing interests thanks to its computational efficiency. We refer to [**1**, **10**, **11**, **16**] for some other results on the OSM for inverse scattering problems.

The paper is organized as follows. The direct problem and its basic setup are presented in Section 2. We discuss in Section 3 the integro-differential formulation for the direct problem and its well-posedness. A spectral Galerkin method, its convergence analysis are presented in Section 4. Numerical examples of the direct solver are presented in Section 5. The inverse problem and the OSM are studied in Section 6.

2. The problem setup

We are concerned with the scattering problem for a bi-anisotropic medium which is 2π-periodic in x_1, x_2 and bounded in x_3 (here x_1, x_2, x_3 are the three coordinates of a vector $\mathbf{x} = (x_1, x_2, x_3)^\top$ in $\mathbb{R}^3$). Let $D_0 \subset \mathbb{R}^3$ be the interior of the medium. The problem is governed by the time-harmonic Maxwell's equations with frequency $\omega > 0$

$$\operatorname{curl} \mathbf{E} + i\omega \mathbf{B} = 0, \quad \operatorname{curl} \mathbf{H} - i\omega \mathbf{D} = 0, \quad \text{in } \mathbb{R}^3, \tag{2.1}$$

where the electric field $\mathbf{E}$, the magnetic field $\mathbf{H}$, the electric flux density $\mathbf{D}$ and the magnetic flux density $\mathbf{B}$ are three dimensional vector-valued functions, along with the constitutive relations

$$\mathbf{B} = \mu \mathbf{H} + \xi \sqrt{\varepsilon_0 \mu_0}\, \mathbf{E}, \quad \mathbf{D} = \varepsilon \mathbf{E} + \overline{\xi} \sqrt{\varepsilon_0 \mu_0}\, \mathbf{H} \tag{2.2}$$

where the permittivity ε and the permeability μ are 3×3 matrix-valued bounded functions. We assume that these functions are 2π-periodic in x_1 and x_2 and that they are respectively equal $\mu_0 I_3$ and $\varepsilon_0 I_3$ in $\mathbb{R}^3 \setminus \overline{D_0}$ (I_3 is the identity matrix). The function ξ is related to the magnetoelectric effect (see [**18**]) which is also a 3×3 matrix-valued bounded function. This function is also assumed to be 2π-periodic in x_1 and x_2 and to be zero in $\mathbb{R}^3 \setminus \overline{D_0}$.

We introduce the relative quantities

$$\varepsilon_r := \frac{\varepsilon}{\varepsilon_0}, \quad \mu_r := \frac{\mu}{\mu_0}$$

and the scaled quantities (with the same notations as the original ones)

$$\mathbf{E} := \sqrt{\varepsilon_0}\, \mathbf{E}, \quad \mathbf{H} := \sqrt{\mu_0}\, \mathbf{H}.$$

Using these new quantities and plugging (2.2) into (2.1) we obtain

$$\operatorname{curl} \mathbf{E} - ik(\mu_r \mathbf{H} + \xi \mathbf{E}) = 0, \quad \operatorname{curl} \mathbf{H} + ik(\varepsilon_r \mathbf{E} + \overline{\xi} \mathbf{H}) = 0 \tag{2.3}$$

where $k := \omega\sqrt{\varepsilon_0 \mu_0}$ is the wave number.

Assume that μ_r is invertible almost everywhere in $\mathbb{R}^3$, by the first equation in (2.3) we can write

$$\mathbf{H} = -\frac{i}{k}\mu_r^{-1} \operatorname{curl} \mathbf{E} - \mu_r^{-1}\xi \mathbf{E}. \tag{2.4}$$

Plugging this into the second one and rearranging the resulting equation we obtain

$$\operatorname{curl}(\mu_r^{-1} \operatorname{curl} \mathbf{E}) + ik\left[\overline{\xi}\mu_r^{-1} \operatorname{curl} \mathbf{E} - \operatorname{curl}(\mu_r^{-1}\xi\mathbf{E})\right] - k^2(\varepsilon_r - \overline{\xi}\mu_r^{-1}\xi)\mathbf{E} = 0. \tag{2.5}$$

Let $\mathbf{d} = (d_1, d_2, d_3) \in \mathbb{R}^3$ be the wave vector satisfying $d_3 \neq 0$ and $|\mathbf{d}| = k$, we consider the plane electromagnetic wave given by

$$\mathbf{E}^{in} = \frac{1}{\omega\varepsilon_0}(\mathbf{s} \times \mathbf{d}) e^{i\mathbf{d}\cdot\mathbf{x}}, \quad \mathbf{H}^{in} = \mathbf{s} e^{i\mathbf{d}\cdot\mathbf{x}} \tag{2.6}$$

where $\mathbf{s}$ is the polarization vector satisfying $\mathbf{s} \cdot \mathbf{d} = 0$. The interaction between this plane wave and the medium gives rise to the scattered wave $(\mathbf{E}^{sc}, \mathbf{H}^{sc})$, and the total wave $(\mathbf{E}, \mathbf{H})$ given by $\mathbf{E} = \mathbf{E}^{in} + \mathbf{E}^{sc}$, $\mathbf{H} = \mathbf{H}^{in} + \mathbf{H}^{sc}$ satisfies (2.4)–(2.5). From now on we will be working with $\mathbf{E}$ only since $\mathbf{E}$ and $\mathbf{H}$ are intimately related by (2.4).

For simplicity we denote $\mathbf{u} := \mathbf{E}^{sc}$. Note that by construction, $\mathbf{E}^{in}$ satisfies

$$\operatorname{curl} \operatorname{curl} \mathbf{E}^{in} - k^2 \mathbf{E}^{in} = 0,$$

thus (2.5) can be rewritten as

$$\begin{aligned} \operatorname{curl}^2 \mathbf{u} - k^2 \mathbf{u} = k^2 &\left[(P - \overline{\xi}\mu_r^{-1}\xi)(\mathbf{E}^{in} + \mathbf{u}) + \frac{i}{k}\overline{\xi}\mu_r^{-1}(\operatorname{curl} \mathbf{E}^{in} + \operatorname{curl} \mathbf{u})\right] \\ &+ \operatorname{curl}\left[Q(\operatorname{curl} \mathbf{E}^{in} + \operatorname{curl} \mathbf{u}) - ik\mu_r^{-1}\xi(\mathbf{E}^{in} + \mathbf{u})\right] \end{aligned} \tag{2.7}$$

where $P := \varepsilon_r - I$ and $Q := I - \mu_r^{-1}$ are the contrasts. It is clear that outside of $\overline{D_0}$, ε_r and μ_r equal the identity matrix, so $P = Q = 0$. Thus we have

$$\operatorname{supp}(P) \cup \operatorname{supp}(Q) \subset \overline{D_0}.$$

We now define that for $\boldsymbol{\alpha} = (\alpha_1, \alpha_2, 0) \in \mathbb{R}^3$, a function $\mathbf{v} : \mathbb{R}^3 \to \mathbb{R}^3$ is called *$\boldsymbol{\alpha}$-quasiperiodic* if for any $n = (n_1, n_2, 0) \in \mathbb{Z}^3$

$$\mathbf{v}(x_1 + n_1 2\pi, x_2 + n_2 2\pi, x_3) = e^{2\pi i \alpha \cdot n} \mathbf{v}(x_1, x_2, x_3), \quad \mathbf{x} \in \mathbb{R}^3.$$

Let $\boldsymbol{\alpha} = (\alpha_1, \alpha_2, 0) = (d_1, d_2, 0)$ and $\boldsymbol{\alpha}_m = (\alpha_1 + m_1, \alpha_2 + m_2, 0)$ for $m \in \mathbb{Z}^2$. Then it can be seen that the plane wave $\mathbf{E}^{in}$ is $\boldsymbol{\alpha}$-quasiperiodic. It is well-known that we will seek for the unique solution $\mathbf{u}$ of the direct problem as an $\boldsymbol{\alpha}$-quasiperiodic function.

Next we prescribe a radiation condition for $\mathbf{u}$. Recall that the medium D_0 is bounded in x_3, thus let $h > 0$ be such that

$$h > \sup\{|x_3| : \mathbf{x} \in D_0\} \tag{2.8}$$

and for $m \in \mathbb{Z}^2$ define

$$\beta_m = \begin{cases} \sqrt{k^2 - |\boldsymbol{\alpha}_m|^2}, & |\boldsymbol{\alpha}_m| \le k \\ i\sqrt{|\boldsymbol{\alpha}_m|^2 - k^2}, & |\boldsymbol{\alpha}_m| > k \end{cases},$$

then $\mathbf{u}$ is called radiating if it can be expressed as the following Rayleigh expansion

$$\mathbf{u}(\mathbf{x}) = \sum_{m \in \mathbb{Z}^2} \widehat{\mathbf{u}}_m^{\pm} e^{i(\boldsymbol{\alpha}_m \cdot \mathbf{x} \pm \beta_m (x_3 \mp h))} \quad \text{for } x_3 \gtrless \pm h \tag{2.9}$$

where

$$\widehat{\mathbf{u}}_m^{\pm} = \frac{1}{4\pi^2} \int_{-\pi}^{\pi} \int_{-\pi}^{\pi} e^{-i\boldsymbol{\alpha}_m \cdot \mathbf{x}} \mathbf{u}(x_1, x_2, \pm h) \mathrm{d}x_1 \mathrm{d}x_2, \quad m \in \mathbb{Z}^2.$$

Note that all but finitely many terms in (2.9) are exponentially decaying, which helps us easily deduce absolute convergence of the series for every $\mathbf{x}$. Moreover, we can choose $\boldsymbol{\alpha}$ and k such that β_m are nonzero for all $m \in \mathbb{Z}^2$, and this will be the case for the remaining part of this paper.

We now introduce notation and assumptions related to the spatial domains and function spaces that we will be using throughout this paper. First, due to periodicity of all the introduced quantities and the condition (2.9), we can restrict our problem from $\mathbb{R}^3$ to one period

$$\Omega = (-\pi, \pi)^2 \times \mathbb{R}.$$

Also, for any $r > 0$, we define a truncation of Ω as

$$\Omega_r = (-\pi, \pi)^2 \times (-r, r).$$

Next we introduce some Sobolev spaces for vector-valued functions

$$\begin{aligned} L^2(\mathcal{O})^3 &= \{\mathbf{w} = (w_1, w_2, w_3)^T : w_1, w_2, w_3 \in L^2(\mathcal{O})\}, \\ H(\mathrm{curl}\,, \mathcal{O}) &= \{\mathbf{w} \in L^2(\mathcal{O}, \mathbb{C}^3) : \mathrm{curl}\,\mathbf{w} \in L^2(\mathcal{O}, \mathbb{C}^3)\}, \\ H_{\mathrm{loc}}(\mathrm{curl}\,, \mathcal{O}) &= \{\mathbf{w} : \mathbf{w}|_V \in H(\mathrm{curl}\,, V) \text{ for all } V \subset \mathcal{O} \text{ open, bounded}\}, \\ H_{\boldsymbol{\alpha}}(\mathrm{curl}\,, \mathcal{O}) &= \{\mathbf{w} \in H(\mathrm{curl}\,, \mathcal{O}) : \mathbf{w} = \mathbf{W}|_{\mathcal{O}} \text{ for some } \boldsymbol{\alpha}\text{-quasiperiodic} \\ &\qquad \text{function } \mathbf{W} \in H_{\mathrm{loc}}(\mathrm{curl}\,, \mathbb{R}^3)\}, \\ H_{\boldsymbol{\alpha},\mathrm{loc}}(\mathrm{curl}\,, \mathcal{O}) &= \{\mathbf{w} \in H_{\mathrm{loc}}(\mathrm{curl}\,, \mathcal{O}) : \mathbf{w} = \mathbf{W}|_{\mathcal{O}} \text{ for some } \boldsymbol{\alpha}\text{-quasiperiodic} \\ &\qquad \text{function } \mathbf{W} \in H_{\mathrm{loc}}(\mathrm{curl}\,, \mathbb{R}^3)\}, \end{aligned}$$

along with $L^\infty(\mathbb{R}^3)^{3\times 3}$ which is the space of 3×3 matrix-valued functions whose components are functions in $L^\infty(\mathbb{R}^3)$. We define the following norm on this space

$$\|\gamma\|_\infty = \max_{i,j=\overline{1,3}} \|\gamma_{ij}\|_\infty$$

for $\gamma \in L^\infty(\mathbb{R}^3)^{3\times 3}$.

Let $D = D_0 \cap \Omega$. It is clear that D is open and bounded. Finally, the following assumption holds true throughout the theoretical analysis in this paper.

ASSUMPTION 2.1. Assume that D has Lipschitz boundary; ε_r, μ_r, μ_r^{-1}, ξ are in $L^\infty(\mathbb{R}^3)^{3\times 3}$, symmetric almost everywhere in $\mathbb{R}^3$ and ξ is real-valued. Moreover,

(1) There exist $c_1, c_2 > 0$ such that for all $\mathbf{z} \in \mathbb{C}^3$

$$\mathrm{Re}\,(\mu_r^{-1})\mathbf{z} \cdot \overline{\mathbf{z}} \ge c_1 |\mathbf{z}|^2, \quad \mathrm{Re}\,(\varepsilon_r - \xi \mu_r^{-1} \xi)\mathbf{z} \cdot \overline{\mathbf{z}} \ge c_2 |\mathbf{z}|^2$$

a.e in $\mathbb{R}^3$ and that

$$\| |\mu_r^{-1} \xi|_F \|_\infty < c_1 c_2.$$

(2) There exist $c_3, c_4 > 0$ such that for all $\mathbf{z} \in \mathbb{C}^3$

$$-\mathrm{Im}\,(\mu_r^{-1})\mathbf{z}\cdot\bar{\mathbf{z}} \geq c_3|\mathbf{z}|^2, \qquad \mathrm{Im}\,(\varepsilon_r - \xi\mu_r^{-1}\xi)\mathbf{z}\cdot\bar{\mathbf{z}} \geq c_4|\mathbf{z}|^2$$

a.e in $\mathbb{R}^3$ and that

$$\frac{1}{2}\left(\|\mu_r^{-1}\xi\|_\infty^2 + 1\right) \leq \min\{c_3, c_4\}.$$

Here the norm $\||\cdot|_F\|_\infty$ is defined as, for a 3×3 matrix-valued function $A = (a_{ij})$,

$$\||A|_F\|_\infty = \left\|\left(\sum_{i,j=1}^{3}|a_{ij}|^2\right)^{1/2}\right\|_\infty.$$

3. Integro-differential formulation

In this section, we will reformulate the direct problem into an integro-differential equation. Define the operators

$$\mathcal{S}\mathbf{u} = (P - \xi\mu_r^{-1}\xi)\mathbf{u} + \frac{i}{k}\xi\mu_r^{-1}\mathrm{curl}\,\mathbf{u}, \quad \mathcal{T}\mathbf{u} = Q\mathrm{curl}\,\mathbf{u} - ik\mu_r^{-1}\xi\mathbf{u}$$

and let $\mathbf{f} = \mathcal{S}\mathbf{E}^{in}$, $\mathbf{g} = \mathcal{T}\mathbf{E}^{in}$. Then it is clear that $\mathcal{S}\mathbf{u}$, $\mathcal{T}\mathbf{u}$, $\mathbf{f}$ and $\mathbf{g}$ are compactly supported in D and (2.7) can be written as

$$\mathrm{curl}^2\mathbf{u} - k^2\mathbf{u} = k^2(\mathcal{S}\mathbf{u} + \mathbf{f}) + \mathrm{curl}\,(\mathcal{T}\mathbf{u} + \mathbf{g}). \tag{3.1}$$

Note that (3.1) needs to be solved in the variational sense, therefore we will be seeking solutions in $H_{\boldsymbol{\alpha},\mathrm{loc}}(\mathrm{curl}\,,\Omega)$ and $\mathbf{u}$ is called a solution of (3.1) in the variational sense if it satisfies

$$\int_\Omega(\mathrm{curl}\,\mathbf{u}\cdot\mathrm{curl}\,\overline{\mathbf{v}} - k^2\mathbf{u}\cdot\overline{\mathbf{v}})\,d\mathbf{x} = k^2\int_D(\mathcal{S}\mathbf{u}+\mathbf{f})\cdot\overline{\mathbf{v}}\,d\mathbf{x} + \int_D(\mathcal{T}\mathbf{u}+\mathbf{g})\cdot\mathrm{curl}\,\overline{\mathbf{v}}\,d\mathbf{x} \tag{3.2}$$

for all $\mathbf{v} \in H_{\boldsymbol{\alpha}}(\mathrm{curl}\,,\Omega)$ with compact support.

Let G be the $\boldsymbol{\alpha}$-quasiperiodic Green's function of the Helmholtz equation, then G has the Fourier expansion (see [**4**])

$$G(\mathbf{x}) = \frac{i}{8\pi^2}\sum_{m\in\mathbb{Z}^2}\frac{1}{\beta_m}e^{i(\boldsymbol{\alpha}_m\cdot\mathbf{x}+\beta_m|x_3|)}$$

for $\mathbf{x} \in \mathbb{R}^3$, $\mathbf{x} \neq (2\pi n_1, 2\pi n_2, 0)$, $n_1, n_2 \in \mathbb{Z}$.

The variational problem (3.2) along with the radiation condition (2.9) is equivalent to the following integro-differential equation

$$\mathbf{u} - \mathcal{A}\mathcal{S}\mathbf{u} - \mathcal{B}\mathcal{T}\mathbf{u} = \mathcal{A}\mathbf{f} + \mathcal{B}\mathbf{g} \quad \text{in } \Omega \tag{3.3}$$

where

$$\mathcal{A}\mathbf{h}(\mathbf{x}) := (k^2 + \nabla\mathrm{div})\int_D G(\mathbf{x}-\mathbf{y})\mathbf{h}(\mathbf{y})d\mathbf{y},$$

$$\mathcal{B}\mathbf{h}(\mathbf{x}) := \mathrm{curl}\int_D G(\mathbf{x}-\mathbf{y})\mathbf{h}(\mathbf{y})d\mathbf{y}$$

in the sense that, if a radiating $\mathbf{u} \in H_{\boldsymbol{\alpha},\mathrm{loc}}(\mathrm{curl}\,,\Omega)$ solves (3.2) then $\mathbf{u}|_D$ solves (3.3). Conversely, if $\mathbf{u} \in H_{\boldsymbol{\alpha}}(\mathrm{curl}\,,D)$ solves (3.3) in then $\mathbf{u}$ can be extended by the right-hand side of (3.3) to a radiating solution of (3.2) in $H_{\boldsymbol{\alpha},\mathrm{loc}}(\mathrm{curl}\,,\Omega)$. The proof for this equivalence is similar to that in [**14**].

The operator $I-\mathcal{AS}-\mathcal{BT}$ on the left-hand side of (3.3) satisfies a Garding-type estimate, and thus the Fredholm property. Note that $\overline{D} \subset \Omega_h$, so we only need to know $\mathbf{u}$ on Ω_h and then extend it to the whole Ω using integration. On the solution space $H_{\boldsymbol{\alpha}}(\mathrm{curl}\,, \Omega_h)$, we define the inner product

$$(\cdot,\cdot)_{H_{\boldsymbol{\alpha}}(\mathrm{curl}\,,\Omega_h)} = k^2(\cdot,\cdot)_{L^2(\Omega_h)^3} + (\mathrm{curl}\,\cdot, \mathrm{curl}\,\cdot)_{L^2(\Omega_h)^3}$$

where $(\cdot,\cdot)_{L^2(\Omega_h)^3}$ is the usual inner product in $L^2(\Omega_h)^3$.

Under Assumption 2.1, the following Fredholm property can be proved in a similar manner as in [**23**].

THEOREM 3.1. *There exist a constant $C > 0$ and a compact operator K on $H_{\boldsymbol{\alpha}}(\mathrm{curl}\,, \Omega_h)$ such that*

$$\mathrm{Re}\,(\mathbf{u} - \mathcal{AS}\mathbf{u} - \mathcal{BT}\mathbf{u}, \mathbf{u})_{H_{\boldsymbol{\alpha}}(\mathrm{curl}\,,\Omega_h)} \geq C\|\mathbf{u}\|^2_{H_{\boldsymbol{\alpha}}(\mathrm{curl}\,,\Omega_h)} + \mathrm{Re}\,(K\mathbf{u}, \mathbf{u})_{H_{\boldsymbol{\alpha}}(\mathrm{curl}\,,\Omega_h)}.$$

Since the Fredholm property holds, existence of solution to (3.3) is equivalent to uniqueness, therefore we show that it has a unique solution.

THEOREM 3.2. *The integro-differential equation* (3.3) *has a unique solution in* $H_{\boldsymbol{\alpha}}(\mathrm{curl}\,, \Omega_h)$.

PROOF. Suppose (3.3) has two solutions $\mathbf{u}_1, \mathbf{u}_2$ in $H_{\boldsymbol{\alpha}}(\mathrm{curl}\,, \Omega_h)$. Let $\mathbf{u} = \mathbf{u}_1 - \mathbf{u}_2$, we have

$$\mathbf{u} - \mathcal{AS}\mathbf{u} - \mathcal{BT}\mathbf{u} = 0 \tag{3.4}$$

in Ω_h. In particular, $\mathbf{u}|_D$ solves the (3.4) in D. Therefore $\mathbf{u}|_D$ can be extended by the right-hand side of (3.4) to a radiating solution $\tilde{\mathbf{u}} \in H_{\boldsymbol{\alpha},\mathrm{loc}}(\mathrm{curl}\,, \Omega)$ of the variational problem

(3.5)
$$\int_\Omega (\mathrm{curl}\,\tilde{\mathbf{u}} \cdot \mathrm{curl}\,\overline{\mathbf{v}} - k^2\tilde{\mathbf{u}} \cdot \overline{\mathbf{v}})\, d\mathbf{x} = \int_D (k^2(P - \xi\mu_r^{-1}\xi)\tilde{\mathbf{u}} + ik\xi\mu_r^{-1}\mathrm{curl}\,\tilde{\mathbf{u}}) \cdot \overline{\mathbf{v}}\, d\mathbf{x} + \int_D (Q\mathrm{curl}\,\tilde{\mathbf{u}} - ik\mu_r^{-1}\xi\tilde{\mathbf{u}}) \cdot \mathrm{curl}\,\overline{\mathbf{v}}\, d\mathbf{x}$$

for all $\mathbf{v} \in H_{\boldsymbol{\alpha}}(\mathrm{curl}\,, \Omega)$ with compact support. Let $\phi \in C_0^\infty(\mathbb{R})$ be a smooth cut-off function such that $\phi(t) = 1$ if $|t| \leq h$ and $\phi(t) = 0$ if $|t| \geq 2h$. Using the test function $\mathbf{v}(\mathbf{x}) = \phi(x_3)\tilde{\mathbf{u}}(\mathbf{x})$ in (3.5) yields

(3.6)
$$\int_{\Omega_{2h}} (\mathrm{curl}\,\tilde{\mathbf{u}} \cdot \mathrm{curl}\,\overline{\phi\tilde{\mathbf{u}}} - k^2\tilde{\mathbf{u}} \cdot \overline{\phi\tilde{\mathbf{u}}})\, d\mathbf{x} = \int_D (k^2(P - \xi\mu_r^{-1}\xi)\tilde{\mathbf{u}} + ik\xi\mu_r^{-1}\mathrm{curl}\,\tilde{\mathbf{u}}) \cdot \overline{\tilde{\mathbf{u}}}\, d\mathbf{x} + \int_D (Q\mathrm{curl}\,\tilde{\mathbf{u}} - ik\mu_r^{-1}\xi\tilde{\mathbf{u}}) \cdot \mathrm{curl}\,\overline{\tilde{\mathbf{u}}}\, d\mathbf{x}.$$

Since $\tilde{\mathbf{u}}$ is divergence-free and solves $\operatorname{curl}^2\tilde{\mathbf{u}} - k^2\tilde{\mathbf{u}} = 0$ on $\Omega_{2h} \setminus \Omega_h$, we have

$$\begin{aligned}
&\int_{\Omega_{2h}} (\operatorname{curl}\tilde{\mathbf{u}} \cdot \operatorname{curl}\overline{\phi\tilde{\mathbf{u}}} - k^2\tilde{\mathbf{u}} \cdot \overline{\phi\tilde{\mathbf{u}}})\, d\mathbf{x} \\
&= \int_{\Omega_h} (|\operatorname{curl}\tilde{\mathbf{u}}|^2 - k^2|\tilde{\mathbf{u}}|^2)\, d\mathbf{x} + \left(\int_{\{x_3=h\}} - \int_{\{x_3=-h\}}\right) (e_3 \times \operatorname{curl}\tilde{\mathbf{u}}) \cdot \overline{\tilde{\mathbf{u}}}\, ds(\mathbf{x}) \\
&= \int_{\Omega_h} (|\operatorname{curl}\tilde{\mathbf{u}}|^2 - k^2|\tilde{\mathbf{u}}|^2)\, d\mathbf{x} \\
&\quad + \left(\int_{\{x_3=h\}} - \int_{\{x_3=-h\}}\right) \left(\tilde{u}_3 \frac{\partial\overline{\tilde{u}_3}}{\partial x_3} - \overline{\tilde{u}_1}\frac{\partial \tilde{u}_1}{\partial x_3} - \overline{\tilde{u}_2}\frac{\partial \tilde{u}_2}{\partial x_3}\right) ds(\mathbf{x}).
\end{aligned}$$

Using radiation condition (2.9) on the boundary term then taking the imaginary part of both sides gives

$$\operatorname{Im} \int_{\Omega_{2h}} (\operatorname{curl}\tilde{\mathbf{u}} \cdot \operatorname{curl}\overline{\phi\tilde{\mathbf{u}}} - k^2\tilde{\mathbf{u}} \cdot \overline{\phi\tilde{\mathbf{u}}})\, d\mathbf{x} = -4\pi^2 \sum_{m:\beta_m>0} \beta_m(|\hat{u}_m^+|^2 + |\hat{u}_m^-|^2) \le 0.$$

On the other hand, by Assumption 2.1,

$$\begin{aligned}
\operatorname{Im} \Bigg(\int_D &(k^2(P - \xi\mu_r^{-1}\xi)\tilde{\mathbf{u}} + ik\xi\mu_r^{-1}\operatorname{curl}\tilde{\mathbf{u}}) \cdot \overline{\tilde{\mathbf{u}}}\, d\mathbf{x} \\
&+ \int_D (Q\operatorname{curl}\tilde{\mathbf{u}} - ik\mu_r^{-1}\xi\tilde{\mathbf{u}}) \cdot \operatorname{curl}\overline{\tilde{\mathbf{u}}}\, d\mathbf{x}\Bigg) \\
&\ge k^2c_2\|\tilde{\mathbf{u}}\|^2 + c_1\|\operatorname{curl}\tilde{\mathbf{u}}\|^2 + k\operatorname{Re}(\xi\mu_r^{-1}\operatorname{curl}\tilde{\mathbf{u}}, \tilde{\mathbf{u}}) - k\operatorname{Re}(\mu_r^{-1}\xi\tilde{\mathbf{u}}, \operatorname{curl}\tilde{\mathbf{u}}),
\end{aligned}$$

here $\|\cdot\|$ and $(\cdot,\cdot)$ denote the $L^2(D)$ norm and $L^2(D)$ inner product, respectively. Using Cauchy-Schwarz inequality and the fact that μ_r^{-1} and ξ are symmetric we have

$$\begin{aligned}
k\operatorname{Re}(\xi\mu_r^{-1}\operatorname{curl}\tilde{\mathbf{u}}, \tilde{\mathbf{u}}) &\ge -\|\xi\mu_r^{-1}\operatorname{curl}\tilde{\mathbf{u}}\|\|k\tilde{\mathbf{u}}\| \\
&\ge -\frac{1}{2}\left(\||\mu_r^{-1}\xi|_F\|^2_{L^\infty(D)}\|\operatorname{curl}\tilde{\mathbf{u}}\|^2 + k^2\|\tilde{\mathbf{u}}\|^2\right)
\end{aligned}$$

and

$$\begin{aligned}
-k\operatorname{Re}(\mu_r^{-1}\xi\tilde{\mathbf{u}}, \operatorname{curl}\tilde{\mathbf{u}}) &\ge -\|k\mu_r^{-1}\xi\tilde{\mathbf{u}}\|\|\operatorname{curl}\tilde{\mathbf{u}}\| \\
&\ge -\frac{1}{2}\left(k^2\||\mu_r^{-1}\xi|_F\|^2_{L^\infty(D)}\|\tilde{\mathbf{u}}\|^2 + \|\operatorname{curl}\tilde{\mathbf{u}}\|^2\right).
\end{aligned}$$

Therefore

$$\begin{aligned}
\operatorname{Im} \Bigg(\int_D &(k^2(P - \xi\mu_r^{-1}\xi)\tilde{\mathbf{u}} + ik\xi\mu_r^{-1}\operatorname{curl}\tilde{\mathbf{u}}) \cdot \overline{\tilde{\mathbf{u}}}\, d\mathbf{x} \\
&+ \int_D (Q\operatorname{curl}\tilde{\mathbf{u}} - ik\mu_r^{-1}\xi\tilde{\mathbf{u}}) \cdot \operatorname{curl}\overline{\tilde{\mathbf{u}}}\, d\mathbf{x}\Bigg) \\
&\qquad\qquad \ge C_1\|\operatorname{curl}\tilde{\mathbf{u}}\|^2 + k^2C_2\|\tilde{\mathbf{u}}\|^2,
\end{aligned}$$

where

$$\begin{aligned}
C_1 &= c_1 - \frac{1}{2}(\||\mu_r^{-1}\xi|_F\|^2_{L^\infty(D)} + 1) > 0, \\
C_2 &= c_2 - \frac{1}{2}(\||\mu_r^{-1}\xi|_F\|^2_{L^\infty(D)} + 1) > 0.
\end{aligned}$$

Combining the estimates for the two sides of (3.6) we have

$$C_1\|\text{curl}\,\tilde{\mathbf{u}}\|^2 + k^2 C_2\|\tilde{\mathbf{u}}\|^2 \leq 0,$$

and thus $\tilde{\mathbf{u}} = 0$ in D or $\mathbf{u}|_D = 0$. Due to (3.4), $\mathbf{u} = 0$ in $H_{\boldsymbol{\alpha}}(\text{curl}\,,\Omega_h)$, which means $\mathbf{u}_1 = \mathbf{u}_2$ in $H_{\boldsymbol{\alpha}}(\text{curl}\,,\Omega_h)$. □

4. Periodization of the integro-differential equation

We will be focusing on solving

$$\mathbf{u} - \mathcal{A}\mathcal{S}\mathbf{u} - \mathcal{B}\mathcal{T}\mathbf{u} = \mathcal{A}\mathbf{f} + \mathcal{B}\mathbf{g}, \quad \mathbf{u} \in H_{\boldsymbol{\alpha}}(\text{curl}\,,\Omega_h). \tag{4.1}$$

The idea is to transform (4.1) into a periodic equation so that we can utilize all the tools of periodic integral equations, especially the fast Fourier transform (FFT), while ensuring that the solution to this periodic equation can be used to find $\mathbf{u}$ that solves (4.1).

For $R > 2h$ let

$$\mathcal{K}_R(\mathbf{x}) = G(\mathbf{x}), \quad \mathbf{x} \in \Omega_R \setminus \{0\}$$

and then extend $\mathcal{K}_R$ $2R$-periodically in x_3 and $\boldsymbol{\alpha}$-quasiperiodically in x_1 and x_2 to the entire $\mathbb{R}^3$. Since G is already $\boldsymbol{\alpha}$-quasiperiodic and smooth everywhere but at its singularities, extending along the x_1 and x_2 axes will not affect the smoothness of $\mathcal{K}_R$. However doing so along the x_3 axis does as G is not periodic in x_3. Thus next we smoothen $\mathcal{K}_R$ by defining

$$\mathcal{K}(\mathbf{x}) = \mathcal{K}_R(\mathbf{x})\chi(x_3), \quad \mathbf{x} \in \mathbb{R}^3 \setminus \{(n_1 2\pi, n_2 2\pi, n_3 2R) : n_1, n_2, n_3 \in \mathbb{Z}\}$$

where $\chi : \mathbb{R} \to [0,1]$ is a $2R$-periodic smooth function satisfying

$$\chi(t) = \begin{cases} 1, & -2h \leq t \leq 2h \\ 0, & t = \pm R \end{cases} \quad \text{and} \quad \chi'(\pm R) = \chi''(\pm R) = \chi'''(\pm R) = 0.$$

The explicit formula for the Fourier coefficients of $\mathcal{K}_R$ can be found in [**21**]. This helps us avoid the evaluation of $\mathcal{K}_R$ in the spectral domain. We also extend P, Q, ξ from Ω_R to $\mathbb{R}^3$ 2π-periodically in x_1, x_2 and $2R$-periodically in x_3, and extend $\mathbf{f}$, $\mathbf{g}$ from Ω_R to $\mathbb{R}^3$ $\boldsymbol{\alpha}$-quasiperiodically in x_1, x_2 and $2R$-periodically in x_3. We obtain the following periodic integro-differential equation

$$\mathbf{u} = (k^2 + \nabla\text{div})\int_D \mathcal{K}_R(\mathbf{x},\mathbf{y})(\mathcal{S}\mathbf{u}(\mathbf{y}) + \mathbf{f}(\mathbf{y}))d\mathbf{y} + \text{curl}\int_D \mathcal{K}_R(\mathbf{x},\mathbf{y})(\mathcal{T}\mathbf{u}(\mathbf{y}) + \mathbf{g}(\mathbf{y}))d\mathbf{y} \tag{4.2}$$

Next we will analyze the periodic integral operators in the equation. To this end, for $\lambda \geq 0$, we consider periodic Sobolev spaces $H^{\lambda}_{\boldsymbol{\alpha},p}(\Omega_R)$ and $H^{\lambda}_{\boldsymbol{\alpha},p}(\text{curl}\,,\Omega_R)$ as spaces consisting of functions in $L^2(\Omega_R)^3$ that satisfy

$$\|\mathbf{w}\|^2_{H^{\lambda}_{\boldsymbol{\alpha},p}(\Omega_R)} = \sum_{\mathbf{j}\in\mathbb{Z}^3}(1 + |\mathbf{j}|^2)^{\lambda}|\widehat{\mathbf{w}}(\mathbf{j})|^2 < \infty,$$

$$\|\mathbf{w}\|^2_{H^{\lambda}_{\boldsymbol{\alpha},p}(\text{curl}\,,\Omega_R)} = \sum_{\mathbf{j}\in\mathbb{Z}^3}(1 + |\mathbf{j}|^2)^{\lambda}\left(k^2|\widehat{\mathbf{w}}(\mathbf{j})|^2 + \left|\tilde{\mathbf{j}} \times \widehat{\mathbf{w}}(\mathbf{j})\right|^2\right) < \infty,$$

respectively, where $\tilde{\mathbf{j}} = \left(\alpha_1 + j_1, \alpha_2 + j_2, \frac{\pi j_3}{R}\right)$. Note that $H^0_{\boldsymbol{\alpha},p}(\Omega_R) = L^2(\Omega_R)^3$ and $H^0_{\boldsymbol{\alpha},p}(\text{curl}\,,\Omega_R) = H_{\boldsymbol{\alpha},p}(\text{curl}\,,\Omega_R)$. Here the orthonormal basis for $L^2(\Omega_R)^3$ is defined as

$$\phi_{\mathbf{j}}(\mathbf{x}) = \frac{1}{\sqrt{8R\pi^2}}\exp\left(i\alpha_{\tilde{j}} \cdot \mathbf{x} + i\frac{\pi j_3}{R}x_3\right), \quad \mathbf{j} = (j_1, j_2, j_3) \in \mathbb{Z}^3 \tag{4.3}$$

where $\tilde{j} = (j_1, j_2)$, and for $\mathbf{w} \in L^2(\Omega_R)^3$, the Fourier coefficients $\widehat{\mathbf{w}}(\mathbf{j})$ are given by

$$\widehat{\mathbf{w}}(\mathbf{j}) = \int_{\Omega_r} \mathbf{w}(\mathbf{x})\overline{\phi_{\mathbf{j}}(\mathbf{x})}d\mathbf{x}. \tag{4.4}$$

In fact

$$\|\mathbf{w}\|^2_{H^\lambda_{\boldsymbol{\alpha},p}(\mathrm{curl}\,,\Omega_R)} = k^2\|\mathbf{w}\|^2_{H^\lambda_{\boldsymbol{\alpha},p}(\Omega_R)} + \|\mathrm{curl}\,\mathbf{w}\|^2_{H^\lambda_{\boldsymbol{\alpha},p}(\Omega_R)}.$$

Note that the extension of a function $\mathbf{w}$ in $H_{\boldsymbol{\alpha},p}(\mathrm{curl}\,,\Omega_R)$ to the entire $\mathbb{R}^3$ using the Fourier expansion

$$\mathbf{w}(\mathbf{x}) = \sum_{\mathbf{j}\in\mathbb{Z}^3} \widehat{\mathbf{w}}(\mathbf{j})\phi_{\mathbf{j}}(\mathbf{x}), \quad \mathbf{x} \in \mathbb{R}^3$$

is $\boldsymbol{\alpha}$-quasiperiodic in x_1, x_2 and $2R$-periodic in x_3.

Let $H^\lambda_{\boldsymbol{\alpha},p}(\Omega_R)^{3\times 3}$ be the space of 3×3 matrix-valued functions whose columns are functions in $H^\lambda_{\boldsymbol{\alpha},p}(\Omega_R)$. Now we prescribe an assumption on regularity of the parameters as well as the solution $\mathbf{u}$ of (4.2).

ASSUMPTION 4.1. Assume that $\varepsilon_r, \mu_r, \mu_r^{-1}, \xi \in H^{\eta+\frac{1}{2}+\epsilon}_{\boldsymbol{\alpha},p}(\Omega_R)^{3\times 3}$ for some $\epsilon > 0$ and $\eta > \frac{1}{2}$. Moreover, $\mathbf{u} \in H^\lambda_{\boldsymbol{\alpha},p}(\mathrm{curl}\,,\Omega_R)$ for some λ such that $0 \le \lambda < \eta - \frac{1}{2}$.

Next, for $\mathbf{h} \in H^\lambda_{\boldsymbol{\alpha},p}(\Omega_R)$, we define

$$\mathcal{A}_p\mathbf{h}(\mathbf{x}) := (k^2 + \nabla\mathrm{div})\int_D \mathcal{K}(\mathbf{x}-\mathbf{y})\mathbf{h}(\mathbf{y})d\mathbf{y},$$
$$\mathcal{B}_p\mathbf{h}(\mathbf{x}) := \mathrm{curl}\,\int_D \mathcal{K}(\mathbf{x}-\mathbf{y})\mathbf{h}(\mathbf{y})d\mathbf{y}$$

as periodized versions of $\mathcal{A}$ and $\mathcal{B}$ respectively.

LEMMA 4.2. *$\mathcal{A}_p$ and $\mathcal{B}_p$ are bounded linear operators from $H^\lambda_{\boldsymbol{\alpha},p}(\Omega_R)$ to $H^\lambda_{\boldsymbol{\alpha},p}(\mathrm{curl}\,,\Omega_R)$.*

PROOF. The linearity is obvious. From the Fourier coefficients of $\mathcal{K}$ in **[21]** there exist $C_1, C_2 > 0$ such that $|\widehat{\mathcal{K}}(\mathbf{j})| \le C_1|\mathbf{j}|^{-2}$ and $|\tilde{\mathbf{j}}| \le C_2|\mathbf{j}|$ for all $\mathbf{j} \in \mathbb{Z}^3 \setminus \{0\}$. Let $\mathbf{v} \in H^\lambda_{\boldsymbol{\alpha},p}(\Omega_R)$,

$$\begin{aligned}
\|\mathcal{A}_p\mathbf{v}\|^2_{H^\lambda_{\boldsymbol{\alpha},p}(\mathrm{curl}\,,\Omega_R)} &= \sum_{\mathbf{j}\in\mathbb{Z}^3}(1+|\mathbf{j}|^2)^\lambda\left(k^2\left|\widehat{\mathcal{A}_p\mathbf{v}}(\mathbf{j})\right|^2 + \left|\tilde{\mathbf{j}}\times\widehat{\mathcal{A}_p\mathbf{v}}(\mathbf{j})\right|^2\right)\\
&= \sum_{\mathbf{j}\in\mathbb{Z}^3}(1+|\mathbf{j}|^2)^\lambda\left(k^2\left|k^2\widehat{\mathcal{K}}(\mathbf{j})\widehat{\mathbf{v}}(\mathbf{j}) + \tilde{\mathbf{j}}\left(\tilde{\mathbf{j}}\cdot\widehat{\mathcal{K}}(\mathbf{j})\widehat{\mathbf{v}}(\mathbf{j})\right)\right|^2 + \left|\tilde{\mathbf{j}}\times k^2\widehat{\mathcal{K}}(\mathbf{j})\widehat{\mathbf{v}}(\mathbf{j})\right|^2\right)\\
&\le k^2(2k^4 + 2|\boldsymbol{\alpha}|^4 + k^2|\boldsymbol{\alpha}|^2)|\widehat{\mathcal{K}}(0)|^2|\widehat{\mathbf{v}}(0)|^2\\
&\quad + C_1^2k^2\left(2k^4 + 2C_2^4 + C_2^2k^2\right)\sum_{\mathbf{j}\in\mathbb{Z}^3\setminus\{0\}}(1+|\mathbf{j}|^2)^\lambda|\widehat{\mathbf{v}}(\mathbf{j})|^2.
\end{aligned}$$

Hence let $C^2 = k^2\max\left\{(2k^4 + 2|\boldsymbol{\alpha}|^4 + k^2|\boldsymbol{\alpha}|^2)|\widehat{\mathcal{K}}(0)|^2, C_1^2\left(2k^4 + 2C_2^4 + C_2^2k^2\right)\right\}$ we have

$$\|\mathcal{A}_p\mathbf{v}\|_{H^\lambda_{\boldsymbol{\alpha},p}(\mathrm{curl}\,,\Omega_R)} \le C\|\mathbf{v}\|_{H^\lambda_{\boldsymbol{\alpha},p}(\Omega_R)}.$$

On the other hand

$$\|\mathcal{B}_p\mathbf{v}\|^2_{H^\lambda_{\boldsymbol{\alpha},p}(\mathrm{curl}\,,\Omega_R)} = \sum_{\mathbf{j}\in\mathbb{Z}^3}(1+|\mathbf{j}|^2)^\lambda\left(k^2\left|\widehat{\mathcal{B}_p\mathbf{v}}(\mathbf{j})\right|^2+\left|\tilde{\mathbf{j}}\times\widehat{\mathcal{B}_p\mathbf{v}}(\mathbf{j})\right|^2\right)$$

$$=\sum_{\mathbf{j}\in\mathbb{Z}^3}(1+|\mathbf{j}|^2)^\lambda\left[k^2\left|\tilde{\mathbf{j}}\times\widehat{\mathcal{K}}(\mathbf{j})\widehat{\mathbf{v}}(\mathbf{j})\right|^2+\left|\tilde{\mathbf{j}}\times\left(\tilde{\mathbf{j}}\times\widehat{\mathcal{K}}(\mathbf{j})\widehat{\mathbf{v}}(\mathbf{j})\right|^2\right)\right]$$

$$\le (k^2|\boldsymbol{\alpha}|^2+|\boldsymbol{\alpha}|^4)|\widehat{\mathcal{K}}(0)|^2|\widehat{\mathbf{v}}(0)|^2+C_1^2\left(C_2^2k^2+C_2^4\right)\sum_{\mathbf{j}\in\mathbb{Z}^3\setminus\{0\}}(1+|\mathbf{j}|^2)^\lambda|\widehat{\mathbf{v}}(\mathbf{j})|^2.$$

Let $C'^2=\max\left\{(k^2|\boldsymbol{\alpha}|^2+|\boldsymbol{\alpha}|^4)|\widehat{\mathcal{K}}(0)|^2, C_1^2\left(C_2^2k^2+C_2^4\right)\right\}$ we have

$$\|\mathcal{B}_p\mathbf{v}\|_{H^\lambda_{\boldsymbol{\alpha},p}(\mathrm{curl}\,,\Omega_R)}\le C'\|\mathbf{v}\|_{H^\lambda_{\boldsymbol{\alpha},p}(\Omega_R)}.$$

□

The operators $\mathcal{A}_p\mathcal{S}$ and $\mathcal{B}_p\mathcal{T}$ have some interesting mapping properties. First the following lemma will be handy.

LEMMA 4.3. *If $A\in H^\gamma_{\boldsymbol{\alpha},p}(\Omega_R)^{3\times3}$ and $\mathbf{v}\in H^\nu_{\boldsymbol{\alpha},p}(\Omega_R)$ where $\gamma>\frac{1}{2}$ and $0\le\nu<\gamma-\frac{1}{2}$ then there exists $C>0$ such that*

$$\|A\mathbf{v}\|_{H^\nu_{\boldsymbol{\alpha},p}(\Omega_R)}\le C\|\mathbf{v}\|_{H^\nu_{\boldsymbol{\alpha},p}(\Omega_R)}.$$

PROOF. Suppose $A=\begin{bmatrix}\mathbf{a}_1 & \mathbf{a}_2 & \mathbf{a}_3\end{bmatrix}$, $v=\begin{bmatrix}v_1 & v_2 & v_3\end{bmatrix}^T$ then

$$\|A\mathbf{v}\|^2_\nu=\left\|\sum_{j=1}^3\mathbf{a}_jv_j\right\|^2_\nu\le 3\sum_{j=1}^3\|\mathbf{a}_jv_j\|^2_\nu. \tag{4.5}$$

For $j=1,2,3$ we have

$$\|\mathbf{a}_jv_j\|^2_{H^\nu_{\boldsymbol{\alpha},p}(\Omega_R)}=\sum_{\mathbf{j}\in\mathbb{Z}^3}(1+|\mathbf{j}|^2)^\nu|\widehat{\mathbf{a}_jv_j}(\mathbf{j})|^2=\sum_{\mathbf{j}\in\mathbb{Z}^3}(1+|\mathbf{j}|^2)^\nu\left|\sum_{\mathbf{l}\in\mathbb{Z}^3}\widehat{\mathbf{a}_j}(\mathbf{j}-\mathbf{l})\widehat{v_j}(\mathbf{l})\right|^2$$

$$\le\sum_{\mathbf{j}\in\mathbb{Z}^3}\left(\sum_{\mathbf{l}\in\mathbb{Z}^3}(1+|\mathbf{j}|^2)^{\frac{\nu}{2}}|\widehat{\mathbf{a}_j}(\mathbf{j}-\mathbf{l})||\widehat{v_j}(\mathbf{l})|\right)^2$$

$$\le 2^{2\nu}\sum_{\mathbf{j}\in\mathbb{Z}^3}\left(\sum_{\mathbf{l}\in\mathbb{Z}^3}\left[(1+|\mathbf{j}-\mathbf{l}|^2)^{\frac{\nu}{2}}+(1+|\mathbf{l}|^2)^{\frac{\nu}{2}}\right]|\widehat{\mathbf{a}_j}(\mathbf{j}-\mathbf{l})||\widehat{v_j}(\mathbf{l})|\right)^2$$

$$\le 2^{2\nu+1}(S_1+S_2)$$

where

$$S_1=\sum_{\mathbf{j}\in\mathbb{Z}^3}\left(\sum_{\mathbf{l}\in\mathbb{Z}^3}(1+|\mathbf{j}-\mathbf{l}|^2)^{\frac{\nu}{2}}|\widehat{\mathbf{a}_j}(\mathbf{j}-\mathbf{l})||\widehat{v_j}(\mathbf{l})|\right)^2,$$

$$S_2=\sum_{\mathbf{j}\in\mathbb{Z}^3}\left(\sum_{\mathbf{l}\in\mathbb{Z}^3}(1+|\mathbf{l}|^2)^{\frac{\nu}{2}}|\widehat{\mathbf{a}_j}(\mathbf{j}-\mathbf{l})||\widehat{v_j}(\mathbf{l})|\right)^2.$$

Applying the Cauchy - Schwarz inequality consecutively gives

$$
\begin{aligned}
S_1 &\le \sum_{\mathbf{j}\in\mathbb{Z}^3}\left(\sum_{\mathbf{l}\in\mathbb{Z}^3}(1+|\mathbf{j}-\mathbf{l}|^2)^{\frac{\nu}{2}}|\widehat{\mathbf{a}_j}(\mathbf{j}-\mathbf{l})||\widehat{v_j}(\mathbf{l})|^2\sum_{\mathbf{l}\in\mathbb{Z}^3}(1+|\mathbf{j}-\mathbf{l}|^2)^{\frac{\nu}{2}}|\widehat{\mathbf{a}_j}(\mathbf{j}-\mathbf{l})|\right)\\
&= \left(\sum_{\mathbf{l}\in\mathbb{Z}^3}(1+|\mathbf{l}|^2)^{\frac{\nu}{2}}|\widehat{\mathbf{a}_j}(\mathbf{l})|\right)\sum_{\mathbf{l}\in\mathbb{Z}^3}|\widehat{v_j}(\mathbf{l})|^2\sum_{\mathbf{j}\in\mathbb{Z}^3}(1+|\mathbf{j}-\mathbf{l}|^2)^{\frac{\nu}{2}}|\widehat{\mathbf{a}_j}(\mathbf{j}-\mathbf{l})|\\
&\le \left(\sum_{\mathbf{l}\in\mathbb{Z}^3}(1+|\mathbf{l}|^2)^{\frac{\nu}{2}}|\widehat{\mathbf{a}_j}(\mathbf{l})|\right)^2\sum_{\mathbf{l}\in\mathbb{Z}^3}(1+|\mathbf{l}|^2)^{\nu}|\widehat{v_j}(\mathbf{l})|^2
\end{aligned}
$$

and

$$
\begin{aligned}
S_2 &\le \sum_{\mathbf{j}\in\mathbb{Z}^3}\left(\sum_{\mathbf{l}\in\mathbb{Z}^3}(1+|\mathbf{l}|^2)^{\nu}|\widehat{\mathbf{a}_j}(\mathbf{j}-\mathbf{l})||\widehat{v_j}(\mathbf{l})|^2\sum_{\mathbf{l}\in\mathbb{Z}^3}|\widehat{\mathbf{a}_j}(\mathbf{j}-\mathbf{l})|\right)\\
&= \left(\sum_{\mathbf{l}\in\mathbb{Z}^3}|\widehat{\mathbf{a}_j}(\mathbf{l})|\right)\sum_{\mathbf{l}\in\mathbb{Z}^3}|\widehat{v_j}(\mathbf{l})|^2\sum_{\mathbf{j}\in\mathbb{Z}^3}(1+|\mathbf{l}|^2)^{\nu}|\widehat{\mathbf{a}_j}(\mathbf{j}-\mathbf{l})||\widehat{v_j}(\mathbf{l})|^2\\
&\le \left(\sum_{\mathbf{l}\in\mathbb{Z}^3}(1+|\mathbf{l}|^2)^{\frac{\nu}{2}}|\widehat{\mathbf{a}_j}(\mathbf{l})|\right)^2\sum_{\mathbf{l}\in\mathbb{Z}^3}(1+|\mathbf{l}|^2)^{\nu}|\widehat{v_j}(\mathbf{l})|^2.
\end{aligned}
$$

Thus

$$
\|\mathbf{a}_j v_j\|^2_{H^{\nu}_{\boldsymbol{\alpha},p}(\Omega_R)} \le 2^{2\nu+2}\left(\sum_{\mathbf{l}\in\mathbb{Z}^3}(1+|\mathbf{l}|^2)^{\frac{\nu}{2}}|\widehat{\mathbf{a}_j}(\mathbf{l})|\right)^2\sum_{\mathbf{l}\in\mathbb{Z}^3}(1+|\mathbf{l}|^2)^{\nu}|\widehat{v_j}(\mathbf{l})|^2.
$$

Also by the Cauchy - Schwarz inequality

$$
\left(\sum_{\mathbf{l}\in\mathbb{Z}^3}(1+|\mathbf{l}|^2)^{\frac{\nu}{2}}|\widehat{\mathbf{a}_j}(\mathbf{l})|\right)^2 \le \sum_{\mathbf{l}\in\mathbb{Z}^3}(1+|\mathbf{l}|^2)^{\nu-\gamma}\sum_{\mathbf{l}\in\mathbb{Z}^3}(1+|\mathbf{l}|^2)^{\gamma}|\widehat{\mathbf{a}_j}(\mathbf{l})|^2 = C_1^2\|\mathbf{a}_j\|^2_{H^{\gamma}_{\boldsymbol{\alpha},p}(\Omega_R)},
$$

where $C_1^2 = \sum_{\mathbf{l}\in\mathbb{Z}^3}(1+|\mathbf{l}|^2)^{\nu-\gamma}$ (this series converges since $\nu-\gamma<-\frac{1}{2}$). Hence

$$
\|\mathbf{a}_j v_j\|^2_{H^{\nu}_{\boldsymbol{\alpha},p}(\Omega_R)} \le 2^{2\nu+2}C_1^2\|\mathbf{a}_j\|^2_{H^{\gamma}_{\boldsymbol{\alpha},p}(\Omega_R)}\|\mathbf{v}\|^2_{H^{\nu}_{\boldsymbol{\alpha},p}(\Omega_R)}. \tag{4.6}
$$

By (4.5) and (4.6)

$$
\|A\mathbf{v}\|^2_{H^{\nu}_{\boldsymbol{\alpha},p}(\Omega_R)} = \sum_{j=1}^{3}\|\mathbf{a}_j v_j\|^2_{H^{\nu}_{\boldsymbol{\alpha},p}(\Omega_R)} \le 3.2^{2\nu+2}C_1^2\max_j\left\{\|\mathbf{a}_j\|^2_{H^{\gamma}_{\boldsymbol{\alpha},p}(\Omega_R)}\right\}\|\mathbf{v}\|^2_{H^{\nu}_{\boldsymbol{\alpha},p}(\Omega_R)}.
$$

Let $C=\sqrt{3}.2^{\nu+1}C_1\max_j\left\{\|\mathbf{a}_j\|_{H^{\gamma}_{\boldsymbol{\alpha},p}(\Omega_R)}\right\}$, we have

$$
\|A\mathbf{v}\|_{H^{\nu}_{\boldsymbol{\alpha},p}(\Omega_R)} \le C\|\mathbf{v}\|_{H^{\nu}_{\boldsymbol{\alpha},p}(\Omega_R)}.
$$

□

LEMMA 4.4. *The matrix-valued functions P, Q, $\xi\mu_r^{-1}$, $\xi\mu_r^{-1}\xi$, $\mu_r^{-1}\xi$ belong to $H^{\eta}_{\boldsymbol{\alpha},p}(\Omega_R)^{3\times3}$. Moreover, the operators $\mathcal{S}$, $\mathcal{T}$ are bounded linear operators from $H^{\lambda}_{\boldsymbol{\alpha},p}(\mathrm{curl}\,,\Omega_R)$ to $H^{\lambda}_{\boldsymbol{\alpha},p}(\Omega_R)$.*

PROOF. For the first part, we only prove for $\xi\mu_r^{-1}\xi$, the other cases are either similar or straightforward. Suppose $\xi = [\boldsymbol{\xi}_1 \quad \boldsymbol{\xi}_2 \quad \boldsymbol{\xi}_3]$, then $\mu_r^{-1}\xi = [\mu_r^{-1}\boldsymbol{\xi}_1 \quad \mu_r^{-1}\boldsymbol{\xi}_2 \quad \mu_r^{-1}\boldsymbol{\xi}_3]$. For $j = 1,2,3$, since $\mu_r^{-1} \in H_{\boldsymbol{\alpha},p}^{\eta+\frac{1}{2}+\epsilon}(\Omega_R)^{3\times 3}$ and $\boldsymbol{\xi}_j \in H_{\boldsymbol{\alpha},p}^{\eta}(\Omega_R)$, by Lemma 4.3 there exists $c > 0$ such that

$$\|\mu_r^{-1}\boldsymbol{\xi}_j\|_{H_{\boldsymbol{\alpha},p}^{\eta}(\Omega_R)} \leq c\|\boldsymbol{\xi}_j\|_{H_{\boldsymbol{\alpha},p}^{\eta}(\Omega_R)} < \infty.$$

Similarly we can show that $\|\bar{\xi}\mu_r^{-1}\boldsymbol{\xi}_j\|_{H_{\boldsymbol{\alpha},p}^{\eta}(\Omega_R)} \leq c'\|\mu_r^{-1}\boldsymbol{\xi}_j\|_{H_{\boldsymbol{\alpha},p}^{\eta}(\Omega_R)} < \infty$ for some $c' > 0$. Thus $\xi\mu_r^{-1}\xi \in H_{\boldsymbol{\alpha},p}^{\eta}(\Omega_R)^{3\times 3}$.

Linearity of $\mathcal{S}$ and $\mathcal{T}$ is clear, we only prove their boundedness. For $\mathbf{u} \in H_{\boldsymbol{\alpha},p}^{\lambda}(\mathrm{curl}\,,\Omega_R)$, $\mathrm{curl}\,\mathbf{u} \in H_{\boldsymbol{\alpha},p}^{\lambda}(\Omega_R)$. Therefore by lemma 4.3

$$\begin{aligned}
\|\mathcal{S}\mathbf{u}\|_{H_{\boldsymbol{\alpha},p}^{\lambda}(\Omega_R)} &\leq \|(P - \xi\mu_r^{-1}\xi)\mathbf{u}\|_{H_{\boldsymbol{\alpha},p}^{\lambda}(\Omega_R)} + \left\|\frac{i}{k}\xi\mu_r^{-1}\mathrm{curl}\,\mathbf{u}\right\|_{H_{\boldsymbol{\alpha},p}^{\lambda}(\Omega_R)} \\
&\leq C_1 k\|\mathbf{u}\|_{H_{\boldsymbol{\alpha},p}^{\lambda}(\Omega_R)} + C_2\|\mathrm{curl}\,\mathbf{u}\|_{H_{\boldsymbol{\alpha},p}^{\lambda}(\Omega_R)} \\
&\leq C\|\mathbf{u}\|_{H_{\boldsymbol{\alpha},p}^{\lambda}(\mathrm{curl}\,,\Omega_R)}
\end{aligned}$$

and

$$\begin{aligned}
\|\mathcal{T}\mathbf{u}\|_{H_{\boldsymbol{\alpha},p}^{\lambda}(\Omega_R)} &\leq \|Q\mathrm{curl}\,\mathbf{u}\|_{H_{\boldsymbol{\alpha},p}^{\lambda}(\Omega_R)} + \|ik\mu_r^{-1}\xi\mathbf{u}\|_{H_{\boldsymbol{\alpha},p}^{\lambda}(\Omega_R)} \\
&\leq C_1'\|\mathrm{curl}\,\mathbf{u}\|_{H_{\boldsymbol{\alpha},p}^{\lambda}(\Omega_R)} + C_2' k\|\mathbf{u}\|_{H_{\boldsymbol{\alpha},p}^{\lambda}(\Omega_R)} \\
&\leq C'\|\mathbf{u}\|_{H_{\boldsymbol{\alpha},p}^{\lambda}(\mathrm{curl}\,,\Omega_R)}
\end{aligned}$$

where $C_1, C_1', C_2, C_2', C, C'$ are positive numbers. □

The following theorem gives us the the relation between the periodic equation (4.2) and the original one as well as its Fredholm property.

THEOREM 4.5. *If* $\mathbf{u} \in H_{\boldsymbol{\alpha},p}(\mathrm{curl}\,,\Omega_R)$ *is a solution to* (4.2) *then* $\mathbf{u}|_{\Omega_h} \in H_{\boldsymbol{\alpha}}(\mathrm{curl}\,,\Omega_h)$ *solves* (4.1). *Moreover, there exist a constant* $C > 0$ *and a compact operator* K *on* $H_{\boldsymbol{\alpha},p}(\mathrm{curl}\,,\Omega_R)$ *such that*

$$\mathrm{Re}\,(\mathbf{u}-\mathcal{A}_p\mathcal{S}\mathbf{u}-\mathcal{B}_p\mathcal{T}\mathbf{u},\mathbf{u})_{H_{\boldsymbol{\alpha},p}(\mathrm{curl}\,,\Omega_R)} \geq C\|\mathbf{u}\|^2_{H_{\boldsymbol{\alpha},p}(\mathrm{curl}\,,\Omega_R)} + \mathrm{Re}\,(K\mathbf{u},\mathbf{u})_{H_{\boldsymbol{\alpha},p}(\mathrm{curl}\,,\Omega_R)}.$$

The proof can be done using the same reasoning as in Theorem 5.2 and Theorem 5.3 in [**21**].

5. A spectral Galerkin method

For $N \in \mathbb{N}$, define

$$\mathbb{Z}_N^3 = \left\{\mathbf{j} = [j_1 \quad j_2 \quad j_3]^T \in \mathbb{Z}^3 : -\frac{N}{2} < j_1, j_2, j_3 \leq \frac{N}{2}\right\}.$$

Now we consider

$$\mathcal{T}_N = \mathrm{span}\left\{\phi_{\mathbf{j}} : \mathbf{j} \in \mathbb{Z}_N^3\right\}$$

which is a finite dimensional subspace of $H_{\boldsymbol{\alpha},p}(\mathrm{curl}\,,\Omega_R)$ and let $P_N : H_{\boldsymbol{\alpha},p}(\mathrm{curl}\,,\Omega_R) \to \mathcal{T}_N$ be the orthogonal projection onto $\mathcal{T}_N$, i.e

$$P_N\mathbf{u} = \sum_{\mathbf{j}\in\mathbb{Z}_N^3} \widehat{\mathbf{u}}(\mathbf{j})\phi_{\mathbf{j}}(\mathbf{x}).$$

We approximate (4.2) by the following Galerkin equation

$$(5.1)\quad \begin{aligned}&(\mathbf{u}_N - \mathcal{A}_p\mathcal{S}\mathbf{u}_N - \mathcal{B}_p\mathcal{T}\mathbf{u}_N, \mathbf{w}_N)_{H_{\boldsymbol{\alpha},p}(\mathrm{curl}\,,\Omega_R)} = (\mathcal{A}_p\mathbf{f} + \mathcal{B}_p\mathbf{g}, \mathbf{w}_N)_{H_{\boldsymbol{\alpha},p}(\mathrm{curl}\,,\Omega_R)},\\ &\mathbf{u}_N \in \mathcal{T}_N\end{aligned}$$

for all $\mathbf{w}_N \in \mathcal{T}_N$, which is equivalent to

$$(5.2)\quad \mathbf{u}_N - P_N\mathcal{A}_p\mathcal{S}\mathbf{u}_N - P_N\mathcal{B}_p\mathcal{T}\mathbf{u}_N = P_N\mathcal{A}_p\mathbf{f} + P_N\mathcal{B}_p\mathbf{g}, \quad \mathbf{u}_N \in \mathcal{T}_N.$$

We will see that convergence is achieved without smoothness of the parameters, but the smoother they are the higher order of convergence we will get.

LEMMA 5.1. *There exists $C_\lambda > 0$ such that for all $\mathbf{w} \in H^\lambda_{\boldsymbol{\alpha},p}(\mathrm{curl}\,,\Omega_R)$*

$$\|P_N\mathbf{w} - \mathbf{w}\|_{H_{\boldsymbol{\alpha},p}(\mathrm{curl}\,,\Omega_R)} \le C_\lambda N^{-\lambda}\|\mathbf{w}\|_{H^\lambda_{\boldsymbol{\alpha},p}(\mathrm{curl}\,,\Omega_R)}.$$

PROOF. For $\mathbf{w} \in H^\lambda_{\boldsymbol{\alpha},p}(\mathrm{curl}\,,\Omega_R)$

$$\begin{aligned}\|P_N\mathbf{w} - \mathbf{w}\|^2_{H_{\boldsymbol{\alpha},p}(\mathrm{curl}\,,\Omega_R)} &= \sum_{\mathbf{j}\notin\mathbb{Z}^3_N}\left(k^2|\widehat{\mathbf{w}}(\mathbf{j})|^2 + \left|\tilde{\mathbf{j}}\times\widehat{\mathbf{w}}(\mathbf{j})\right|^2\right)\\ &= \sum_{\mathbf{j}\notin\mathbb{Z}^3_N}\frac{1}{(1+|\mathbf{j}|^2)^\lambda}(1+|\mathbf{j}|^2)^\lambda\left(k^2|\widehat{\mathbf{w}}(\mathbf{j})|^2 + \left|\tilde{\mathbf{j}}\times\widehat{\mathbf{w}}(\mathbf{j})\right|^2\right)\\ &\le \frac{1}{\left(1+\frac{3N^2}{4}\right)^\lambda}\sum_{\mathbf{j}\notin\mathbb{Z}^3_N}(1+|\mathbf{j}|^2)^\lambda\left(k^2|\widehat{\mathbf{w}}(\mathbf{j})|^2 + \left|\tilde{\mathbf{j}}\times\widehat{\mathbf{w}}(\mathbf{j})\right|^2\right)\\ &\le \left(\frac{4}{7}\right)^\lambda N^{-2\lambda}\|\mathbf{w}\|^2_{H^\lambda_{\boldsymbol{\alpha},p}(\mathrm{curl}\,,\Omega_R)}.\end{aligned}$$

□

Recall that the equation (4.2) satisfies a Garding estimate, we have the following convergence theorem for the Galerkin method.

THEOREM 5.2. *There exists $N_0 > 0$ such that for all $N \ge N_0$, the equation* (5.2) *has a unique solution $\mathbf{u}_N$ and the following error estimate holds*

$$\|\mathbf{u}_N - \mathbf{u}\|_{H_{\boldsymbol{\alpha},p}(\mathrm{curl}\,,\Omega_R)} \le C_\lambda N^{-\lambda}\|\mathbf{u}\|_{H^\lambda_{\boldsymbol{\alpha},p}(\mathrm{curl}\,,\Omega_R)}$$

where $C_\lambda > 0$ is a constant depending on λ and $\mathbf{u}$ is the solution to (4.2).

PROOF. Since (4.2) satisfies the Garding estimate in Theorem 4.5, by Theorem 4.2.9 in [**31**], there exists N_0 and a positive constant C independent of N such that for all $N \ge N_0$,

$$\|\mathbf{u}_N - \mathbf{u}\|_{H_{\boldsymbol{\alpha},p}(\mathrm{curl}\,,\Omega_R)} \le C\inf_{\mathbf{w}_N\in\mathcal{T}_N}\|\mathbf{w}_N - \mathbf{u}\|_{H_{\boldsymbol{\alpha},p}(\mathrm{curl}\,,\Omega_R)}.$$

Moreover, according to Lemma 5.1,

$$\inf_{\mathbf{w}_N\in\mathcal{T}_N}\|\mathbf{w}_N - \mathbf{u}\|_{H_{\boldsymbol{\alpha},p}(\mathrm{curl}\,,\Omega_R)} \le \|P_N\mathbf{u} - \mathbf{u}\|_{H_{\boldsymbol{\alpha},p}(\mathrm{curl}\,,\Omega_R)} \le C_\lambda N^{-\lambda}\|\mathbf{u}\|_{H^\lambda_{\boldsymbol{\alpha},p}(\mathrm{curl}\,,\Omega_R)},$$

therefore the desired estimate holds. □

By this theorem, we can conclude that the spectral Galerkin method converges with arbitrarily high order provided $\mathbf{u}$ is sufficiently smooth. The discretization of equation (5.2) as well as the process of calculating the Fourier coefficients of the involved quantities can be done similarly as in [**21**], therefore are omitted here.

6. Numerical examples

We now present numerical results for the Galerkin method. The computational domain is the rectangular box $(-\pi,\pi)^2 \times (-1,1)$ $(h = 1)$ which is partitioned uniformly in each dimension into N^3 grid points for some $N \in \mathbb{N}$. The wave number $k = 3\pi/2$. We choose the following $\boldsymbol{\alpha}$-quasiperiodic plane wave as used as the incident electric field

$$\mathbf{E}^{in}(\mathbf{x}) = \begin{bmatrix} 1 \\ 1 \\ 0 \end{bmatrix} e^{ikx_3}.$$

We ran the simulation in MATLAB and observed the scattered electric field $\mathbf{u}$ at $\mathbf{x} = (0,0,1.5)$ when N varies from 20 to 100 with step size 10. For each value of N, we calculated the relative error between $\mathbf{u}$ of the current step and that of the previous one, that means

$$\text{error}_j = \frac{|\mathbf{u}_{N_j}(\mathbf{x}) - \mathbf{u}_{N_{j-1}}(\mathbf{x})|}{|\mathbf{u}_{N_{j-1}}(\mathbf{x})|}, \quad j = 1,2,\ldots,8.$$

Here $N_0 = 20$ and $N_j = N_{j-1} + 10$. We consider three test cases in the numerical simulations.

6.1. Small smooth spheres. The first case is a medium consisting of a small sphere in each period with the geometry given by

$$D = \left\{\mathbf{x} \in \Omega_h : x_1^2 + x_2^2 + x_3^2 < 0.2^2\right\}$$

and, for $\mathbf{x} \in \Omega_h$, the material parameters are

$$\begin{aligned} \varepsilon_r(\mathbf{x}) &= I + \gamma_0(\mathbf{x})\text{diag}\left\{0.2, 0.4, 0.1\right\}, \\ \mu_r^{-1}(\mathbf{x}) &= I + \gamma_0(\mathbf{x})\text{diag}\left\{0.3, 0.1, 0.2\right\}, \\ \xi(\mathbf{x}) &= \gamma_0(\mathbf{x})\text{diag}\left\{0.01, 0.02, 0.05)\right\}, \end{aligned}$$

where $\gamma_0 : \mathbb{R}^3 \to [0,1]$ is a smooth function that equals 1 at the center of the sphere and 0 outside it. More specifically, for $\mathbf{x} \in D$,

$$\gamma_0(\mathbf{x}) = \exp\left(1 - \frac{0.2^2}{0.2^2 - x_1^2 - x_2^2 - x_3^2}\right).$$

The three-dimensional image for the geometry as well as the relative errors for increasing values of N are shown in Figure 1. The error decreases as N increases and the difference between $N = 90$ and $N = 100$ is very slim.

6.2. Smooth rectangular boxes. Next we consider a medium that consists of one rectangular box in each period. The geometry is given by

$$D = \{\mathbf{x} \in \Omega_h : |x_1| < 0.5, |x_2| < 0.5, |x_3| < 0.2\}$$

and, for $\mathbf{x} \in \Omega_h$, the material parameters are

$$\begin{aligned} \varepsilon_r(\mathbf{x}) &= I + \gamma_1(x_1)\gamma_2(x_2)\gamma_3(x_3)\text{diag}\left\{0.2, 0.4, 0.1\right\}, \\ \mu_r^{-1}(\mathbf{x}) &= I + \gamma_1(x_1)\gamma_2(x_2)\gamma_3(x_3)\text{diag}\left\{0.3, 0.1, 0.2\right\}, \\ \xi(\mathbf{x}) &= \gamma_1(x_1)\gamma_2(x_2)\gamma_3(x_3)\text{diag}\left\{0.01, 0.02, 0.05)\right\}, \end{aligned}$$

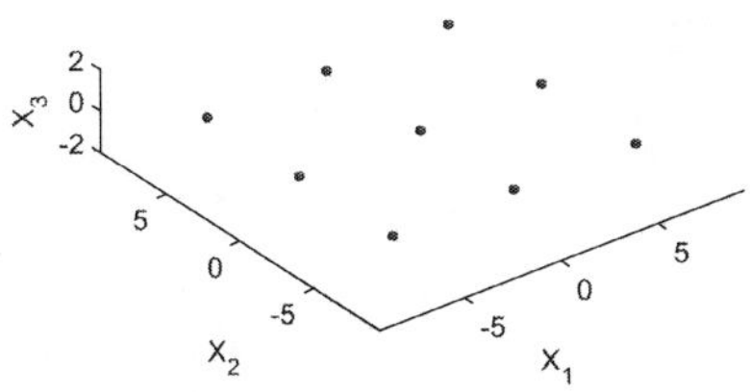

(A) Periodic structure (9 periods)

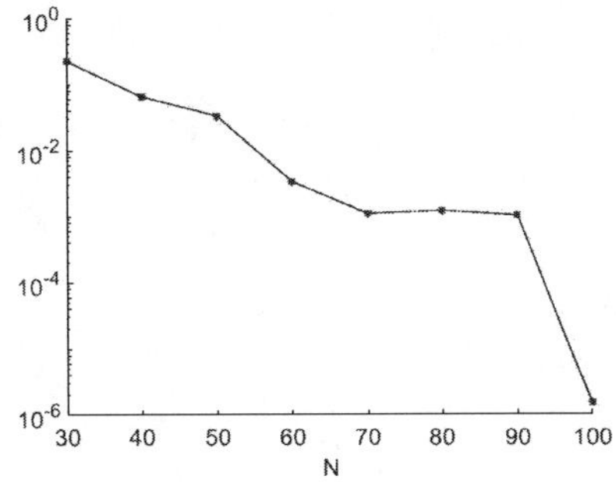

(B) Relative errors of $\mathbf{u}$ at $(0, 0, 1.5)$

FIGURE 1. Solving the direct scattering problem for periodically aligned spheres.

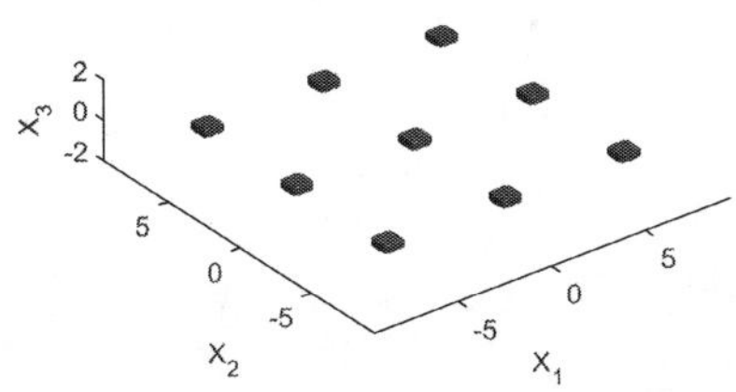

(A) Periodic structure (9 periods)

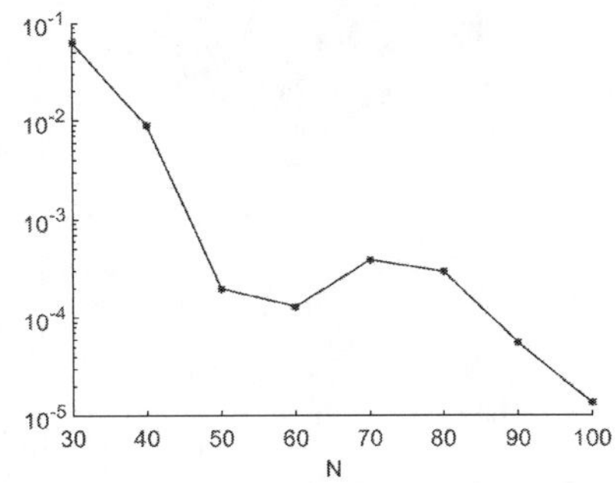

(B) Relative errors of $\mathbf{u}$ at $(0, 0, 1.5)$

FIGURE 2. Solving the direct scattering problem for periodically aligned boxes.

where $\gamma_1, \gamma_2, \gamma_3 : \mathbb{R} \to [0, 1]$ are smooth functions such that their product equals 1 at the center of the box and 0 outside it. For $\mathbf{x} \in D$, they are given by

$$\gamma_1(x_1) = \exp\left(1 - \frac{0.5^2}{0.5^2 - x_1^2}\right)$$
$$\gamma_2(x_2) = \exp\left(1 - \frac{0.5^2}{0.5^2 - x_2^2}\right)$$
$$\gamma_3(x_3) = \exp\left(1 - \frac{0.2^2}{0.2^2 - x_3^2}\right).$$

Figure 2 shows the true geometry and relative errors for the rectangular box. Overall, the error decreases from $N = 30$ to $N = 100$. There is an abnormal rise in error from $N = 60$ to $N = 70$, but the error is already very small at this point so this might be due to roundoff errors.

6.3. Smooth flat strip. The last scattering medium is a strip that spans the entire period horizontally. The geometry is

$$D = \{\mathbf{x} \in \Omega_h : |x_3| < 0.4\}$$

and, for $\mathbf{x} \in \Omega_h$, the material parameters are

$$\varepsilon_r(\mathbf{x}) = I + \gamma_4(\mathbf{x})\text{diag}\{0.2, 0.4, 0.1\},$$
$$\mu_r^{-1}(\mathbf{x}) = I + \gamma_4(\mathbf{x})\text{diag}\{0.3, 0.1, 0.2\},$$
$$\xi(\mathbf{x}) = \gamma_4(\mathbf{x})\text{diag}\{0.01, 0.02, 0.05)\},$$

where

$$\gamma_4(\mathbf{x}) = \begin{cases} \exp\left(1 - \frac{0.4^2}{0.4^2 - x_3^2}\right) & \text{if } \mathbf{x} \in D, \\ 0 & \text{if } \mathbf{x} \in \Omega_h \setminus D. \end{cases}$$

In this case, the error gradually decreases as N increases, see Figure 3.

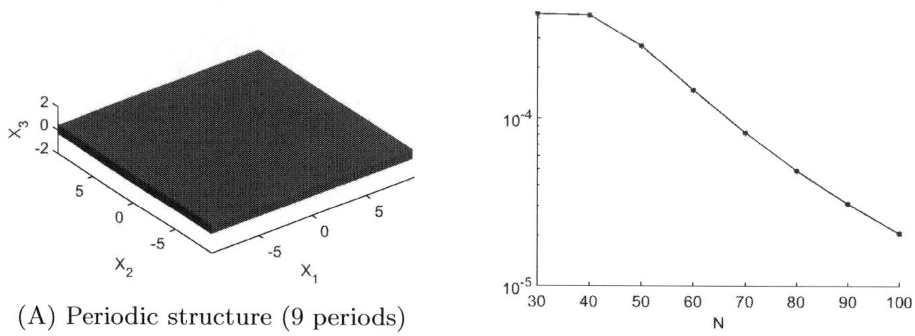

(A) Periodic structure (9 periods)

(B) Relative errors of $\mathbf{u}$ at $(0, 0, 1.5)$

FIGURE 3. Solving the direct scattering problem for a flat strip

In summary, the spectral method converges numerically for all examples. The periodization of the integro-differential equation along with the use of the fast Fourier transform helps significantly improve computational time and accuracy since we can avoid direct computation of the $\boldsymbol{\alpha}$-quasiperiodic Green's function.

7. The inverse problem

We consider the inverse problem of reconstructing the geometry of periodic scattering objects using data generated by solving the direct problem. For $s > h$, let

$$\Gamma_{\pm s} = (-\pi, \pi)^2 \times \{\pm s\}$$

be the set where data is measured. Moreover, we consider two incident electric fields $\mathbf{E}^{in}_{\pm}(\mathbf{x})$ which are $\boldsymbol{\alpha}$-quasiperiodic plane waves of the form

$$\mathbf{E}^{in}_{\pm}(\mathbf{x}) = \mathbf{q}e^{i(\alpha_1 x_1 + \alpha_2 x_2 \pm \beta x_3)}$$

where $\beta = \sqrt{k^2 - \alpha_1^2 - \alpha_2^2}$ and $\mathbf{q} \in \mathbb{R}^3$ is the polarization vector satisfying $\mathbf{q} \cdot \begin{bmatrix}\alpha_1 & \alpha_2 & \beta\end{bmatrix}^T = 0$. Note that this choice of the polarization vector $\mathbf{q}$ is appropriate according to 2.6, in fact, it is proportional to $\mathbf{s} \times \mathbf{d}$ when $\mathbf{s} = \mathbf{q} \times \mathbf{d}$. These two incident fields can be understood as one propagating upward and one propagating downward.

Inverse problem. Given the scattered electric field $\mathbf{u}_{\pm}(\mathbf{x})$ for $\mathbf{x} \in \Gamma_s = \Gamma_{+s} \cup \Gamma_{-s}$ corresponding to $\mathbf{E}^{in}_{\pm}(\mathbf{x})$, respectively. Determine the geometry D of the scattering object.

7.1. The imaging functional. For $\mathbf{p} \in \mathbb{R}^3$, $\mathbf{z} \in \Omega_h$, and $p > 0$, define

$$\mathcal{I}(\mathbf{z}) = \left| \int_{\Gamma_s} \mathbf{u}_+(\mathbf{x}) \cdot \mathbf{p}\overline{G}(\mathbf{x}, \mathbf{z})\, ds(\mathbf{x}) \right|^p + \left| \int_{\Gamma_s} \mathbf{u}_-(\mathbf{x}) \cdot \mathbf{p}\overline{G}(\mathbf{x}, \mathbf{z})\, ds(\mathbf{x}) \right|^p . \tag{7.1}$$

The definition of $\mathcal{I}(\mathbf{z})$ as in (7.1) is inspired by that of the orthogonality sampling method solving the inverse scattering problem for bi-anisotropic bounded objects in [**23**]. Based on our numerical experiments, $\mathcal{I}(\mathbf{z})$ also behaves as an imaging functional for the inverse problem in this paper, meaning it attains large values when $\mathbf{z} \in D$ and is close to 0 when $\mathbf{z} \notin D$. Therefore, we use it as a tool to solve the inverse problem numerically. The analysis of the imaging functional will be addressed in the future.

7.2. Numerical results. We use the same set of parameters as in the numerical study for the direct problem, which means $h = 1$, $k = 3\pi/2$, $\alpha_1 = \alpha_2 = 0$, $\beta = k$. The two incident waves are

$$\mathbf{E}^{in}_{\pm}(\mathbf{x}) = \begin{bmatrix} 1 \\ 1 \\ 0 \end{bmatrix} e^{\pm ikx_3}.$$

The scattered electric fields are measured on $\Gamma_{\pm 1.5} = (-\pi, \pi)^2 \times \{\pm 1.5\}$, each contains 32×32 points of measurement. The sampling domain $(-\pi, \pi)^2 \times (-1, 1)$ is uniformly partitioned into a $64 \times 64 \times 64$ grid. The direct problem is solved on the same domain but with grid density $32 \times 32 \times 32$. We also added 50% of artificial noise to the data obtained from the direct solver.

We used a discretized version of the imaging functional $\mathcal{I}(\mathbf{z})$ for the reconstruction

$$\begin{aligned}
\mathcal{I}(\mathbf{z}) = &\Bigg| \sum_{j_1=1}^{32} \sum_{j_2=1}^{32} \mathbf{u}_+(x_1^{j_1}, x_2^{j_2}, +1.5) \cdot \mathbf{p}\overline{G}\left((x_1^{j_1}, x_2^{j_2}, +1.5), \mathbf{z}\right) \\
&+ \sum_{j_1=1}^{32} \sum_{j_2=1}^{32} \mathbf{u}_+(x_1^{j_1}, x_2^{j_2}, -1.5) \cdot \mathbf{p}\overline{G}\left((x_1^{j_1}, x_2^{j_2}, -1.5), \mathbf{z}\right) \Bigg|^p \\
&+ \Bigg| \sum_{j_1=1}^{32} \sum_{j_2=1}^{32} \mathbf{u}_-(x_1^{j_1}, x_2^{j_2}, +1.5) \cdot \mathbf{p}\overline{G}\left((x_1^{j_1}, x_2^{j_2}, +1.5), \mathbf{z}\right) \\
&+ \sum_{j_1=1}^{32} \sum_{j_2=1}^{32} \mathbf{u}_-(x_1^{j_1}, x_2^{j_2}, -1.5) \cdot \mathbf{p}\overline{G}\left((x_1^{j_1}, x_2^{j_2}, -1.5), \mathbf{z}\right) \Bigg|^p ,
\end{aligned}$$

where $\mathbf{p} = \begin{bmatrix} 1 & 1 & 1 \end{bmatrix}^T$ and the exponent $p = 4$. We then normalized the values of $\mathcal{I}(\mathbf{z})$ by dividing it by the largest value over all $\mathbf{z} \in \Omega_h$, thus we ignored all the constant factors in its discretization including the grid step size. We consider two periodic scattering media. The first is a medium consisting of a small smooth sphere in each period identical to that in the direct problem. Cross-sections of the true and reconstructed geometries are shown in Figure 4.

For this small smooth object, the imaging functional was able to give a good reconstruction even though we only used two incident waves. The second medium

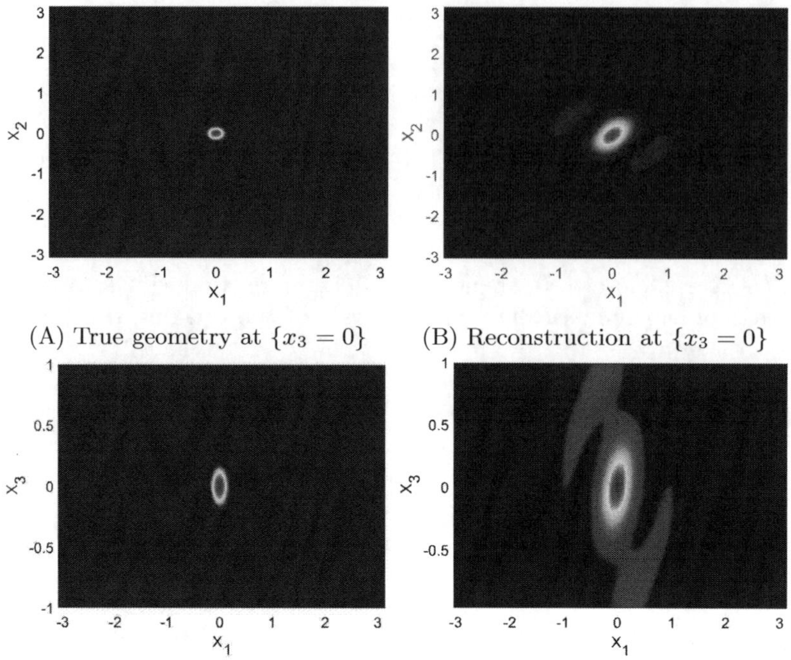

(A) True geometry at $\{x_3 = 0\}$ (B) Reconstruction at $\{x_3 = 0\}$

(C) True geometry at $\{x_2 = 0\}$ (D) Reconstruction at $\{x_2 = 0\}$

FIGURE 4. Reconstruction of a small sphere (in one period) using two incident fields.

consists of a non-smooth ring-like object in each period. Each ring is in fact a relatively thin hollow cylinder with small height. The geometry is given by

$$D = \{x \in \Omega_h : 0.8^2 < x_1^2 + x_2^2 < 1, |x_3| < 0.1\}.$$

For $\mathbf{x} \in D$, the material parameters are

$$\begin{aligned} \varepsilon_r(\mathbf{x}) &= I + \operatorname{diag}\{0.2, 0.4, 0.1\}, \\ \mu_r^{-1}(\mathbf{x}) &= I + \operatorname{diag}\{0.3, 0.1, 0.2\}, \\ \xi(\mathbf{x}) &= \operatorname{diag}\{0.01, 0.02, 0.05)\}, \end{aligned}$$

For $\mathbf{x} \in \Omega_h \setminus D$, $\varepsilon_r = \mu_r^{-1} = I$ and $\xi = 0$. The true and reconstructed cross-sections are shown in Figure 5.

This ring is an interesting object but challenging to reconstruct when there are only a small number of incident waves used. Regardless, the imaging functional gives us some decent information about the shape of the ring, especially in the $\{x_3 = 0\}$ cross-section.

The reconstruction method using $\mathcal{I}(\mathbf{z})$ is viable for reconstructing relatively small and thin objects when data is provided for two incident waves. The most promising aspects of the method are its simple implementation and stability. In its discretized form, calculating the imaging functional consists of only dot products and summations. As for stability, it can deal with very high levels of noise compared to some other methods, as we can see from the above examples.

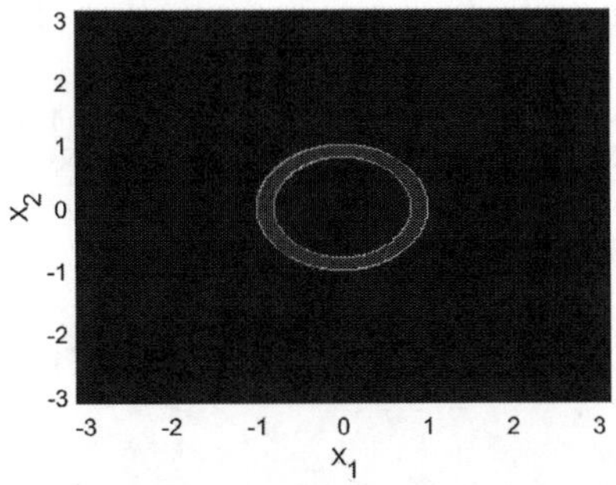

(A) True geometry at $\{x_3 = 0\}$

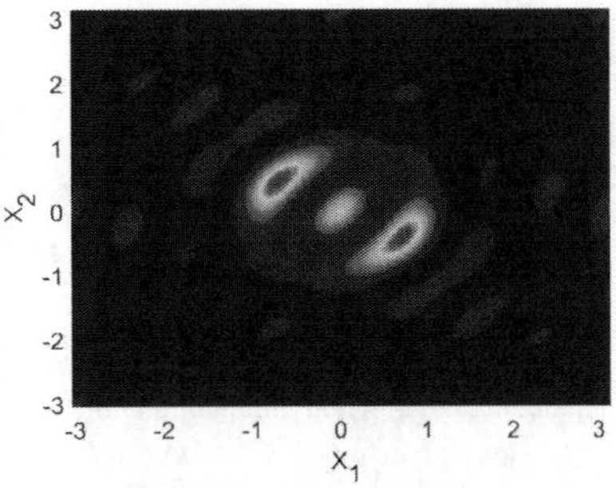

(B) Reconstruction at $\{x_3 = 0\}$

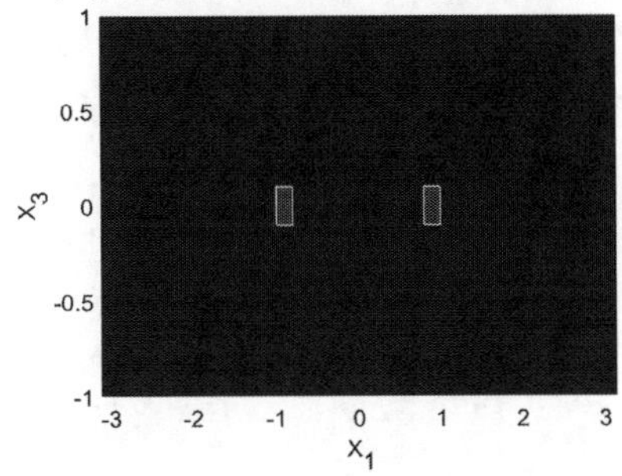

(C) True geometry at $\{x_2 = 0\}$

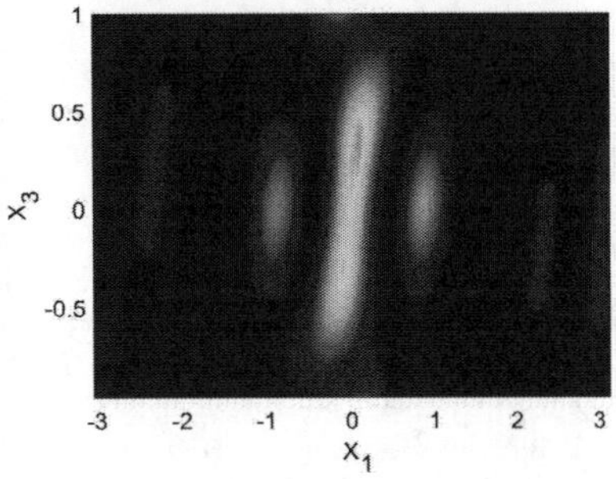

(D) Reconstruction at at $\{x_2 = 0\}$

FIGURE 5. Reconstruction of a ring (in one period) with two incident fields.

References

[1] Chi Young Ahn, Seongje Chae, and Won-Kwang Park, *Fast identification of short, sound-soft open arcs by the orthogonality sampling method in the limited-aperture inverse scattering problem*, Appl. Math. Lett. **109** (2020), 106556, 8, DOI 10.1016/j.aml.2020.106556. MR4109343

[2] Habib Ammari and Gang Bao, *Maxwell's equations in periodic chiral structures*, Math. Nachr. **251** (2003), 3–18, DOI 10.1002/mana.200310026. MR1960800

[3] Habib Ammari and Gang Bao, *Coupling of finite element and boundary element methods for the scattering by periodic chiral structures*, J. Comput. Math. **26** (2008), no. 3, 261–283. MR2421880

[4] Tilo Arens, *Scattering by biperiodic layered media: The integral equation approach*, Habilitation Thesis, Universität Karlsruhe, 2010.

[5] Christodoulos Athanasiadis, George Costakis, and Ioannis G. Stratis, *Electromagnetic scattering by a homogeneous chiral obstacle in a chiral environment*, IMA J. Appl. Math. **64** (2000), no. 3, 245–258, DOI 10.1093/imamat/64.3.245. MR1764940

[6] C. Athanasiadis and E. Kardasi, *Inverse electromagnetic scattering by a perfect conductor in a chiral environment*, J. Inverse Ill-Posed Probl. **16** (2008), no. 1, 1–18, DOI 10.1515/jiip.2008.001. MR2387647

[7] Patrick Ciarlet Jr. and Guillaume Legendre, *Well-posedness of the Drude-Born-Fedorov model for chiral media*, Math. Models Methods Appl. Sci. **17** (2007), no. 3, 461–484, DOI 10.1142/S0218202507001991. MR2311927

[8] Pierre-Henri Cocquet, Pierre-Alain Mazet, and Vincent Mouysset, *Sur l'existence et l'unicité des solutions pour des équations de Drude-Born-Fedorov homogénéisées en domaine borné et applications aux métamatériaux* (French, with English and French summaries), C. R. Math. Acad. Sci. Paris **349** (2011), no. 1-2, 99–103, DOI 10.1016/j.crma.2010.11.019. MR2755706

[9] K. M. Flood and D. L. Jaggard, *Band-gap structure for periodic chiral media*, J. Opt. Soc. Am. A **13** (1996), 1395.

[10] Roland Griesmaier, *Multi-frequency orthogonality sampling for inverse obstacle scattering problems*, Inverse Problems **27** (2011), no. 8, 085005, 23, DOI 10.1088/0266-5611/27/8/085005. MR2819947

[11] Isaac Harris and Dinh-Liem Nguyen, *Orthogonality sampling method for the electromagnetic inverse scattering problem*, SIAM J. Sci. Comput. **42** (2020), no. 3, B722–B737, DOI 10.1137/19M129783X. MR4101374

[12] S. Heumann, *The factorization method for inverse scattering from chiral media*, Ph.D. thesis, Karlsruher Institut für Technologie, 2012.

[13] D. L. Jaggard, N. Engheta, M. W. Kowarz, P. Pelet, J. C. Liu, and Y. Kim, *Periodic chiral structures*, IEEE Trans. Antennas Propagat. **37** (1989), 1447–1452.

[14] Andreas Kirsch, *An integral equation approach and the interior transmission problem for Maxwell's equations*, Inverse Probl. Imaging **1** (2007), no. 1, 159–179, DOI 10.3934/ipi.2007.1.159. MR2262751

[15] C. E. Kriegler, M. S. Rill, S. Linden, and M. Wegener, *Bianisotropic photonic metamaterials*, IEEE J. Sel. Topics Quantum Electron. **16** (2010), 367–375.

[16] Thu Le, Dinh-Liem Nguyen, Hayden Schmidt, and Trung Truong, *Imaging of 3D objects with experimental data using orthogonality sampling methods*, Inverse Problems **38** (2022), no. 2, Paper No. 025007, 20, DOI 10.1088/1361-6420/ac3d85. MR4374010

[17] Shumin Li, *An inverse problem for Maxwell's equations in bi-isotropic media*, SIAM J. Math. Anal. **37** (2005), no. 4, 1027–1043, DOI 10.1137/S003614100444366X. MR2192286

[18] T. G. Mackay and A. Lakhtakia, *Electromagnetic anisotropy and bi-anisotropy: A field guide*, World Scientific, Singapore, 2010.

[19] Stephen R. McDowall, *Boundary determination of material parameters from electromagnetic boundary information*, Inverse Problems **13** (1997), no. 1, 153–163, DOI 10.1088/0266-5611/13/1/012. MR1435874

[20] Stephen R. McDowall, *An electromagnetic inverse problem in chiral media*, Trans. Amer. Math. Soc. **352** (2000), no. 7, 2993–3013, DOI 10.1090/S0002-9947-00-02518-6. MR1675214

[21] Dinh-Liem Nguyen, *A volume integral equation method for periodic scattering problems for anisotropic Maxwell's equations*, Appl. Numer. Math. **98** (2015), 59–78, DOI 10.1016/j.apnum.2015.08.005. MR3400433

[22] Dinh-Liem Nguyen, *The factorization method for the Drude-Born-Fedorov model for periodic chiral structures*, Inverse Probl. Imaging **10** (2016), no. 2, 519–547, DOI 10.3934/ipi.2016010. MR3507905

[23] Fenglong Qu and Jiaqing Yang, *Unique determination of inverse electromagnetic scattering by a two-layered cavity*, Inverse Problems **35** (2019), no. 12, 125010, 16, DOI 10.1088/1361-6420/ab2b02. MR4039939

[24] Dinh-Liem Nguyen, Thanh-Nhan Nguyen, and Minh-Phuong Tran, *A Galerkin approximation for integro-differential equations in electromagnetic scattering from a chiral medium*, Appl. Anal. **96** (2017), no. 1, 159–172, DOI 10.1080/00036811.2016.1212335. MR3581693

[25] Dinh-Liem Nguyen and Trung Truong, *Imaging of bi-anisotropic periodic structures from electromagnetic near-field data*, J. Inverse Ill-Posed Probl. **30** (2022), no. 2, 205–219, DOI 10.1515/jiip-2020-0114. MR4401764

[26] Serge Nicaise, *Time domain study of the Drude-Born-Fedorov model for a class of heterogeneous chiral materials*, Math. Methods Appl. Sci. **36** (2013), no. 7, 794–813, DOI 10.1002/mma.2627. MR3056937

[27] J. Pendry, *A Chiral Route to Negative Refraction*, Science **19** (2004), 1353–1355.

[28] R. Potthast, *A study on orthogonality sampling*, Inverse Problems **26** (2010), 074015.

[29] Roland Potthast and Ioannis G. Stratis, *On the domain derivative for scattering by impenetrable obstacles in chiral media*, IMA J. Appl. Math. **68** (2003), no. 6, 621–635, DOI 10.1093/imamat/68.6.621. MR2019987

[30] G. F. Roach, I. G. Stratis, and A. N. Yannacopoulos, *Mathematical analysis of deterministic and stochastic problems in complex media electromagnetics*, Princeton Series in Applied Mathematics, Princeton University Press, Princeton, NJ, 2012. Appendix D by George Dassios. MR2896222

[31] Stefan A. Sauter and Christoph Schwab, *Boundary element methods*, Springer Series in Computational Mathematics, vol. 39, Springer-Verlag, Berlin, 2011. Translated and expanded from the 2004 German original, DOI 10.1007/978-3-540-68093-2. MR2743235

Department of Mathematics, Kansas State University, Manhattan, Kansas 66506
Email address: `dlnguyen@ksu.edu`

Department of Mathematics, Kansas State University, Manhattan, Kansas 66506
Email address: `trungt@ksu.edu`

Contemporary Mathematics
Volume **784**, 2023
https://doi.org/10.1090/conm/784/15755

Reconstructing a space-dependent source term via the quasi-reversibility method

Loc H. Nguyen and Huong T.T. Vu

ABSTRACT. The aim of this paper is to solve an important inverse source problem which arises from the well-known inverse scattering problem. We propose to truncate the Fourier series of the solution to the governing equation with respect to a special basis of L^2. By this, we obtain a system of linear elliptic equations. Solutions to this system are the Fourier coefficients of the solution to the governing equation. After computing these Fourier coefficients, we can directly find the desired source function. Numerical examples are presented.

1. Introduction

Let $d \geq 1$ be the spatial dimension. Fix a wavenumber $k > 0$. Let $[\underline{\theta}, \overline{\theta}]$ be an interval of angles. Let $u_0 = u_0(\mathbf{x}, \theta)$, $(\mathbf{x}, \theta) \in \mathbb{R}^d \times [\underline{\theta}, \overline{\theta}]$, be the incident plane wave that illuminates the medium at an angle θ. This choice of u_0 as the incident wave arises from the well-known inverse scattering problem, stated later. Let $u(\mathbf{x}, \theta)$, $(\mathbf{x}, \theta) \in \mathbb{R}^d \times [\underline{\theta}, \overline{\theta}]$, be the wave function that is governed by the Helmholtz equation and the Sommerfeld radiation condition

$$\begin{cases} \Delta u(\mathbf{x}, \theta) + k^2 c_0(\mathbf{x}) u(\mathbf{x}, \theta) = -k^2 p(\mathbf{x}) u_0(\mathbf{x}, \theta) & \mathbf{x} \in \mathbb{R}^d, \\ \partial_{|\mathbf{x}|} u(\mathbf{x}, \theta) - \mathrm{i}\, k u(\mathbf{x}, \theta) = o(|\mathbf{x}|^{\frac{1-d}{2}}) & |\mathbf{x}| \to \infty. \end{cases} \tag{1.1}$$

Here, $c_0 : \mathbb{R}^d \to [1, \infty)$ and $u_0 : \mathbb{R}^d \times [\underline{\theta}, \overline{\theta}] \to \mathbb{C}$ are given functions. The function c_0 represents the background dielectric constant of the medium.

Let Ω be an open and bounded domain of $\mathbb{R}^d$ with a smooth boundary. Assume that the source function p is compactly contained in Ω. We are interested in the following problem.

PROBLEM 1.1 (Inverse source problem). *Determine the source function $p(\mathbf{x})$ for all $\mathbf{x} \in \Omega$ from the following boundary measurements*

$$f(\mathbf{x}, \theta) = u(\mathbf{x}, \theta) \quad \text{and} \quad g(\mathbf{x}, \theta) = \partial_\nu u(\mathbf{x}, \theta) \tag{1.2}$$

for all $(\mathbf{x}, \theta) \in \partial\Omega \times [\underline{\theta}, \overline{\theta}]$. Here, $\nu(\mathbf{x})$ is the unit normal vector to $\partial\Omega$ at $\mathbf{x}$.

2020 *Mathematics Subject Classification.* Primary 35R30; Secondary 78A46.

Key words and phrases. Inverse source problem, truncated Fourier series, approximation, Carleman estimate, convergence.

LHN was partially supported by National Science Foundation grant DMS-2208159, by funds provided by the Faculty Research Grant program at UNC Charlotte Fund No. 111272, and by the CLAS small grant provided by the College of Liberal Arts & Sciences, UNC Charlotte.

This inverse source problem is the linearization of the nonlinear inverse scattering problem. Since the inverse scattering problem has many real-world applications; mostly in exploring some inaccessible regions from external measurement; for e.g, bio-medical imaging, nano-sciences, security, seismic exploration. Therefore, Problem 1.1 is significant in those fields. The solver for Problem 1.1 proposed in this paper consists of three steps.

(1) We derive a system of linear elliptic equations. Solution of this system is a vector involving the first N Fourier coefficients of the function $w = -u/(k^2 u_0)$, $N \in \mathbb{N}$, with respect to a special basis of L^2 introduced in [**35**].
(2) We apply the quasi-reversibility method developed in [**49**] to solve the system obtained in Step 1.
(3) Directly compute the desired source from the solution obtained in Step 2.

The inverse source problem and some of its versions were studied intensively. We cite to [**7–9, 24, 27, 28**] for the uniqueness, stability and numerical methods to solve inverse source problems for the case when the medium is homogeneous; i.e. c_0 = constant. The numerical reconstruction methods in those publications are based on the least squares optimization method. Good quality reconstructions due to those approaches are achieved only when the wave number k is large. We observe that when k is large, the data is very sensitive with noise. This is because of the high oscillation of the data. Unlike this, the reconstructive method in this paper and the method in [**49**] do not require the data at high frequency. Therefore, we can reconstruct the source with reasonable value of k. The difficulty about the noise is overcome.

As mentioned in step 2 above, we only compute the first N Fourier coefficients of the function $w = -u/(k^2 u_0)$. That means we only solve Problem 1.1 in an "Galerkin' approximation context. Rigorously verifying the convergence of this approximation as $N \to \infty$ is extremely challenging. We assume that this approximation is valid. In contrast, this numerical approach is very effective for many kinds of inverse problems; see e.g. [**36, 41, 47, 48, 50**]. This is the reason we employ this technique again in this paper.

The paper is organized as follows. In Section 2, we provide in details the formulation of Problem 1.1. In Section 3, we establish the approximation context and derive a system of PDEs that plays an important role in our algorithm. In Section 4, we recall the quasi-reversibility method to solve the system obtained in Section 3. We present numerical study in Section 5. Section 6 is for concluding remarks.

2. The significance of Problem 1.1

Problem 1.1 is the linearization of the nonlinear inverse scattering problem, which has many real-world applications. We will list some of the important applications later. Let $c : \mathbb{R}^d \to [1, \infty)$ be the spatially distributed dielectric constant of the medium. For each angle $\theta \in [\underline{\theta}, \overline{\theta}]$, define

$$\xi(\theta) = (\cos\theta, \sin\theta, 0, \dots, 0) \in \mathbb{S}^{d-1}$$

that represents the direction of the angle θ in the x_1x_2-plane. Here, $\mathbb{S}^{d-1} = \{\mathbf{x} \in \mathbb{R}^d, |\mathbf{x}| = 1\}$ is the unit sphere in $\mathbb{R}^d$. We use the plane wave, so-called the incident plane wave, of the form

$$v_0(\mathbf{x}, \theta) = e^{\mathrm{i}\, k\mathbf{x}\cdot\xi(\theta)} \quad \text{for all } (\mathbf{x}, \theta) \in \mathbb{R}^d \times [\underline{\theta}, \overline{\theta}], k > 0, \tag{2.1}$$

to illuminate the medium. The incident wave propagates in space and scatters. The resulting total wave $v(\mathbf{x}, \theta)$ is governed by the Helmholtz equation

$$\Delta v(\mathbf{x}, \theta) + k^2 c(\mathbf{x}) v(\mathbf{x}, \theta) = 0 \quad \text{for all } \mathbf{x} \in \mathbb{R}^d, \theta \in [\underline{\theta}, \overline{\theta}] \tag{2.2}$$

and the Sommerfeld radiation condition

$$\partial_{|\mathbf{x}|} v_{\mathrm{sc}}(\mathbf{x}, \theta) - \mathrm{i}\, k v_{\mathrm{sc}}(\mathbf{x}, \theta) = o(|\mathbf{x}|^{\frac{1-d}{2}})$$

where

$$v_{\mathrm{sc}}(\mathbf{x}, \theta) = v(\mathbf{x}, \theta) - v_0(\mathbf{x}, \theta) \quad \text{for all } (\mathbf{x}, \theta) \in \mathbb{R}^d \times [\underline{\theta}, \overline{\theta}] \tag{2.3}$$

is the scattering wave.

By (2.1), the angle between the x_1 axis and the propagation of incident wave in the x_1x_2 plane is θ. Roughly speaking, θ can be considered as the angle of view. Since we consider the case of complete angles of view, we choose $[\underline{\theta}, \overline{\theta}] = [0, 2\pi]$. The study of Problem 1.1 in the case of partial angle of views will be studied later.

The inverse scattering problem is formulated as follows.

PROBLEM 2.1 (The inverse scattering problem). *Let Ω be a bounded and open domain of $\mathbb{R}^d$ with smooth boundary. Given the boundary measurement $v_{\mathrm{sc}}(\mathbf{x}, \theta)$ and $\partial_\nu v_{\mathrm{sc}}(\mathbf{x}, \theta)$ for all $(\mathbf{x}, \theta) \in \partial\Omega \times [\underline{\theta}, \overline{\theta}]$, determine the spatially distributed dielectric constant $c(\mathbf{x})$ for all $\mathbf{x} \in \Omega$,*

Why is the inverse scattering problem interesting and significant? According to the formulation of Problem 2.1, we want to compute the information of the spatially distributed dielectric constant of the medium from the external measurement of the wave field. The knowledge of the reconstructed spatially distributed dielectric constant of a medium directly provides significant information about unknown objects inside that medium; for e.g, position, shape, size and physical properties. Typical examples of those objects are anti-personnel explosive devices buried under the ground, cancerous tumors inside living tissues, and nano structures. Therefore, solving the inverse scattering problem has important applications in bio-medical imaging, nondestructive testing, radar, security, optical physics, seismic exploration, nano science. Hence, the inverse scattering problem and the related ones have been intensively studied. We list here several approaches: the imaging techniques based on sampling and the factorization methods [**1, 3, 14, 16, 20, 26, 32, 42, 43, 45, 52**], the methods based on optimization [**4, 15, 17, 23, 25, 53**], the methods based on Born series [**12, 19, 22, 33, 37, 44**], the method based on linearization [**5, 6, 10, 18**], and the convexification method [**29–31**]. See [**21**] for a more complete list.

The current paper contributes to the field by solving the linearization of the inverse scattering problem. Assume that the dielectric constant c is a perturbation of a background function c_0. For simplicity, in this section, we choose c_0 to be identically 1. That means, the function c is of the form

$$c(\mathbf{x}) = 1 + \eta p(\mathbf{x}) \quad \text{for all } \mathbf{x} \in \mathbb{R}^d \tag{2.4}$$

where $0 < \eta \ll 1$ and p is a function that indicate the unknown inclusion. Therefore, the wave function v satisfies

$$\Delta v(\mathbf{x}, \theta) + k^2(1 + \eta p(\mathbf{x})) v(\mathbf{x}, \theta) = 0 \quad \text{for all } (\mathbf{x}, \theta) \in \mathbb{R}^d \times [\underline{\theta}, \overline{\theta}]. \tag{2.5}$$

It is not hard to verify that

$$\Delta v_0(\mathbf{x},\theta) + k^2 v_0(\mathbf{x},\theta) = 0 \quad \text{for all } (\mathbf{x},\theta) \in \mathbb{R}^d \times [\underline{\theta},\overline{\theta}]. \tag{2.6}$$

Let $u = v_{\rm sc}/\eta$. It follows from (2.5) and (2.6) that

$$\Delta u(\mathbf{x},\theta) + k^2 u(\mathbf{x},\theta) = -k^2 p(\mathbf{x}) v(\mathbf{x},\theta) \quad \text{for all } (\mathbf{x},\theta) \in \mathbb{R}^d \times [\underline{\theta},\overline{\theta}]. \tag{2.7}$$

It is obvious that $\lim_{\eta\to 0} v(\mathbf{x},\theta) = v_0(\mathbf{x},\theta)$. Thus, since η is a small number, equation (2.7) can be approximated as

$$\Delta u(\mathbf{x},\theta) + k^2 u(\mathbf{x},\theta) = -k^2 p(\mathbf{x}) v_0(\mathbf{x},\theta) \quad \text{for all } (\mathbf{x},\theta) \in \mathbb{R}^d \times [\underline{\theta},\overline{\theta}]. \tag{2.8}$$

By choosing u_0 as the incident wave v_0, we derive Problem 1.1. While the argument above only give an example for the motivation to solve Problem 1.1, our method to solve inverse problem is not limited in the context of the inverse scattering problem. That means, the proposed method in this paper can be applied in the general case when u_0 is not necessary to be the incident wave v_0.

3. An approximation context

Define

$$w(\mathbf{x},\theta) = -\frac{u(\mathbf{x},\theta)}{k^2 u_0(\mathbf{x},\theta)} \quad \text{or} \quad u(\mathbf{x},\theta) = -k^2 w(\mathbf{x},\theta) u_0(\mathbf{x},\theta) \tag{3.1}$$

for all $(\mathbf{x},k) \in \Omega \times [\underline{\theta},\overline{\theta}]$. Since $\Delta u(\mathbf{x},\theta) = -k^2 \Delta[w(\mathbf{x},\theta)u_0(\mathbf{x},\theta)]$, we have for all $(\mathbf{x},\theta) \in \Omega \times [\underline{\theta},\overline{\theta}]$

$$\Delta u(\mathbf{x},\theta) = -k^2\big[w(\mathbf{x},\theta)\Delta u_0(\mathbf{x},\theta) + u_0(\mathbf{x},\theta)\Delta w(\mathbf{x},\theta) + 2\nabla w(\mathbf{x},\theta)\cdot\nabla u_0(\mathbf{x},\theta)\big]. \tag{3.2}$$

On the other hand, it follows from the Helmholtz equation in (1.1) and the second identity in (3.1) that

$$\begin{aligned}\Delta u(\mathbf{x},\theta) &= -k^2\big[c_0(\mathbf{x})u(\mathbf{x},\theta) + p(\mathbf{x})u_0(\mathbf{x},\theta)\big]\\ &= -k^2\big[-k^2 c_0(\mathbf{x}) w(\mathbf{x},\theta) u_0(\mathbf{x},\theta) + p(\mathbf{x}) u_0(\mathbf{x},\theta)\big]\end{aligned}$$

for all $(\mathbf{x},\theta) \in \Omega \times [\underline{\theta},\overline{\theta}]$. This and (3.2) imply

$$\begin{aligned} w(\mathbf{x},\theta)\Delta u_0(\mathbf{x},\theta) + u_0(\mathbf{x},\theta)\Delta w(\mathbf{x},\theta) + 2\nabla w(\mathbf{x},\theta)\cdot\nabla u_0(\mathbf{x},\theta)&\\ = -k^2 c_0(\mathbf{x}) w(\mathbf{x},\theta) u_0(\mathbf{x},\theta) + p(\mathbf{x}) u_0(\mathbf{x},\theta)&\end{aligned}$$

or all $(\mathbf{x},\theta) \in \Omega \times [\underline{\theta},\overline{\theta}]$. Therefore,

$$\Delta w(\mathbf{x},\theta) + \Big[\frac{\Delta u_0(\mathbf{x},\theta)}{u_0(\mathbf{x},\theta)} + k^2 c_0(\mathbf{x})\Big] w(\mathbf{x},\theta) + 2\nabla w(\mathbf{x},\theta)\cdot\frac{\nabla u_0(\mathbf{x},\theta)}{u_0(\mathbf{x},\theta)} = p(\mathbf{x}) \tag{3.3}$$

for all $(\mathbf{x},\theta) \in \Omega \times [\underline{\theta},\overline{\theta}]$. To eliminate the unknown p, we differentiate (3.3) with respect to θ. We obtain

$$\begin{aligned}\Delta\partial_\theta w(\mathbf{x},\theta) + \partial_\theta\Big[\frac{\Delta u_0(\mathbf{x},\theta)}{u_0(\mathbf{x},\theta)} + k^2 c_0(\mathbf{x})\Big] w(\mathbf{x},\theta)&\\ + \Big[\frac{\Delta u_0(\mathbf{x},\theta)}{u_0(\mathbf{x},\theta)} + k^2 c_0(\mathbf{x})\Big]\partial_\theta w(\mathbf{x},\theta) + 2\nabla\partial_\theta w(\mathbf{x},\theta)\cdot\frac{\nabla u_0(\mathbf{x},\theta)}{u_0(\mathbf{x},\theta)}&\\ + 2\nabla w(\mathbf{x},\theta)\cdot\partial_\theta\Big[\frac{\nabla u_0(\mathbf{x},\theta)}{u_0(\mathbf{x},\theta)}\Big] = 0&\end{aligned} \tag{3.4}$$

for all $(\mathbf{x}, \theta) \in \Omega \times [\underline{\theta}, \overline{\theta}]$. Solving the third order equation (3.4) is extremely challenging. An analytic and numerical approach to solve it is not developed yet. Suggested by [**49**], we only consider an approximation context due to a truncation of the Fourier series of the solution $w(\mathbf{x}, \theta)$ with respect to the special basis. This basis was first introduced in [**35**]. We briefly recall the construction of this basis. For each $m \geq 1$, define $\phi_m(\theta) = (\theta - \theta_0)^{m-1} \exp(\theta - \theta_0)$ where $\theta_0 = (\underline{\theta} + \overline{\theta})/2$. The sequence $\{\phi_m\}_{m=1}^{\infty}$ is complete in $L^2(\underline{\theta}, \overline{\theta})$. Applying the Gram-Schmidt orthonormalization procedure to the sequence $\{\phi_m\}_{m=1}^{\infty}$, we obtain an orthonormal basis in $L^2(\underline{\theta}, \overline{\theta})$, denoted by $\{\Psi_m\}_{m=1}^{\infty}$. It is not hard to verify that for each m, the function $\Psi_m(\theta)$ has the form $\Psi_m(\theta) = P_{m-1}(\theta - \theta_0) \exp(\theta - \theta_0)$, where P_{m-1} is a polynomial of the degree $(m-1)$. The following result plays an important role in our analysis.

PROPOSITION 3.1 (see [**35**]). *For $m, n \geq 1$, we have*

$$s_{mn} = \int_{\underline{\theta}}^{\overline{\theta}} \Psi_m(\theta) \Psi_n'(\theta) d\theta = \begin{cases} 1 & \text{if } n = m, \\ 0 & \text{if } n < m. \end{cases} \tag{3.5}$$

Consequently, let $N > 1$ be an integer. Then, the $N \times N$ matrix

$$S = (s_{mn})_{m,n=1}^{N} \tag{3.6}$$

has determinant 1 and is invertible.

Recall the function w defined in (3.1). For each $(\mathbf{x}, \theta) \in \Omega \times [\underline{\theta}, \overline{\theta}]$, we write

$$w(\mathbf{x}, \theta) = \sum_{n=1}^{\infty} w_n(\mathbf{x}) \Psi_n(\theta) \simeq \sum_{n=1}^{N} w_n(\mathbf{x}) \Psi_n(\theta) \tag{3.7}$$

where

$$w_n(\mathbf{x}) = \int_{\underline{\theta}}^{\overline{\theta}} w(\mathbf{x}, \theta) \Psi_n(\theta) d\theta, \quad n = 1, 2, \ldots \tag{3.8}$$

REMARK 3.1 (The choice of N). *The cut-off number N in (3.7) is chosen numerically as follows. For each N, we define*

$$\varphi(N) = \Big\| w(\mathbf{x}, \theta) - \sum_{n=1}^{N} w_n(\mathbf{x}) \Psi_n(\theta) \Big\|_{L^{\infty}(\Gamma \times [\underline{\theta}, \overline{\theta}])}$$

where Γ is a subset of $\partial\Omega$. Since the function u is given on $\partial\Omega \times [\underline{\theta}, \overline{\theta}])$, using (3.1) and (3.8), we can compute $\varphi(N)$ directly. We then choose N such that $\varphi(N)$ is sufficiently small. In our computation, $N = 35$. With this choice of N, we have $\varphi(N) < 5 \times 10^{-3}$. See Figure 1 for an illustration.

From now on, we assume that the approximation in (3.7) is valid. Plugging the truncating formula (3.7) into (3.4), we obtain

$$
\begin{aligned}
(3.9)\quad &\sum_{n=1}^{N} \Delta w_n(\mathbf{x})\Psi_n'(\theta) + \partial_\theta\Big[\frac{\Delta u_0(\mathbf{x},\theta)}{u_0(\mathbf{x},\theta)} + k^2 c_0(\mathbf{x})\Big]\sum_{n=1}^{N} w_n(\mathbf{x})\Psi_n(\theta) \\
&+ \Big[\frac{\Delta u_0(\mathbf{x},\theta)}{u_0(\mathbf{x},\theta)} + k^2 c_0(\mathbf{x})\Big]\sum_{n=1}^{N} w_n(\mathbf{x})\Psi_n'(\theta) + 2\sum_{n=1}^{N}\nabla w_n(\mathbf{x})\Psi_n'(\theta)\cdot\frac{\nabla u_0(\mathbf{x},\theta)}{u_0(\mathbf{x},\theta)} \\
&\qquad + 2\sum_{n=1}^{N}\nabla w_n(\mathbf{x})\Psi_n(\theta)\cdot\partial_\theta\Big[\frac{\nabla u_0(\mathbf{x},\theta)}{u_0(\mathbf{x},\theta)}\Big] = 0
\end{aligned}
$$

for all $(\mathbf{x},\theta) \in \Omega\times[\underline{\theta},\overline{\theta}]$. For each $m \in \{1,\dots,N\}$, we multiply $\Psi_m(\theta)$ to both sides of (3.9). Then, we integrate the resulting equation with respect to θ on $[\underline{\theta},\overline{\theta}]$. We get

$$
(3.10)\qquad \sum_{n=1}^{N} s_{mn}\Delta w_n(\mathbf{x}) + \sum_{n=1}^{N} a_{mn}(\mathbf{x})w_n(\mathbf{x}) + \sum_{n=1}^{N}\mathbf{b}_{mn}(\mathbf{x})\cdot\nabla w_n(\mathbf{x}) = 0
$$

for all $\mathbf{x}\in\Omega$. Here,

$$
\begin{aligned}
s_{mn} &= \int_{\underline{\theta}}^{\overline{\theta}} \Psi_n'(\theta)\Psi_m(\theta)d\theta,\\
a_{mn}(\mathbf{x}) &= \int_{\underline{\theta}}^{\overline{\theta}} \Big[\partial_\theta\Big(\frac{\Delta u_0(\mathbf{x},\theta)}{u_0(\mathbf{x},\theta)} + k^2c_0(\mathbf{x})\Big)\Psi_n(\theta)\\
&\qquad\qquad + \Big(\frac{\Delta u_0(\mathbf{x},\theta)}{u_0(\mathbf{x},\theta)} + k^2c_0(\mathbf{x})\Big)\Psi_n'(\theta)\Big]\Psi_m(\theta)d\theta,\\
\mathbf{b}_{mn}(\mathbf{x}) &= 2\int_{\underline{\theta}}^{\overline{\theta}} \Big[\frac{\nabla u_0(\mathbf{x},\theta)}{u_0(\mathbf{x},\theta)}\Psi_n'(\theta) + \partial_\theta\Big(\frac{\nabla u_0(\mathbf{x},\theta)}{u_0(\mathbf{x},\theta)}\Big)\Psi_n(\theta)\Big]\Psi_m(\theta)d\theta
\end{aligned}
$$

for all $m,n\in\{1,\dots,N\}$. Equation (3.10), with $m\in\{1,\dots,N\}$, forms a linear system of linear elliptic equations for the vector $W = (w_n)_{n=1}^N$.

We next derive the boundary conditions for W. Due to (1.2) and (3.8), for all $\mathbf{x}\in\partial\Omega$ and $n\in\{1,\dots,N\}$, we have

$$
(3.11)\qquad w_n(\mathbf{x}) = \int_{\underline{\theta}}^{\overline{\theta}} \frac{u(\mathbf{x},\theta)}{-k^2u_0(\mathbf{x},\theta)}\Psi_n(\theta)d\theta = \int_{\underline{\theta}}^{\overline{\theta}} \frac{f(\mathbf{x},\theta)}{-k^2u_0(\mathbf{x},\theta)}\Psi_n(\theta)d\theta
$$

and

$$
\begin{aligned}
\partial_\nu w_n(\mathbf{x}) &= \int_{\underline{\theta}}^{\overline{\theta}} \partial_\nu\Big(\frac{u(\mathbf{x},\theta)}{-k^2u_0(\mathbf{x},\theta)}\Big)\Psi_n(\theta)d\theta\\
&= \int_{\underline{\theta}}^{\overline{\theta}} \frac{\partial_\nu u(\mathbf{x},\theta)u_0(\mathbf{x},\theta) - u(\mathbf{x},\theta)\partial_\nu u_0(\mathbf{x},\theta)}{-k^2u_0^2(\mathbf{x},\theta)}\Psi_n(\theta)d\theta\\
(3.12)\qquad &= \int_{\underline{\theta}}^{\overline{\theta}} \frac{g(\mathbf{x},\theta)u_0(\mathbf{x},\theta) - f(\mathbf{x},\theta)\partial_\nu u_0(\mathbf{x},\theta)}{-k^2u_0^2(\mathbf{x},\theta)}\Psi_n(\theta)d\theta.
\end{aligned}
$$

Due to (3.10), (3.11) and (3.12), the vector W satisfies

$$
\begin{cases} \Delta W(\mathbf{x}) + S^{-1}A(\mathbf{x})W(\mathbf{x}) + S^{-1}\mathbf{B}(\mathbf{x}) \cdot \nabla W(\mathbf{x}) = 0 & \mathbf{x} \in \Omega, \\ W(\mathbf{x}) = F(\mathbf{x}) & \mathbf{x} \in \partial\Omega, \\ \partial_\nu W(\mathbf{x}) = G(\mathbf{x}) & \mathbf{x} \in \partial\Omega \end{cases} \tag{3.13}
$$

where $S = (s_{mn})_{m,n=1}^N$, $A(\mathbf{x}) = (a_{mn}(\mathbf{x}))_{m,n=1}^N$, $\mathbf{B}(\mathbf{x}) = (\mathbf{b}_{mn}(\mathbf{x}))_{m,n=1}^N$ and

$$
F(\mathbf{x}) = \Big(\int_{\underline{\theta}}^{\overline{\theta}} \frac{f(\mathbf{x}, \theta)}{-k^2 u_0(\mathbf{x}, \theta)} \Psi_n(\theta) d\theta \Big)_{n=1}^N, \tag{3.14}
$$

$$
G(\mathbf{x}) = \Big(\int_{\underline{\theta}}^{\overline{\theta}} \frac{g(\mathbf{x}, \theta)u_0(\mathbf{x}, \theta) - f(\mathbf{x}, \theta)\partial_\nu u_0(\mathbf{x}, k)}{-k^2 u_0^2(\mathbf{x}, \theta)} \Psi_n(\theta) d\theta \Big)_{n=1}^N. \tag{3.15}
$$

The invertibility of S is an important property of the basis $\{\Psi_m\}_{m\geq 1}$. See Proposition 3.1 and the proof in [**35**]. Using other basis, for e.g the popular trigonometric basis, is not suitable because the corresponding S is not invertible. Solving Problem 1.1 becomes the problem of finding a function W satisfying (3.13).

REMARK 3.2. *Using the truncation in* (3.7) *to derive* (3.13) *is inspired by the Garlekin approximation in the frequency domain. Studying the behavior of this approximation context as $N \to \infty$ is extremely challenging. This is out of the scope of the paper. Although* (3.13) *is not exact, the approximation is good enough for us to obtain out of expectation numerical results. This phenomenon is true for a long list of inverse problems. Here is an incomplete list of inverse problems that were solved numerically by using similar truncation approaches. We refer the reader to* [**36**] *for the X-ray tomography problem,* [**51**] *for an inverse source problem for the full radiative transfer equation,* [**29, 31, 40**] *for the inverse scattering problem,* [**47, 50**] *for a coefficient inverse problem for parabolic equations,* [**39**] *for an inverse source problem for nonlinear parabolic equations;* [**41**] *for an inverse source problem for hyperbolic equations. Especially, we refer the reader to* [**49**] *for an algorithm to solve an inverse source problem that is a particular case of Problem* 1.1. *The inverse source problem in* [**49**] *only for the case when u_0 does only on the wave number k while in this paper, u_0 is allowed to depend on the spatial variable* $\mathbf{x}$.

Solving Problem 1.1 becomes finding a vector valued function W that satisfies (3.13). Since (3.13) is over-determined, we apply the quasi-reversibility method to solve it. It is worth mentioning that the quasi-reversibility method was first introduced in [**38**]. Then, it was used very often in solving over-determined boundary value problems, see e.g., [**11, 13, 34, 41, 46, 49**]. Having the solution $W = (w_1, \dots, w_N) \in H^2(\Omega)^N$ of (3.13) in hand, we can compute $w(\mathbf{x}, \theta)$ for all $(\mathbf{x}, \theta) \in \overline{\Omega} \times [\underline{\theta}, \overline{\theta}]$ via (3.7). Then, we can compute the source function p by using (3.3). This procedure is summarized in Algorithm 1.

The quasi-reversibility method used in Step 2 of Algorithm 1 will be presented in Section 4.

4. The quasi-reversibility method

For the convenience in the analysis, we consider in this section the boundary data $W = F$ and $\partial_\nu W = G$ respectively as the "indirect data" of Problem 1.1. In fact, having the data of the inverse problem, see (1.2), in hand, we can use the explicit formulas in (3.14) and (3.15) to compute F and G. Let $\delta > 0$ be a noise level. Let F^δ and G^δ and F^* and G^* be the noisy and noiseless versions of F and

Algorithm 1 A numerical method to solve Problem 1.1

1: Define the basis $\{\Psi\}_{n\geq 1}$ (see [**35**]) and choose a cut-off number N as in Remark 3.1.
2: Solve (3.13) by the quasi-reversibility method (see Section 4) for a vector value function $W^{\rm comp} \in H^2(\Omega)^N$.
3: Compute

$$w^{\rm comp}(\mathbf{x}, \theta) = \sum_{n=1}^{N} w_n(\mathbf{x})\Psi_n(\theta) \quad \text{for all } (\mathbf{x}, \theta) \in \overline{\Omega} \times [\underline{\theta}, \overline{\theta}].$$

4: The source function is given by

$$p^{\rm comp}(\mathbf{x}) = \frac{-1}{\overline{\theta} - \underline{\theta}} \int_{\underline{\theta}}^{\overline{\theta}} \Big[\Delta w^{\rm comp}(\mathbf{x}, \theta) + \Big[\frac{\Delta u_0(\mathbf{x}, \theta)}{u_0(\mathbf{x}, \theta)} + k^2 c_0(\mathbf{x})\Big] w^{\rm comp}(\mathbf{x}, \theta) + 2\nabla w^{\rm comp}(\mathbf{x}, \theta) \cdot \frac{\nabla u_0(\mathbf{x}, \theta)}{u_0(\mathbf{x}, \theta)}\Big] d\theta$$

for all $\mathbf{x} \in \Omega$.

G respectively. The corresponding noiseless versions are denoted by F^* and G^*. By noise, we mean that there exists an "error" function $\mathcal{E} \in H^2(\Omega)^N$ such that

$$\begin{cases} \mathcal{E}|_{\partial\Omega} = F^\delta - F^*, \\ \partial_\nu \mathcal{E}|_{\partial\Omega} = G^\delta - G^*, \\ \|\mathcal{E}\|_{H^2(\Omega)} \leq \delta. \end{cases} \tag{4.1}$$

REMARK 4.1. *The existence of the error function in* (4.1) *imply that the noise can be regularized; for e.g.* [**2**, Section 5] *for the noise model such that it can be smooth out. This condition is significant only for the proof of the convergence theorem (see Theorem* 4.1 *and its proof in* [**49**]*). However, in numerical study, we do not have to smooth the noise. In fact, we compute the noisy data* F^δ *and* G^δ *using formulas* (3.14) *and*(3.15) *with* f *and* g *replaced by* f^δ *and* f^δ *respectively. Here,*

$$f^\delta = f^*(1 + \delta rand), \quad g^\delta = g^*(1 + \delta rand) \tag{4.2}$$

where rand is a function taking uniformly distributed random numbers in $[-1, 1]$ *and* f^* *and* g^* *are the noiseless versions of* f *and* g *respectively.*

Define the set of admissible solution

$$H = \{V \in H^2(\Omega) : V|_{\partial\Omega} = F^\delta \text{and } \partial_\nu V|_{\partial\Omega} = G^\delta\}. \tag{4.3}$$

We assume that the set $H \neq \emptyset$. Due to the presence of noise in the given data, problem (3.13) is over-determined. It is natural to solve it using the quasi-reversibility method, which is similar to the least-squares optimization together with a Tikhonov regularization term. That means, we minimize the functional

$$J(W) = \int_\Omega \left|\Delta W + S^{-1}A(\mathbf{x})W(\mathbf{x}) + S^{-1}\mathbf{B}(\mathbf{x}) \cdot \nabla W(\mathbf{x})\right|^2 d\mathbf{x} + \epsilon\|W\|^2_{H^2(\Omega)^N}$$

subject to the boundary conditions $W|_{\partial\Omega} = F^\delta$ and $\partial_\nu W|_{\partial\Omega} = G^\delta$.

REMARK 4.2. *The presence of the Tikhonov regularization term is significant. In the theoretical part, it makes the functional J coercive, which is important for the existence of a minimizer of J. In practice, we cannot obtain good numerical result without the presence of this Tikhonov regularization term. In our numerical study in Section 4, we choose $\epsilon = 10^{-5}$. This value of ϵ was chosen by a trial and error process. We manually try many values of ϵ for test 1 until we obtain good numerical result. Then, we use this value for all other tests and for several noise levels $\delta \in [5\%, 50\%]$.*

We have the theorem.

THEOREM 4.1. *Let F^δ and G^δ be the noisy boundary data for W. Here, $\delta > 0$ is the noise level in the sense of (4.1). Assume that the set H defined in (4.3) is nonempty. Then, for any $\epsilon > 0$, the functional J has a unique minimizer in H. Denote by $W_{\min}$ the obtained minimizer. Moreover, let W^* be true solution to (3.13). Then, the following estimate holds true*

$$\|W^\delta - W^*\|^2_{H^1(\Omega)^N} \leq C\Big(\delta^2 + \epsilon\|W^*\|^2_{H^2(\Omega)^N}\Big). \tag{4.4}$$

For brevity, we do not present the proof of Theorem 4.1. We refer the reader to [**49**, Theorem 3.1 and Theorem 5.1] for the proof of this theorem. We also refer to [**47**, Proposition 4.1 and Theorem 4.1] for the proof of a similar theorem when J involves the boundary integrals of the data.

REMARK 4.3. *The quasi-reversibility method we employ here is based on least squares optimization with a special Tikhonov regularization term. In general, to solve the over-determined boundary value problem (3.13), one can use the least squares optimization method with many other choices for the regularization term. It is worth to study the convergence of the method with each of such choice.*

5. Numerical study

For simplicity in implementation, we numerically solve Problem 1.1 in 2D. Let Ω be the square $(-1,1)^2 \subset \mathbb{R}^2$. Since solving the Helmholtz equation in (1.1) on the whole space, to generate the simulated data, is challenging, we solve its approximation on Ω, say

$$\begin{cases} \Delta u(\mathbf{x},\theta) + k^2 c_0(\mathbf{x})u(\mathbf{x},\theta) = -k^2 p(\mathbf{x})u_0(\mathbf{x},\theta) & \mathbf{x} \in \Omega, \\ \partial_{nu} u(\mathbf{x},\theta) - \mathrm{i}\, ku(\mathbf{x},\theta) = 0 & \mathbf{x} \in \partial\Omega. \end{cases} \tag{5.1}$$

This change is acceptable in the sense that it does not effect the analysis in Section 3 because our arguments depend only on the form of the governing partial differential equation while the boundary values of u and the flux $\partial_\nu u$ serve as the given data. Due to the motivation of Problem 1.1 in Section 2, we choose u_0 as the incident wave v_0, defined in (2.1) where $k = 3\pi$. For simplicity, we choose $c_0 \equiv 1$ The range of the angle $[\underline{\theta}, \overline{\theta}] = [0, 2\pi]$. We solve (5.1) by the finite difference method. We arrange a uniform $N_x \times N_x \times N_\theta$ grid of points in $\Omega \times [\underline{\theta}, \overline{\theta}]$ as

$$\Big\{(x_i, y_j, \theta_l) : x_i = -1 + (i-1)d_x, y_j = -1 + (j-1)d_x,$$
$$\theta_l = \underline{\theta} + (l-1)d_\theta, 1 \leq i,j \leq N_x 1 \leq l \leq N_\theta\Big\}$$

where $d_x = 2/(N_x - 1)$ and $d_\theta = (\overline{\theta} - \underline{\theta})/(N_\theta - 1)$. In our computational program $N_x = 80$ and $N_\theta = 250$. Since solving (5.1) by the finite difference method to compute the solution $u(\mathbf{x}, \theta)$ is standard in the scientific community, we do not describe the procedure here. Having $u|_{\partial\Omega}$ and $\partial_\nu u$ in hand, we can compute the indirect data $F|_{\partial\Omega}$ and $G|_{\partial\Omega}$ (3.14) and (3.15) respectively. The noisy data are as in Remark 4.1 and (4.2).

We now present an example of the choice of the cut-off number N in Remark 3.1. We choose $\Gamma = \{(x = 1, y) : |y| \leq 1\} \subset \partial\Omega$. We numerically examine how the function $\sum_{n=1}^{N} w(\mathbf{x}, \theta)$ approximates the function $w(\mathbf{x}, \theta)$ on $\Gamma \times [\underline{\theta}, \overline{\theta}]$ by testing their L^∞ difference

$$\varphi(N) = \Big\| \sum_{n=1}^{N} w_n(\mathbf{x})\Psi_n(\theta) - w(\mathbf{x}) \Big\|_{L^\infty(\Gamma \times [\underline{\theta}, \overline{\theta}])}.$$

The graphs of $\varphi(N)$ for $N \in \{15, 25, 35\}$ are displayed in Figure 1. It is evidence that $\varphi(N = 35)$ is sufficiently small. We choose $N = 35$ in all of our numerical tests.

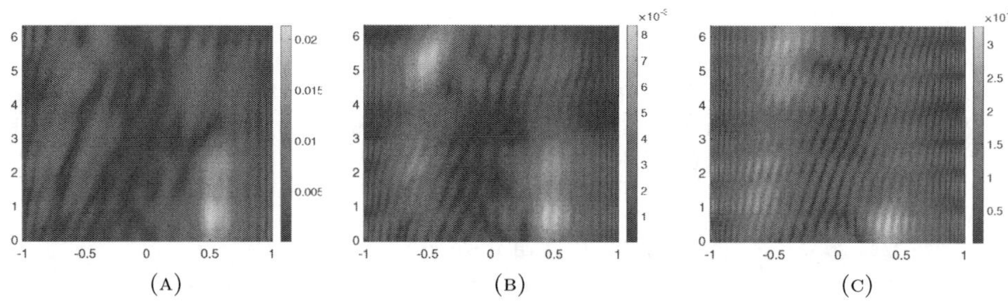

FIGURE 1. The graph of the function $\varphi(N) = \|w(\mathbf{x}, \theta) - \sum_{n=1}^{N} w_n(\mathbf{x})\Psi_n(\theta)\|_{L^\infty(\Gamma \times [\underline{\theta}, \overline{\theta}])}$ where $\Gamma = \{(x = R, y) : |y| < R\}$ is a subset of $\partial\Omega$. In (A), $N = 15$ and $\varphi(15) = 21.3 \times 10^{-3}$ In (B), $N = 25$, $\varphi(N) = 8.34 \times 10^{-3}$ In (C), $N = 35$, $\varphi(N) = 3.26 \times 10^{-3}$. The data is taken from Test 4 below. In these graphs, the horizontal axis indicates the range of $y \in [-R, R]$ and the vertical axis represents the angle $\theta \in [0, 2\pi]$.

In Step 2 of Algorithm 1, rather than solving (3.13), we solve the equivalent problem

$$\begin{cases} S\Delta W(\mathbf{x}) + A(\mathbf{x})W(\mathbf{x}) + \mathbf{B}(\mathbf{x}) \cdot \nabla W(\mathbf{x}) = 0 & \mathbf{x} \in \Omega, \\ W(\mathbf{x}) = F(\mathbf{x}) & \mathbf{x} \in \partial\Omega, \\ \partial_\nu W(\mathbf{x}) = G(\mathbf{x}) & \mathbf{x} \in \partial\Omega. \end{cases} \tag{5.2}$$

We suggest this change to remove the inverse of the matrix S in the system of PDE. The main reason is that the Gram-Schmidt procedure is unstable, so is S^{-1}. The unstability of S^{-1} might lead to some unnecessary difficulties in numerical experiments. The implementation of the quasi-reversibility method to solve (5.2) is very similar to that in [**47, 50**]. We do not repeat it here. The implementation of other Steps of Algorithm 1 are straightforward.

We next show four numerical tests.

Test 1. We test the case when the support of the true source function takes the form

$$p_{\text{true}}(x,y) = \begin{cases} 2 & \text{if } \max(|x-y|/0.8, |x+y|/0.35) < 1, \\ 0 & \text{otherwise} \end{cases} \quad \text{for all } (x,y) \in \Omega.$$

The graphs of the true source function and its reconstructions are displayed in Figure 2. In this test, the support of the true source function p looks like a rectangle. It is evident that this rectangle is detected successfully. On the other hand, the maximum values of the source are computed quite accurate. When $\delta = 5\%$, $\frac{|\max_{\mathbf{x}\in\overline{\Omega}} p_{\text{comp}} - \max_{\mathbf{x}\in\overline{\Omega}} p_{\text{true}}|}{|\max_{\mathbf{x}\in\overline{\Omega}} p_{\text{true}}|} = 6.72\%$. When $\delta = 10\%$, this error is 6.95%.

Test 2. We next test the case when the value and the support of the true source are larger than those in Test 1. The true source function is given by

$$p_{\text{true}}(x,y) = \begin{cases} 4 & \text{if } \max\{|x+y|, |x-y|\} < 0.6 \\ 0 & \text{otherwise} \end{cases} \quad \text{for all } (x,y) \in \Omega.$$

The graphs of the true source function and its reconstructions are displayed in Figure 3. The support of true source function p_{true} is the square centered at the origin, rotated 45°. We can see that reconstruction this square is out of expectation. The maximum values the source function inside the square is quite correctly computed. When $\delta = 10\%$, $\frac{|\max_{\mathbf{x}\in\overline{\Omega}} p_{\text{comp}} - \max_{\mathbf{x}\in\overline{\Omega}} p_{\text{true}}|}{|\max_{\mathbf{x}\in\overline{\Omega}} p_{\text{true}}|} = 5.13\%$. When $\delta = 30\%$, this error is 5.44%.

Test 3. We next test the case when the graph of the true source has a void. The true source function in this test is

$$p_{\text{true}}(x,y) = \begin{cases} 4 & \text{if } 0.4^2 < x^2 + y^2 < 0.8^2 \\ 0 & \text{otherwise} \end{cases} \quad \text{for all } (x,y) \in \Omega.$$

The graphs of the true source function and its reconstructions are displayed in Figure 4. The support of true source function p_{true} is the ring centered at the origin with outer radius 0.8 and inner radius 0.4. This test is interesting due to the presence of the void. We can see that reconstruction this ring is acceptable. The maximum values the source function inside the square is quite correctly computed. When $\delta = 5\%$, $\frac{|\max_{\mathbf{x}\in\overline{\Omega}} p_{\text{comp}} - \max_{\mathbf{x}\in\overline{\Omega}} p_{\text{true}}|}{|\max_{\mathbf{x}\in\overline{\Omega}} p_{\text{true}}|} = 6.35\%$. When $\delta = 50\%$, this error is 8.13%.

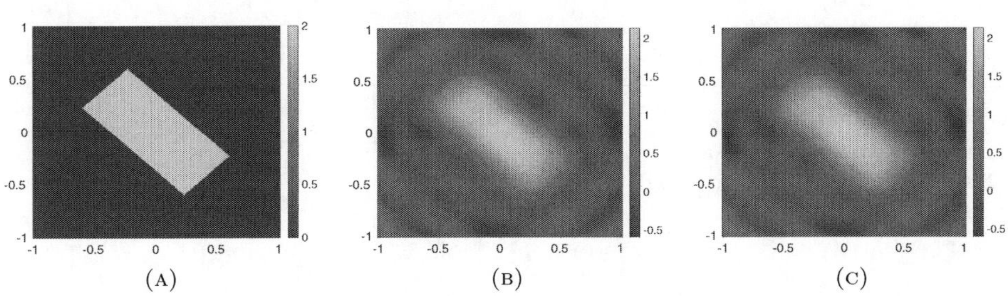

FIGURE 2. Test 1. The true and reconstruction of the source functions p. (A) The function p_{true}. (B) The function p_{comp} computed from data containing $\delta = 5\%$ noise. (C) The function p_{comp} computed from data containing $\delta = 10\%$ noise.

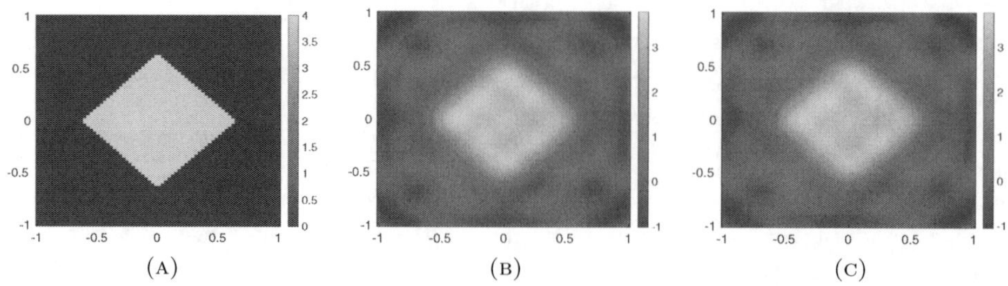

FIGURE 3. Test 2. The true and reconstruction of the source functions p. (A) The function p_{true}. (B) The function p_{comp} computed from data containing $\delta = 10\%$ noise. (C) The function p_{comp} computed from data containing $\delta = 30\%$ noise.

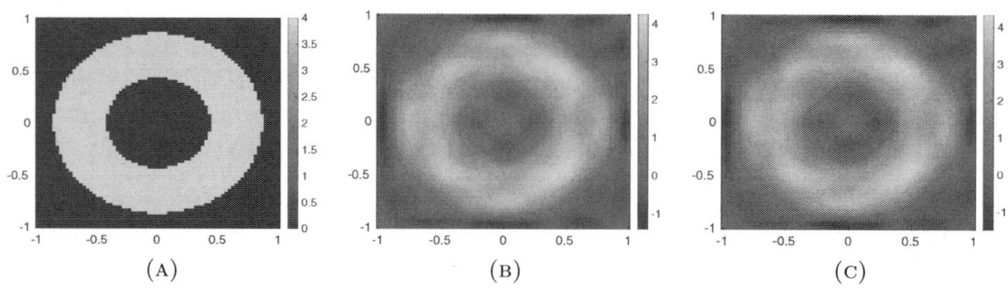

FIGURE 4. Test 3. The true and reconstruction of the source functions p. (A) The function p_{true}. (B) The function p_{comp} computed from data containing $\delta = 5\%$ noise. (C) The function p_{comp} computed from data containing $\delta = 50\%$ noise.

Test 4. We next consider a more interesting case. The function p_{true} is the characteristic function of the letter Y. Although this true source function has complicated structure, we are able to well compute it. The graphs of the true source function and its reconstructions are displayed in Figure 5. We can see that reconstruction the

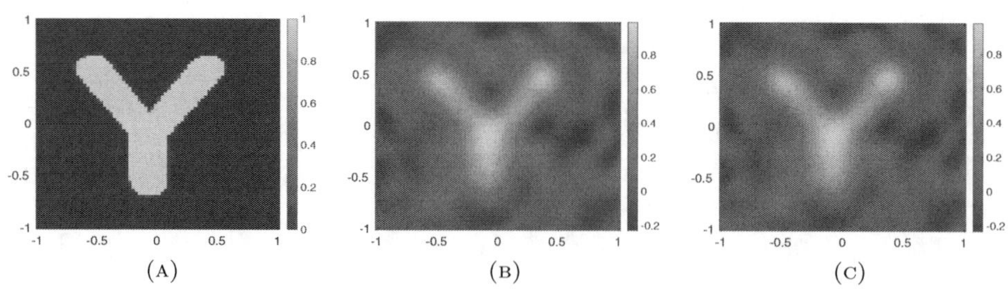

FIGURE 5. Test 4. The true and reconstruction of the source functions p. (A) The function p_{true}. (B) The function p_{comp} computed from data containing $\delta = 10\%$ noise. (C) The function p_{comp} computed from data containing $\delta = 50\%$ noise.

letter Y is acceptable. The maximum values the source function inside the square is quite correctly computed. When $\delta = 10\%$, $\frac{|\max_{\mathbf{x}\in\overline{\Omega}} p_{\text{comp}} - \max_{\mathbf{x}\in\overline{\Omega}} p_{\text{true}}|}{|\max_{\mathbf{x}\in\overline{\Omega}} p_{\text{true}}|} = 0.83\%$. When $\delta = 50\%$, this error is 1.92%.

REMARK 5.1. *These numerical examples numerically show that Algorithm* 1 *is robust. Especially, it is stable with respect to the noise. We can obtain good numerical results even when the noise level is up to* 50%.

6. Concluding remarks

In this paper, we solve an inverse source problem. This inverse source problem is the linearization of the highly nonlinear and severely ill-posed inverse scattering problem. In order to solve the inverse source problem, in the first step, we derive a system of linear elliptic PDEs in which the source function is absent. We solve this system by the quasi-reversibility method. We choose the quasi-reversibility method because its convergence as noise tends to 0 was proved. The efficiency of our method is confirmed by some interesting numerical examples.

References

[1] Habib Ammari, Yat Tin Chow, and Jun Zou, *Phased and phaseless domain reconstructions in the inverse scattering problem via scattering coefficients*, SIAM J. Appl. Math. **76** (2016), no. 3, 1000–1030, DOI 10.1137/15M1043959. MR3505314

[2] Habib Ammari, Josselin Garnier, Wenjia Jing, and Loc Hoang Nguyen, *Quantitative thermo-acoustic imaging: an exact reconstruction formula*, J. Differential Equations **254** (2013), no. 3, 1375–1395, DOI 10.1016/j.jde.2012.10.019. MR2997376

[3] Habib Ammari and Hyeonbae Kang, *Reconstruction of small inhomogeneities from boundary measurements*, Lecture Notes in Mathematics, vol. 1846, Springer-Verlag, Berlin, 2004, DOI 10.1007/b98245. MR2168949

[4] A. B. Bakushinsky and M. Yu. Kokurin, *Iterative methods for approximate solution of inverse problems*, Mathematics and Its Applications (New York), vol. 577, Springer, Dordrecht, 2004. MR2133802

[5] Gang Bao and Peijun Li, *Inverse medium scattering for the Helmholtz equation at fixed frequency*, Inverse Problems **21** (2005), no. 5, 1621–1641, DOI 10.1088/0266-5611/21/5/007. MR2173413

[6] Gang Bao and Peijun Li, *Inverse medium scattering problems for electromagnetic waves*, SIAM J. Appl. Math. **65** (2005), no. 6, 2049–2066, DOI 10.1137/040607435. MR2177738

[7] Gang Bao, Junshan Lin, and Faouzi Triki, *A multi-frequency inverse source problem*, J. Differential Equations **249** (2010), no. 12, 3443–3465, DOI 10.1016/j.jde.2010.08.013. MR2737437

[8] Gang Bao, Junshan Lin, and Faouzi Triki, *An inverse source problem with multiple frequency data* (English, with English and French summaries), C. R. Math. Acad. Sci. Paris **349** (2011), no. 15-16, 855–859, DOI 10.1016/j.crma.2011.07.009. MR2835891

[9] Gang Bao, Junshan Lin, and Faouzi Triki, *Numerical solution of the inverse source problem for the Helmholtz equation with multiple frequency data*, Mathematical and statistical methods for imaging, Contemp. Math., vol. 548, Amer. Math. Soc., Providence, RI, 2011, pp. 45–60, DOI 10.1090/conm/548/10835. MR2868487

[10] Gang Bao, Peijun Li, Junshan Lin, and Faouzi Triki, *Inverse scattering problems with multi-frequencies*, Inverse Problems **31** (2015), no. 9, 093001, 21, DOI 10.1088/0266-5611/31/9/093001. MR3401991

[11] Eliane Bécache, Laurent Bourgeois, Lucas Franceschini, and Jérémi Dardé, *Application of mixed formulations of quasi-reversibility to solve ill-posed problems for heat and wave equations: the 1D case*, Inverse Probl. Imaging **9** (2015), no. 4, 971–1002, DOI 10.3934/ipi.2015.9.971. MR3461700

[12] Norman Bleistein, *Mathematical methods for wave phenomena*, Computer Science and Applied Mathematics, Academic Press, Inc., Orlando, FL, 1984. MR755514

[13] L. Bourgeois, *Convergence rates for the quasi-reversibility method to solve the Cauchy problem for Laplace's equation*, Inverse Problems **22** (2006), no. 2, 413–430, DOI 10.1088/0266-5611/22/2/002. MR2216406

[14] Tan Bui-Thanh and Omar Ghattas, *Analysis of the Hessian for inverse scattering problems: I. Inverse shape scattering of acoustic waves*, Inverse Problems **28** (2012), no. 5, 055001, 32, DOI 10.1088/0266-5611/28/5/055001. MR2915289

[15] Tan Bui-Thanh and Omar Ghattas, *Analysis of the Hessian for inverse scattering problems: II. Inverse medium scattering of acoustic waves*, Inverse Problems **28** (2012), no. 5, 055002, 17, DOI 10.1088/0266-5611/28/5/055002. MR2915290

[16] Martin Burger and Stanley J. Osher, *A survey on level set methods for inverse problems and optimal design*, European J. Appl. Math. **16** (2005), no. 2, 263–301, DOI 10.1017/S0956792505006182. MR2155293

[17] G. Chavent, *Nonlinear least squares for inverse problems*, Scientific Computation, Springer, New York, 2009. Theoretical foundations and step-by-step guide for applications. MR2554448

[18] Yu Chen, *Inverse scattering via Heisenberg's uncertainty principle*, Inverse Problems **13** (1997), no. 2, 253–282, DOI 10.1088/0266-5611/13/2/005. MR1445918

[19] W. Chew. *Waves and Fields in Inhomogeneous Media.* Van Nostrand Reinhold, New York, 1990.

[20] David Colton and Andreas Kirsch, *A simple method for solving inverse scattering problems in the resonance region*, Inverse Problems **12** (1996), no. 4, 383–393, DOI 10.1088/0266-5611/12/4/003. MR1402098

[21] David Colton and Rainer Kress, *Inverse acoustic and electromagnetic scattering theory*, 3rd ed., Applied Mathematical Sciences, vol. 93, Springer, New York, 2013, DOI 10.1007/978-1-4614-4942-3. MR2986407

[22] Anthony J. Devaney, *Mathematical foundations of imaging, tomography and wavefield inversion*, Cambridge University Press, Cambridge, 2012, DOI 10.1017/CBO9781139047838. MR2975719

[23] Heinz W. Engl, Martin Hanke, and Andreas Neubauer, *Regularization of inverse problems*, Mathematics and its Applications, vol. 375, Kluwer Academic Publishers Group, Dordrecht, 1996. MR1408680

[24] Mozhgan Nora Entekhabi and Victor Isakov, *On increasing stability in the two dimensional inverse source scattering problem with many frequencies*, Inverse Problems **34** (2018), no. 5, 055005, 14, DOI 10.1088/1361-6420/aab465. MR3788155

[25] Alexander V. Goncharsky and Sergey Y. Romanov, *Supercomputer technologies in inverse problems of ultrasound tomography*, Inverse Problems **29** (2013), no. 7, 075004, 22, DOI 10.1088/0266-5611/29/7/075004. MR3073872

[26] Isaac Harris and Dinh-Liem Nguyen, *Orthogonality sampling method for the electromagnetic inverse scattering problem*, SIAM J. Sci. Comput. **42** (2020), no. 3, B722–B737, DOI 10.1137/19M129783X. MR4101374

[27] Victor Isakov and Shuai Lu, *Increasing stability in the inverse source problem with attenuation and many frequencies*, SIAM J. Appl. Math. **78** (2018), no. 1, 1–18, DOI 10.1137/17M1112704. MR3740331

[28] Victor Isakov and Shuai Lu, *Inverse source problems without (pseudo) convexity assumptions*, Inverse Probl. Imaging **12** (2018), no. 4, 955–970, DOI 10.3934/ipi.2018040. MR3824734

[29] Vo Anh Khoa, Grant W. Bidney, Michael V. Klibanov, Loc H. Nguyen, Lam H. Nguyen, Anders J. Sullivan, and Vasily N. Astratov, *Convexification and experimental data for a 3D inverse scattering problem with the moving point source*, Inverse Problems **36** (2020), no. 8, 085007, 34, DOI 10.1088/1361-6420/ab95aa. MR4151408

[30] Vo Anh Khoa, Grant W. Bidney, Michael V. Klibanov, Loc H. Nguyen, Lam H. Nguyen, Anders J. Sullivan, and Vasily N. Astratov, *An inverse problem of a simultaneous reconstruction of the dielectric constant and conductivity from experimental backscattering data*, Inverse Probl. Sci. Eng. **29** (2021), no. 5, 712–735, DOI 10.1080/17415977.2020.1802447. MR4258641

[31] Vo Anh Khoa, Michael Victor Klibanov, and Loc Hoang Nguyen, *Convexification for a three-dimensional inverse scattering problem with the moving point source*, SIAM J. Imaging Sci. **13** (2020), no. 2, 871–904, DOI 10.1137/19M1303101. MR4097812

[32] Andreas Kirsch, *Characterization of the shape of a scattering obstacle using the spectral data of the far field operator*, Inverse Problems **14** (1998), no. 6, 1489–1512, DOI 10.1088/0266-5611/14/6/009. MR1662460

[33] Andreas Kirsch, *Remarks on the Born approximation and the factorization method*, Appl. Anal. **96** (2017), no. 1, 70–84, DOI 10.1080/00036811.2016.1188286. MR3581689

[34] Michael V. Klibanov, *Carleman estimates for global uniqueness, stability and numerical methods for coefficient inverse problems*, J. Inverse Ill-Posed Probl. **21** (2013), no. 4, 477–560, DOI 10.1515/jip-2012-0072. MR3089754

[35] Michael V. Klibanov, *Convexification of restricted Dirichlet-to-Neumann map*, J. Inverse Ill-Posed Probl. **25** (2017), no. 5, 669–685, DOI 10.1515/jiip-2017-0067. MR3707100

[36] Micheal V. Klibanov and Loc H. Nguyen, *PDE-based numerical method for a limited angle X-ray tomography*, Inverse Problems **35** (2019), no. 4, 045009, 32, DOI 10.1088/1361-6420/ab0133. MR3939496

[37] Karl-J. Langenberg, *Applied inverse problems for acoustic, electromagnetic and elastic wave scattering*, Basic methods of tomography and inverse problems (Bréau-sans-Nappe, 1985), Malvern Phys. Ser., Hilger, Bristol, 1987, pp. 125–467. MR976839

[38] R. Lattès and J.-L. Lions, *The method of quasi-reversibility. Applications to partial differential equations*, Modern Analytic and Computational Methods in Science and Mathematics, No. 18, American Elsevier Publishing Co., Inc., New York, 1969. Translated from the French edition and edited by Richard Bellman. MR0243746

[39] Thuy Thi Thu Le and Loc Hoang Nguyen, *A convergent numerical method to recover the initial condition of nonlinear parabolic equations from lateral Cauchy data*, J. Inverse Ill-Posed Probl. **30** (2022), no. 2, 265–286, DOI 10.1515/jiip-2020-0028. MR4401759

[40] Thuy T. Le and Loc H. Nguyen, *The gradient descent method for the convexification to solve boundary value problems of quasi-linear PDEs and a coefficient inverse problem*, J. Sci. Comput. **91** (2022), no. 3, Paper No. 74, 23, DOI 10.1007/s10915-022-01846-3. MR4414399

[41] Thuy T. Le, Loc H. Nguyen, Thi-Phong Nguyen, and William Powell, *The quasi-reversibility method to numerically solve an inverse source problem for hyperbolic equations*, J. Sci. Comput. **87** (2021), no. 3, Paper No. 90, 23, DOI 10.1007/s10915-021-01501-3. MR4254715

[42] Jingzhi Li, Hongyu Liu, and Qi Wang, *Enhanced multilevel linear sampling methods for inverse scattering problems.* part A, J. Comput. Phys. **257** (2014), no. part A, 554–571, DOI 10.1016/j.jcp.2013.09.048. MR3129549

[43] Jingzhi Li, Hongyu Liu, and Jun Zou, *Locating multiple multiscale acoustic scatterers*, Multiscale Model. Simul. **12** (2014), no. 3, 927–952, DOI 10.1137/13093409X. MR3226744

[44] Shari Moskow and John C. Schotland, *Convergence and stability of the inverse scattering series for diffuse waves*, Inverse Problems **24** (2008), no. 6, 065005, 16, DOI 10.1088/0266-5611/24/6/065005. MR2456952

[45] Dinh-Liem Nguyen and Trung Truong, *Imaging of bi-anisotropic periodic structures from electromagnetic near-field data*, J. Inverse Ill-Posed Probl. **30** (2022), no. 2, 205–219, DOI 10.1515/jiip-2020-0114. MR4401764

[46] Loc Hoang Nguyen, *An inverse space-dependent source problem for hyperbolic equations and the Lipschitz-like convergence of the quasi-reversibility method*, Inverse Problems **35** (2019), no. 3, 035007, 28, DOI 10.1088/1361-6420/aafe8f. MR3916467

[47] Loc Hoang Nguyen, *A new algorithm to determine the creation or depletion term of parabolic equations from boundary measurements*, Comput. Math. Appl. **80** (2020), no. 10, 2135–2149, DOI 10.1016/j.camwa.2020.09.010. MR4154907

[48] Loc H. Nguyen and Michael V. Klibanov, *Carleman estimates and the contraction principle for an inverse source problem for nonlinear hyperbolic equations*, Inverse Problems **38** (2022), no. 3, Paper No. 035009, 19, DOI 10.1088/1361-6420/ac4d09. MR4385429

[49] Loc H. Nguyen, Qitong Li, and Michael V. Klibanov, *A convergent numerical method for a multi-frequency inverse source problem in inhomogenous media*, Inverse Probl. Imaging **13** (2019), no. 5, 1067–1094, DOI 10.3934/ipi.2019048. MR4027044

[50] Phuong Mai Nguyen and Loc Hoang Nguyen, *A numerical method for an inverse source problem for parabolic equations and its application to a coefficient inverse problem*, J. Inverse Ill-Posed Probl. **28** (2020), no. 3, 323–339, DOI 10.1515/jiip-2019-0026. MR4104322

[51] Alexey V. Smirnov, Michael V. Klibanov, and Loc H. Nguyen, *On an inverse source problem for the full radiative transfer equation with incomplete data*, SIAM J. Sci. Comput. **41** (2019), no. 5, B929–B952, DOI 10.1137/19M1253605. MR4002730

[52] M. Soumekh. *Synthetic Aperture Radar Signal Processing with MATLAB Algorithms (https://www.mathworks.com/matlabcentral/fileexchange/2188-synthetic-aperture-radar-signal-processing-with-matlab-algorithms), MATLAB Central File Exchange.* John Wiley & Sons, 1999.

[53] A. N. Tikhonov, A. V. Goncharsky, V. V. Stepanov, and A. G. Yagola, *Numerical methods for the solution of ill-posed problems*, Mathematics and its Applications, vol. 328, Kluwer Academic Publishers Group, Dordrecht, 1995. Translated from the 1990 Russian original by R. A. M. Hoksbergen and revised by the authors, DOI 10.1007/978-94-015-8480-7. MR1350538

Department of Mathematics and Statistics, University of North Carolina at Charlotte, Charlotte , North Carolina, USA, 28223

Email address: loc.nguyen@uncc.edu

Department of Information Technology, University of Finance-Marketing, Ho Chi Minh City, Vietnam

Email address: vtthuong@ufm.edu.vn

Contemporary Mathematics
Volume **784**, 2023
https://doi.org/10.1090/conm/784/15756

Convergence analysis of Nédélec finite element approximations for a stationary Maxwell's system

Quyen Tran

Dedicated to Professor Dinh Nho Hào on the occasion of his 60th birthday

ABSTRACT. In this paper we analyze the Maxwell system

$$\begin{cases} \nabla \times \left(\sigma^{-1}(\mathbf{x})\nabla \times \mathbf{E}(\mathbf{x})\right) & = \epsilon(\mathbf{x})\mathbf{u}(\mathbf{x}), \quad \mathbf{x} \in \Omega \\ \nabla \cdot (\epsilon(\mathbf{x})\mathbf{E}(\mathbf{x})) & = \rho(\mathbf{x}), \quad \mathbf{x} \in \Omega \\ \mathbf{E}(\mathbf{x}) \times \vec{\mathbf{n}}(\mathbf{x}) & = \mathbf{0}, \quad \mathbf{x} \in \partial\Omega \\ \nabla \cdot (\epsilon(\mathbf{x})\mathbf{u}(\mathbf{x})) & = 0, \quad \mathbf{x} \in \Omega \end{cases}$$

with the Nédélec finite elements. We show the convergence of the finite element approximations as well as establish their error bounds.

1. Introduction

Let Ω be an open, bounded, nonempty domain of $\mathbb{R}^3$ with a connected Lipschitz-continuous boundary $\partial\Omega$. We consider the Maxwell system

$$(1) \qquad \begin{cases} \nabla \times \left(\sigma^{-1}(\mathbf{x})\nabla \times \mathbf{E}(\mathbf{x})\right) & = \epsilon(\mathbf{x})\mathbf{u}(\mathbf{x}), \quad \mathbf{x} \in \Omega \\ \nabla \cdot (\epsilon(\mathbf{x})\mathbf{E}(\mathbf{x})) & = \rho(\mathbf{x}), \quad \mathbf{x} \in \Omega \\ \mathbf{E}(\mathbf{x}) \times \vec{\mathbf{n}}(\mathbf{x}) & = \mathbf{0}, \quad \mathbf{x} \in \partial\Omega, \end{cases}$$

where $\vec{\mathbf{n}} = \vec{\mathbf{n}}(\mathbf{x})$ is the unit outward normal on $\partial\Omega$. In this paper bold typeface is used to indicate a point of $\mathbb{R}^3$, a (three-dimensional) vector-valued function or a Hilbert space of vector-valued functions. For the above system (1) the source term $\mathbf{u}(\mathbf{x}) \in \mathbf{L}^2(\Omega) := \left(L^2(\Omega)\right)^3$, the charge density $\rho(\mathbf{x}) \in L^2(\Omega)$, and the dielectric permittivity $\epsilon(\mathbf{x}) \in L^\infty(\Omega)$, the magnetic permeability $\sigma(\mathbf{x}) \in L^\infty(\Omega)$ are assumed to satisfy

$$(2) \qquad \underline{\epsilon} \le \epsilon(\mathbf{x}) \quad \text{and} \quad \underline{\sigma} \le \sigma(\mathbf{x}) \le \overline{\sigma}$$

a.e. in Ω with given positive constants $\underline{\epsilon}$, $\underline{\sigma}$ and $\overline{\sigma}$. Furthermore, we assume that Gauss' law is applied to the current source, i.e.

$$(3) \qquad \nabla \cdot (\epsilon(\mathbf{x})\mathbf{u}(\mathbf{x})) = 0, \quad \mathbf{x} \in \Omega.$$

2020 *Mathematics Subject Classification.* Primary 35R25, 47A60, 65M12, 65M15.
Key words and phrases. Maxwell system, Nédélec finite element, Convergence analysis, Error bounds.

Maxwell system with the Gauss's law plays a central role in applied sciences, applications including electromagnetic induction heating, magnetohydrodynamics, magnetic levitation and many more which attracted great attention of many scientists, see, e.g., [**2, 5, 6, 9, 11, 12, 14–16**].

Our aim in this paper is to approximate the Maxwell system by the Nédélec finite elements and show the convergence of the approximations as well as establish their error bounds. We recall that an element $\mathbf{E}$ in $\mathbf{H}_0(\mathbf{curl};\Omega)$ is called to be a weak solution of (1) if

$$\begin{cases} \left(\sigma^{-1}\nabla\times\mathbf{E},\nabla\times\mathbf{\Phi}\right)_{\mathbf{L}^2(\Omega)} & = (\epsilon\mathbf{u},\mathbf{\Phi})_{\mathbf{L}^2(\Omega)}, \quad \forall\mathbf{\Phi}\in\mathbf{H}_0(\mathbf{curl};\Omega) \\ (\epsilon\mathbf{E},\nabla\phi)_{\mathbf{L}^2(\Omega)} & = -(\rho,\phi)_{L^2(\Omega)}, \quad \forall\phi\in H_0^1(\Omega) \end{cases} \tag{4}$$

(see Subsection 2.2 for the definition of the related functional spaces).

Let $\mathcal{E}_h$ and V_h be respectively the finite dimensional subspaces of $\mathbf{H}_0(\mathbf{curl};\Omega)$ and $H_0^1(\Omega)$, where h is the mesh size of the Nédélec finite element (see Section 3 for more details). The authors of [**6**] first consider the auxiliary variational equation

$$(\epsilon\nabla\chi_h,\nabla\phi_h)_{\mathbf{L}^2(\Omega)} = -(\rho,\phi_h)_{L^2(\Omega)}, \quad \forall\phi_h\in V_h \tag{5}$$

for $\chi_h\in V_h$. An approximation of the solution $\mathbf{E}\in\mathbf{H}_0(\mathbf{curl};\Omega)$ of (4) can be then interpolated as the element $\tilde{\mathbf{E}}_h\in\mathcal{E}_h$ of the variational equation

$$\begin{aligned} &\left(\sigma^{-1}\nabla\times\tilde{\mathbf{E}}_h,\nabla\times\mathbf{\Phi}_h\right)_{\mathbf{L}^2(\Omega)} + \gamma(h)\left(\epsilon\tilde{\mathbf{E}}_h,\mathbf{\Phi}_h\right)_{\mathbf{L}^2(\Omega)} \\ &\qquad = (\epsilon(\mathbf{u}+\gamma(h)\nabla\chi_h),\mathbf{\Phi}_h)_{\mathbf{L}^2(\Omega)} \end{aligned} \tag{6}$$

for all $\mathbf{\Phi}_h\in\mathcal{E}_h$, where $\gamma(h)\in(0,1]$ depends on h. If $\mathbf{E}\in\mathbf{H}^s(\mathbf{curl};\Omega)$ for some $s\in(1/2,1]$, the following inequality is then proved

$$\|\mathbf{E}-\tilde{\mathbf{E}}_h\|_{\mathbf{H}(\mathbf{curl};\Omega)} \le \mathcal{C}h^s\|\mathbf{E}\|_{\mathbf{H}^s(\mathbf{curl};\Omega)} \tag{7}$$

with an appropriate choice $\gamma(h)$, where the positive constant $\mathcal{C}$ is independent of h (see [**6**, Section 4]).

In this paper we do not employ auxiliary problems which may be expensive for numerical computations. We approximate the solution $\mathbf{E}\in\mathbf{H}_0(\mathbf{curl};\Omega)$ of (4) with its discrete version $\mathbf{E}_h\in\mathcal{E}_h$ via the variational formulation

$$\begin{cases} \left(\sigma^{-1}\nabla\times\mathbf{E}_h,\nabla\times\mathbf{\Phi}_h\right)_{\mathbf{L}^2(\Omega)} & = (\epsilon\mathbf{u},\mathbf{\Phi}_h)_{\mathbf{L}^2(\Omega)} \\ (\epsilon\mathbf{E}_h,\nabla\phi_h)_{\mathbf{L}^2(\Omega)} & = -(\rho,\phi_h)_{L^2(\Omega)} \end{cases} \tag{8}$$

for all $(\mathbf{\Phi}_h,\phi_h)\in\mathcal{E}_h\times V_h$. We then show that

$$\lim_{h\to 0}\|\nabla\times(\mathbf{E}-\mathbf{E}_h)\|_{\mathbf{L}^2(\Omega)} = 0.$$

Furthermore, under additional smoothness conditions of the dielectric permittivity ϵ and the magnetic permeability σ, the estimates

$$\|\mathbf{E}-\mathbf{E}_h\|_{\mathbf{H}(\mathbf{curl};\Omega)} \le Ch^s\|\mathbf{E}\|_{\mathbf{H}^s(\mathbf{curl};\Omega)}$$

$$\|\nabla\cdot\epsilon(\mathbf{E}-\mathbf{E}_h)\|_{H^{-\nu}(\Omega)} \le Ch^{\nu+\nu'-1}\left(\|\nabla\times\mathbf{E}\|_{\mathbf{L}^2(\Omega)} + \|\rho\|_{H^{\nu'-1}(\Omega)}\right)$$

hold true for some constants $\nu,\nu'\in(1/2,1]$ and $s\in(1/2,1]$ with the positive constant C being independent of h (Theorem 3.1). We note that $\mathbf{E}_h$ defined by (8) does not satisfy the equation (6) in general.

The remainder of the paper is organized as follows. In Section 2, we present the related functional spaces and the solution uniqueness of the variational equation

(4). The convergence and error bounds of the Nédélec finite element approximations are discussed in Section 3.

2. Preliminaries

2.1. Mixed variational problem. We start with presenting briefly the existence and uniqueness of a solution to mixed variational problems. Assume that H and Q are Hilbert spaces with their duals H^* and Q^*. Let $a : H \times H \to \mathbb{R}$ and $b : H \times Q \to \mathbb{R}$ be bounded bilinear forms, i.e.

$$|a(x, y)| \leq C_a \|x\|_H \|y\|_H \quad \text{and} \quad |b(x, q)| \leq C_b \|x\|_H \|q\|_Q$$

for all $x, y \in H$ and $q \in Q$ with some positive constants C_a and C_b. Let $h^* \in H^*$ and $q^* \in Q^*$, we consider the following mixed variational problem: find $(h, q) \in H \times Q$ such that

$$\begin{cases} a(h, \tau) + b(\tau, q) &= h^*(\tau), \quad \forall \tau \in H, \\ b(h, \xi) &= q^*(\xi), \quad \forall \xi \in Q. \end{cases} \tag{9}$$

Set

$$V := \{\tau \in H \mid b(\tau, \xi) = 0 \quad \text{for all} \quad \xi \in Q\},$$

we have the following result.

LEMMA 2.1 ([**4, 8**]). *Assume that the bounded bilinear form a is V-elliptic, i.e.*

$$a(\tau, \tau) \geq C_1 \|\tau\|_H^2$$

for all $\tau \in V \subset H$, where the positive constant C_1 is independent of τ. Furthermore, the bounded bilinear form b satisfies the Ladyzhenskaya-Babuška-Brezzi (LBB) condition, i.e. there exists a positive constant C_2 such that

$$\sup_{0 \neq \tau \in H} \frac{b(\tau, \xi)}{\|\tau\|_H} \geq C_2 \|\xi\|_Q$$

for all $\xi \in Q$. Then for each pair $(h^*, q^*) \in H^* \times Q^*$ the system (9) admits a unique solution $(h, q) \in H \times Q$ which satisfies the inequality

$$\|h\|_H + \|q\|_Q \leq C_3 \left(\|h^*\|_{H^*} + \|q^*\|_{Q^*}\right),$$

where the positive constant C_3 depends only on C_1 and C_2.

2.2. Functional spaces. We here present briefly functional spaces utilized in the paper, for more details one can consult [**1, 12**]. As usual, we use the differential operator $\nabla := \left(\frac{\partial}{\partial x^1}, \frac{\partial}{\partial x^2}, \frac{\partial}{\partial x^3}\right)$, so that $\nabla \phi(\mathbf{x}) = \left(\frac{\partial \phi(\mathbf{x})}{\partial x^1}, \frac{\partial \phi(\mathbf{x})}{\partial x^2}, \frac{\partial \phi(\mathbf{x})}{\partial x^3}\right)$ for the scalar function $\phi = \phi(\mathbf{x})$. Furthermore, for the vector-valued function $\mathbf{\Phi} = \mathbf{\Phi}(\mathbf{x}) = \left(\Phi^1(\mathbf{x}), \Phi^2(\mathbf{x}), \Phi^3(\mathbf{x})\right)$ we have

$$\nabla \cdot \mathbf{\Phi} = \frac{\partial \Phi^1(\mathbf{x})}{\partial x^1} + \frac{\partial \Phi^2(\mathbf{x})}{\partial x^2} + \frac{\partial \Phi^3(\mathbf{x})}{\partial x^3} := \mathrm{div}\mathbf{\Phi} \quad \text{and}$$

$$\nabla \times \mathbf{\Phi} = \left(\frac{\partial \Phi^3(\mathbf{x})}{\partial x^2} - \frac{\partial \Phi^2(\mathbf{x})}{\partial x^3}, \frac{\partial \Phi^1(\mathbf{x})}{\partial x^3} - \frac{\partial \Phi^3(\mathbf{x})}{\partial x^1}, \frac{\partial \Phi^2(\mathbf{x})}{\partial x^1} - \frac{\partial \Phi^1(\mathbf{x})}{\partial x^2}\right) := \mathbf{curl}\mathbf{\Phi}.$$

We also mention that $\nabla \times (\nabla \phi) = \mathbf{0}$ and $\nabla \cdot (\nabla \times \mathbf{\Phi}) = 0$.

The Hilbert spaces

$$\mathbf{H}(\mathrm{div};\Omega) := \{\boldsymbol{\Phi} \in \mathbf{L}^2(\Omega) \mid \nabla\cdot\boldsymbol{\Phi} \in L^2(\Omega)\} \quad \text{and}$$
$$\mathbf{H}(\mathbf{curl};\Omega) := \{\boldsymbol{\Phi} \in \mathbf{L}^2(\Omega) \mid \nabla\times\boldsymbol{\Phi} \in \mathbf{L}^2(\Omega)\}$$

are respectively equipped with the inner product

$$(\boldsymbol{\Phi},\boldsymbol{\Psi})_{\mathbf{H}(\mathrm{div};\Omega)} := (\boldsymbol{\Phi},\boldsymbol{\Psi})_{\mathbf{L}^2(\Omega)} + (\nabla\cdot\boldsymbol{\Phi},\nabla\cdot\boldsymbol{\Psi})_{L^2(\Omega)} \quad \text{and}$$
$$(\boldsymbol{\Phi},\boldsymbol{\Psi})_{\mathbf{H}(\mathbf{curl};\Omega)} := (\boldsymbol{\Phi},\boldsymbol{\Psi})_{\mathbf{L}^2(\Omega)} + (\nabla\times\boldsymbol{\Phi},\nabla\times\boldsymbol{\Psi})_{\mathbf{L}^2(\Omega)}.$$

The *normal trace operator* $\gamma_n(\boldsymbol{\Phi}) := \vec{\mathbf{n}}\cdot\boldsymbol{\Phi}_{|\partial\Omega}$ for all $\boldsymbol{\Phi} \in \mathbf{C}^\infty(\overline{\Omega})$ can be extended to a surjective, continuous linear map from $\mathbf{H}(\mathrm{div};\Omega) \to H^{-1/2}(\partial\Omega) := \big(H^{1/2}(\partial\Omega)\big)^*$ such that Green's formula (cf. [**12**, §3])

$$(\nabla\cdot\boldsymbol{\Phi},\phi)_{L^2(\Omega)} = -(\boldsymbol{\Phi},\nabla\phi)_{\mathbf{L}^2(\Omega)} + \langle\gamma_n(\boldsymbol{\Phi}),\phi\rangle_{\left(H^{-1/2}(\partial\Omega),H^{1/2}(\partial\Omega)\right)} \tag{10}$$

holds true for all $\boldsymbol{\Phi} \in \mathbf{H}(\mathrm{div};\Omega)$ and $\phi \in H^1(\Omega)$. The *tangential trace operator* $\gamma_t(\boldsymbol{\Phi}) := \vec{\mathbf{n}}\times\boldsymbol{\Phi}_{|\partial\Omega}$ for all $\boldsymbol{\Phi} \in \mathbf{C}^\infty(\overline{\Omega})$ can be also extended to a continuous linear map from $\mathbf{H}(\mathbf{curl};\Omega) \to \mathbf{H}^{-1/2}(\partial\Omega)$. Further, Green's formula [**12**, Theorem 3.29]

$$(\nabla\times\boldsymbol{\Phi},\boldsymbol{\Psi})_{\mathbf{L}^2(\Omega)} = (\boldsymbol{\Phi},\nabla\times\boldsymbol{\Psi})_{\mathbf{L}^2(\Omega)} + \langle\gamma_t(\boldsymbol{\Phi}),\boldsymbol{\Psi}\rangle_{\left(\mathbf{H}^{-1/2}(\partial\Omega),\mathbf{H}^{1/2}(\partial\Omega)\right)} \tag{11}$$

holds true for all $\boldsymbol{\Phi} \in \mathbf{H}(\mathbf{curl};\Omega)$ and $\boldsymbol{\Psi} \in \mathbf{H}^1(\Omega)$.

We conclude this subsection by the following definition

$$\mathbf{H}_0(\mathrm{div};\Omega) := \overline{\mathbf{C}_c^\infty(\Omega)}^{\mathbf{H}(\mathrm{div};\Omega)} \quad \text{and} \quad \mathbf{H}_0(\mathbf{curl};\Omega) := \overline{\mathbf{C}_c^\infty(\Omega)}^{\mathbf{H}(\mathbf{curl};\Omega)},$$

where the closures are respectively taken with respect to the norm of the space $\mathbf{H}(\mathrm{div};\Omega)$ and $\mathbf{H}(\mathbf{curl};\Omega)$, and $C_c^\infty(\Omega)$ is the space of all infinitely continuously differential functions with compact support in Ω. Notice that

$$\mathbf{H}_0(\mathrm{div};\Omega) := \{\boldsymbol{\Phi} \in \mathbf{H}(\mathrm{div};\Omega) \mid \gamma_n(\boldsymbol{\Phi}) = 0\},$$
$$\mathbf{H}_0(\mathbf{curl};\Omega) := \{\boldsymbol{\Phi} \in \mathbf{H}(\mathbf{curl};\Omega) \mid \gamma_t(\boldsymbol{\Phi}) = \mathbf{0}\}.$$

2.3. The variational formulation of the system (1)**.** An element $\mathbf{E}$ in $\mathbf{H}_0(\mathbf{curl};\Omega)$ is called to be a weak solution of (1) if it satisfies the system (4). The first equation of the system (4) is obtained by multiplying the first equation of (1) with $\boldsymbol{\Phi} \in \mathbf{H}_0(\mathbf{curl};\Omega)$, and then using the identity (11), while the second equation is derived from the second one of (1) and Green's formula (10). We will see that the system (4) admits a unique solution. For this purpose we set

$$a(\mathbf{E},\tau) := \left(\sigma^{-1}\nabla\times\mathbf{E},\nabla\times\tau\right)_{\mathbf{L}^2(\Omega)},$$
$$b(\mathbf{E},\xi) := (\epsilon\mathbf{E},\nabla\xi)_{\mathbf{L}^2(\Omega)},$$

where $(\mathbf{E},\tau,\xi) \in \mathbf{H}_0(\mathbf{curl};\Omega)\times\mathbf{H}_0(\mathbf{curl};\Omega)\times H_0^1(\Omega)$. Denote by

$$\mathbf{V} := \{\tau \in \mathbf{H}_0(\mathbf{curl};\Omega) \mid b(\tau,\xi) = 0 \quad \text{for all} \quad \xi \in H_0^1(\Omega)\}$$
$$= \{\tau \in \mathbf{H}_0(\mathbf{curl};\Omega) \mid \nabla\cdot(\epsilon\tau) = 0\}.$$

By the compactness of the embedding $\mathbf{V} \hookrightarrow \mathbf{L}^2(\Omega)$ (for ϵ piecewise smooth) and

$$\{\tau \in \mathbf{V} \mid \nabla\times\tau = \mathbf{0}\} = \{\tau \in \mathbf{L}^2(\Omega) \mid \nabla\times\tau = \mathbf{0},\ \nabla\cdot(\epsilon\tau) = 0,\ \vec{\mathbf{n}}\times\tau_{|\partial\Omega} = \mathbf{0}\} = \{\mathbf{0}\}$$

(cf. [**7, 16**]), an application of Peetre's lemma (cf. [**10**]) yields that there exists a positive C^Ω such that

$$\|\tau\|_{\mathbf{L}^2(\Omega)} \le C^\Omega\|\nabla\times\tau\|_{\mathbf{L}^2(\Omega)} \tag{12}$$

for all $\tau \in \mathbf{V}$. With the aid of the condition (2), we have

$$\|\mathbf{v}\|^2_{\mathbf{H}(\mathbf{curl};\Omega)} \leq C^{\Omega}_{\overline{\sigma}} \left(\sigma^{-1}\nabla \times \mathbf{v}, \nabla \times \mathbf{v}\right)_{\mathbf{L}^2(\Omega)} \tag{13}$$

for all $\mathbf{v} \in \mathbf{V}$, where the constant $C^{\Omega}_{\overline{\sigma}} > 0$ is independent of $\mathbf{v}$. Therefore, the bounded bilinear form a is $\mathbf{V}$-elliptic. Next, since $\nabla H^1_0(\Omega) := \left\{\nabla\phi \mid \phi \in H^1_0(\Omega)\right\} \subset \mathbf{H}_0(\mathbf{curl};\Omega)$, we get for all $\xi \in H^1_0(\Omega)$ that

$$\begin{aligned}\sup_{0\neq\tau\in\mathbf{H}_0(\mathbf{curl};\Omega)} \frac{b(\tau,\xi)}{\|\tau\|_{\mathbf{H}_0(\mathbf{curl};\Omega)}} &= \sup_{0\neq\tau\in\mathbf{H}_0(\mathbf{curl};\Omega)} \frac{(\epsilon\tau,\nabla\xi)_{\mathbf{L}^2(\Omega)}}{\|\tau\|_{\mathbf{H}_0(\mathbf{curl};\Omega)}} \geq \frac{(\epsilon\nabla\xi,\nabla\xi)_{\mathbf{L}^2(\Omega)}}{\|\nabla\xi\|_{\mathbf{H}_0(\mathbf{curl};\Omega)}}\\ &\geq \frac{\underline{\epsilon}\,(\nabla\xi,\nabla\xi)_{\mathbf{L}^2(\Omega)}}{(\nabla\xi,\nabla\xi)^{1/2}_{\mathbf{L}^2(\Omega)}} = \underline{\epsilon}\,(\nabla\xi,\nabla\xi)^{1/2}_{\mathbf{L}^2(\Omega)} \geq C^{\Omega}_{\underline{\epsilon}}\|\xi\|_{H^1_0(\Omega)}\end{aligned}$$

for all $\xi \in H^1_0(\Omega)$ with $C^{\Omega}_{\underline{\epsilon}} = \frac{\underline{\epsilon}}{\sqrt{1+C^2_{\Omega}}} > 0$, where we used the identity $\nabla \times (\nabla\xi) = \mathbf{0}$ and the Poincaré-Friedrichs constant C_{Ω} which appears in the inequality

$$\|\xi\|_{L^2(\Omega)} \leq C_{\Omega}\|\nabla\xi\|_{L^2(\Omega)} \tag{14}$$

for all $\xi \in H^1_0(\Omega)$. We thus deduce that the bounded bilinear form b satisfies the LBB condition. An application of Lemma 2.1 yields that there exists a unique solution $(\mathbf{E}, q) \in \mathbf{H}_0(\mathbf{curl};\Omega) \times H^1_0(\Omega)$ of the following system

$$\begin{cases}\left(\sigma^{-1}\nabla \times \mathbf{E}, \nabla \times \mathbf{\Phi}\right)_{\mathbf{L}^2(\Omega)} + (\epsilon\mathbf{\Phi},\nabla q)_{\mathbf{L}^2(\Omega)} &= (\epsilon\mathbf{u},\mathbf{\Phi})_{\mathbf{L}^2(\Omega)}\\ (\epsilon\mathbf{E},\nabla\phi)_{\mathbf{L}^2(\Omega)} &= -(\rho,\phi)_{L^2(\Omega)}\end{cases} \tag{15}$$

for all $(\mathbf{\Phi},\phi) \in \mathbf{H}_0(\mathbf{curl};\Omega) \times H^1_0(\Omega)$. We show that $q = 0$. Indeed, by Gauss' law (3), taking $\mathbf{\Phi} = \nabla q \in \nabla H^1_0(\Omega) \subset \mathbf{H}_0(\mathbf{curl};\Omega)$ in the first equation of (15) and using Green's formula (10) we get

$$\underline{\epsilon}\,(\nabla q,\nabla q)_{\mathbf{L}^2(\Omega)} \leq (\epsilon\nabla q,\nabla q)_{\mathbf{L}^2(\Omega)} = (\epsilon\mathbf{u},\nabla q)_{\mathbf{L}^2(\Omega)} = -\left(\nabla\cdot(\epsilon\mathbf{u}), q\right)_{L^2(\Omega)} = 0.$$

Combining this with (14), we arrive at $\|q\|_{H^1(\Omega)} = 0$ or $q = 0$. Consequently, we conclude that the system (4) attains a unique solution $\mathbf{E} \in \mathbf{H}_0(\mathbf{curl};\Omega)$ which satisfies the estimate

$$\|\mathbf{E}\|_{\mathbf{H}(\mathbf{curl};\Omega)} \leq C^{\Omega}_{\underline{\epsilon},\overline{\sigma}}\left(\|\mathbf{u}\|_{\mathbf{L}^2(\Omega)} + \|\rho\|_{L^2(\Omega)}\right). \tag{16}$$

3. Nédélec finite Element

Let $(\mathcal{T}_h)_{h>0}$ be a quasi-uniform family of regular triangulations of the domain Ω with the mesh size h (cf. [**3**]). For the definition of the discretization space of the state function defined by the system (4) let us denote the Nédélec finite element spaces (cf. [**13**])

$$\mathcal{E}_h := \left\{\mathbf{E}_h \in \mathbf{H}_0(\mathbf{curl};\Omega) \mid \mathbf{E}_{h|T} = \mathbf{a}_T + \mathbf{b}_T \times \mathbf{x},\ \forall T \in \mathcal{T}_h \text{ with } \mathbf{a}_T,\mathbf{b}_T \in \mathbb{R}^3\right\},$$

$$V_h := \left\{\phi_h \in H^1_0(\Omega) \mid \phi_{h|T} = a_T + \mathbf{b}_T\cdot\mathbf{x},\ \forall T \in \mathcal{T}_h \text{ with } a_T \in \mathbb{R}, \mathbf{b}_T \in \mathbb{R}^3\right\},$$

where $\nabla V_h \subset \mathcal{E}_h$.

The discrete variational formulation corresponding to the system (4) then reads: find $\mathbf{E}_h \in \mathcal{E}_h$ such that the system (8) is fulfilled.

Since the discrete Poincaré-Friedrichs-type inequality (cf. [**9**, Theorem 4.7], [**12**, §7])

$$\|\mathbf{v}_h\|_{\mathbf{L}^2(\Omega)} \leq C^{\Omega}\|\nabla \times \mathbf{v}_h\|_{\mathbf{L}^2(\Omega)} \tag{17}$$

is satisfied for all discrete ϵ-divergence-free functions, i.e.

$$\mathbf{v}_h \in \mathcal{D}_h^{(\epsilon)} := \left\{ \mathbf{E}_h \in \mathcal{E}_h \;\middle|\; (\epsilon \mathbf{E}_h, \nabla \phi_h)_{\mathbf{L}^2(\Omega)} = 0 \quad \text{for all} \quad \phi_h \in V_h \right\},$$

we have that

$$\|\mathbf{v}_h\|^2_{\mathbf{H}(\mathbf{curl};\Omega)} \le C^{\Omega}_{\sigma}(\sigma^{-1}\nabla \times \mathbf{v}_h, \nabla \times \mathbf{v}_h)_{\mathbf{L}^2(\Omega)} \tag{18}$$

for all $\mathbf{v}_h \in \mathcal{D}_h^{(\epsilon)}$. Therefore, we can conclude that the system (8) has a unique solution $\mathbf{E}_h \in \mathcal{E}_h$ satisfying the estimate

$$\|\mathbf{E}_h\|_{\mathbf{H}(\mathbf{curl};\Omega)} \le C^{\Omega}_{\underline{\epsilon},\overline{\sigma}} \left(\|\mathbf{u}\|_{\mathbf{L}^2(\Omega)} + \|\rho\|_{L^2(\Omega)} \right). \tag{19}$$

For all $s \ge 0$ we denote the Banach space

$$\mathbf{H}^s(\mathbf{curl};\Omega) := \left\{ \mathbf{\Phi} \in \mathbf{H}^s(\Omega) \;\middle|\; \nabla \times \mathbf{\Phi} \in \mathbf{H}^s(\Omega) \right\}$$

equipped with the norm

$$\|\mathbf{\Phi}\|_{\mathbf{H}^s(\mathbf{curl};\Omega)} := \left(\|\mathbf{\Phi}\|^2_{\mathbf{H}^s(\Omega)} + \|\nabla \times \mathbf{\Phi}\|^2_{\mathbf{H}^s(\Omega)} \right)^{1/2}.$$

We are in the position to state the main result of the section on the convergence and error bounds of the Nédélec finite element approximations.

THEOREM 3.1. *Let $\mathbf{E}$ and $\mathbf{E}_h$ be the unique solution to* (4) *and* (8), *respectively. Then:*

(i) There holds the limit

$$\lim_{h \to 0} \|\nabla \times (\mathbf{E} - \mathbf{E}_h)\|_{\mathbf{L}^2(\Omega)} = 0.$$

(ii) In addition $\epsilon, \sigma^{-1} \in W^{1,\infty}(\Omega)$ we get the regularity $\mathbf{E} \in \mathbf{H}^s(\mathbf{curl};\Omega)$ for some $s \in (1/2, 1]$. Furthermore, there exist constants $\nu, \nu' \in (1/2, 1]$ such that the estimates

$$\begin{aligned} \|\mathbf{E} - \mathbf{E}_h\|_{\mathbf{H}(\mathbf{curl};\Omega)} &\le Ch^s \|\mathbf{E}\|_{\mathbf{H}^s(\mathbf{curl};\Omega)} \\ \|\nabla \cdot \epsilon(\mathbf{E} - \mathbf{E}_h)\|_{H^{-\nu}(\Omega)} &\le Ch^{\nu+\nu'-1} \left(\|\nabla \times \mathbf{E}\|_{\mathbf{L}^2(\Omega)} + \|\rho\|_{H^{\nu'-1}(\Omega)} \right) \end{aligned}$$

hold true.

REMARK 3.2. Before going to prove the result, it is useful to recall the following estimates for χ_h and $\tilde{\mathbf{E}}_h$ defined (5) and (6), respectively

$$\begin{aligned} \|\chi_h\|_{H^1(\Omega)} &\le \mathcal{C}\|\rho\|_{L^2(\Omega)} \\ \|\tilde{\mathbf{E}}_h\|_{\mathbf{H}(\mathbf{curl};\Omega)} &\le \mathcal{C} \left(\|\mathbf{u}\|_{\mathbf{L}^2(\Omega)} + \|\rho\|_{L^2(\Omega)} \right) \\ \|\nabla \cdot \epsilon(\mathbf{E} - \tilde{\mathbf{E}}_h)\|_{H^{-\nu}(\Omega)} &\le \mathcal{C}h^{\nu+\nu'-1} \left(\|\nabla \times \mathbf{E}\|_{\mathbf{L}^2(\Omega)} + \|\rho\|_{H^{\nu'-1}(\Omega)} \right) \end{aligned} \tag{20}$$

for the constants $\nu, \nu' \in (1/2, 1]$, $s \in (1/2, 1]$ and $\epsilon \in W^{1,\infty}(\Omega)$ (cf. [**6**, Lemma 4.2]). Furthermore,

$$\|\nabla \cdot (\epsilon \mathbf{v}_h)\|_{H^{-\nu}(\Omega)} \le \mathcal{C}h^{\nu+\nu'-1} \|\nabla \times \mathbf{v}_h\|_{\mathbf{L}^2(\Omega)} \tag{21}$$

for all $\mathbf{v}_h \in \mathcal{D}_h^{(\epsilon)}$ (cf. [**6**, Lemma 3.8]).

PROOF OF THEOREM 3.1. By the first equations of the systems (4) and (8), we deduce

$$\left(\sigma^{-1}\nabla\times(\mathbf{E}-\mathbf{E}_h),\nabla\times\mathbf{\Phi}_h\right)_{\mathbf{L}^2(\Omega)}=0$$

for all $\mathbf{\Phi}_h\in\mathcal{E}_h$ and thus

$$\left(\sigma^{-1}\nabla\times(\mathbf{E}-\mathbf{E}_h),\nabla\times(\mathbf{E}-\mathbf{E}_h)\right)_{\mathbf{L}^2(\Omega)}=\left(\sigma^{-1}\nabla\times(\mathbf{E}-\mathbf{E}_h),\nabla\times(\mathbf{E}-\mathbf{\Phi}_h)\right)_{\mathbf{L}^2(\Omega)}.$$

Using (2), we then have

$$\begin{aligned}\|\nabla\times(\mathbf{E}-\mathbf{E}_h)\|^2_{\mathbf{L}^2(\Omega)}&\le\left\|\sqrt{\frac{\overline{\sigma}}{\sigma}}|\nabla\times(\mathbf{E}-\mathbf{E}_h)|\right\|^2_{\mathbf{L}^2(\Omega)}\\&=\overline{\sigma}\left(\sigma^{-1}\nabla\times(\mathbf{E}-\mathbf{E}_h),\nabla\times(\mathbf{E}-\mathbf{\Phi}_h)\right)_{\mathbf{L}^2(\Omega)}\\&\le\overline{\sigma}\int_\Omega\sigma^{-1}|\nabla\times(\mathbf{E}-\mathbf{E}_h)||\nabla\times(\mathbf{E}-\mathbf{\Phi}_h)|dx\\&\le\frac{\overline{\sigma}}{\underline{\sigma}}\|\nabla\times(\mathbf{E}-\mathbf{E}_h)\|_{\mathbf{L}^2(\Omega)}\|\nabla\times(\mathbf{E}-\mathbf{\Phi}_h)\|_{\mathbf{L}^2(\Omega)}\end{aligned}$$

for all $\mathbf{\Phi}_h\in\mathcal{E}_h$, and thus

$$\|\nabla\times(\mathbf{E}-\mathbf{E}_h)\|_{\mathbf{L}^2(\Omega)}\le\frac{\overline{\sigma}}{\underline{\sigma}}\inf_{\Phi_h\in\mathcal{E}_h}\|\mathbf{E}-\Phi_h\|_{\mathbf{H}(\mathbf{curl};\Omega)}.$$

By the approximation property of the finite element space, we get

$$\begin{aligned}\lim_{h\to0}\|\nabla\times(\mathbf{E}-\mathbf{E}_h)\|_{\mathbf{L}^2(\Omega)}&\le\frac{\overline{\sigma}}{\underline{\sigma}}\lim_{h\to0}\inf_{\Phi_h\in\mathcal{E}_h}\|\mathbf{E}-\Phi_h\|_{\mathbf{H}(\mathbf{curl};\Omega)}\\&=0,\end{aligned}$$

the part (i) is therefore proved.

Next, for all $\mathbf{\Phi}_h\in\mathcal{E}_h$ we have from (6) and (8), respectively that

$$\begin{aligned}\left(\sigma^{-1}\nabla\times\tilde{\mathbf{E}}_h,\nabla\times\mathbf{\Phi}_h\right)_{\mathbf{L}^2(\Omega)}&=(\epsilon\mathbf{u},\mathbf{\Phi}_h)_{\mathbf{L}^2(\Omega)}+\gamma(h)\left(\epsilon(\nabla\chi_h-\tilde{\mathbf{E}}_h),\mathbf{\Phi}_h\right)_{\mathbf{L}^2(\Omega)}\\&=\left(\sigma^{-1}\nabla\times\mathbf{E}_h,\nabla\times\mathbf{\Phi}_h\right)_{\mathbf{L}^2(\Omega)}\\&\quad+\gamma(h)\left(\epsilon(\nabla\chi_h-\tilde{\mathbf{E}}_h),\mathbf{\Phi}_h\right)_{\mathbf{L}^2(\Omega)}\end{aligned}$$

which gives

$$\left(\sigma^{-1}\nabla\times(\tilde{\mathbf{E}}_h-\mathbf{E}_h),\nabla\times\mathbf{\Phi}_h\right)_{\mathbf{L}^2(\Omega)}=\gamma(h)\left(\epsilon(\nabla\chi_h-\tilde{\mathbf{E}}_h),\mathbf{\Phi}_h\right)_{\mathbf{L}^2(\Omega)}.\tag{22}$$

We now notice that

$$\left(\epsilon\tilde{\mathbf{E}}_h,\nabla\phi_h\right)_{\mathbf{L}^2(\Omega)}=-(\rho,\phi_h)_{L^2(\Omega)}\quad\text{for all}\quad\phi_h\in V_h.\tag{23}$$

In fact, taking $\mathbf{\Phi}_h=\nabla\phi_h$ in (6) and the first equation of (8), using the identity $\nabla\times\nabla\phi_h=\mathbf{0}$ and (5), we get

$$\begin{aligned}\gamma(h)\left(\epsilon\tilde{\mathbf{E}}_h,\nabla\phi_h\right)_{\mathbf{L}^2(\Omega)}&=(\epsilon\mathbf{u},\nabla\phi_h)_{\mathbf{L}^2(\Omega)}+\gamma(h)\left(\epsilon\nabla\chi_h,\nabla\phi_h\right)_{\mathbf{L}^2(\Omega)}\\&=\gamma(h)\left(\epsilon\nabla\chi_h,\nabla\phi_h\right)_{\mathbf{L}^2(\Omega)}\\&=-\gamma(h)(\rho,\phi_h)_{L^2(\Omega)}\end{aligned}$$

which yields (23), by $\gamma(h) > 0$. Combining this with the second equation of the system (8), we have $\tilde{\mathbf{E}}_h - \mathbf{E}_h \in \mathcal{D}_h^{(\epsilon)}$ and thus the discrete version of Gauss's law is applied to $\tilde{\mathbf{E}}_h - \mathbf{E}_h$. Now, taking $\mathbf{\Phi}_h = \tilde{\mathbf{E}}_h - \mathbf{E}_h$ in (22) and using (18), we have

$$\begin{aligned}
&\left\|\tilde{\mathbf{E}}_h - \mathbf{E}_h\right\|^2_{\mathbf{H}(\mathbf{curl};\Omega)} \\
&\leq C^{\Omega}_{\overline{\sigma}} \left(\sigma^{-1}\nabla \times (\tilde{\mathbf{E}}_h - \mathbf{E}_h), \nabla \times (\tilde{\mathbf{E}}_h - \mathbf{E}_h)\right)_{\mathbf{L}^2(\Omega)} \\
&= C^{\Omega}_{\overline{\sigma}}\gamma(h) \left(\epsilon(\nabla \chi_h - \tilde{\mathbf{E}}_h), \tilde{\mathbf{E}}_h - \mathbf{E}_h\right)_{\mathbf{L}^2(\Omega)} \\
&\leq C^{\Omega}_{\overline{\sigma}}\gamma(h)\|\epsilon\|_{L^\infty(\Omega)}\left\|\nabla\chi_h - \tilde{\mathbf{E}}_h\right\|_{\mathbf{L}^2(\Omega)}\left\|\tilde{\mathbf{E}}_h - \mathbf{E}_h\right\|_{\mathbf{L}^2(\Omega)} \\
&\leq C^{\Omega}_{\overline{\sigma}}\gamma(h)\|\epsilon\|_{L^\infty(\Omega)} \left(\|\nabla\chi_h\|_{\mathbf{L}^2(\Omega)} + \|\tilde{\mathbf{E}}_h\|_{\mathbf{L}^2(\Omega)}\right) \left\|\tilde{\mathbf{E}}_h - \mathbf{E}_h\right\|_{\mathbf{H}(\mathbf{curl};\Omega)}
\end{aligned}$$

which with the aid of (20) yields

$$\|\tilde{\mathbf{E}}_h - \mathbf{E}_h\|_{\mathbf{H}(\mathbf{curl};\Omega)} \leq \overline{C}\gamma(h), \tag{24}$$

where

$$\overline{C} := 2\mathcal{C}C^{\Omega}_{\overline{\sigma}}\|\epsilon\|_{L^\infty(\Omega)} \left(\|\mathbf{u}\|_{\mathbf{L}^2(\Omega)} + \|\rho\|_{L^2(\Omega)}\right) > 0$$

is independent of h. Since $\sigma^{-1} \in W^{1,\infty}(\Omega)$, it follows that $\mathbf{E} \in \mathbf{H}^s(\mathbf{curl};\Omega)$ (cf. [**6**, Lemma 3.6]). We therefore deduce from the last inequality and (7) that

$$\begin{aligned}
\|\mathbf{E} - \mathbf{E}_h\|_{\mathbf{H}(\mathbf{curl};\Omega)} &\leq \|\mathbf{E} - \tilde{\mathbf{E}}_h\|_{\mathbf{H}(\mathbf{curl};\Omega)} + \|\tilde{\mathbf{E}}_h - \mathbf{E}_h\|_{\mathbf{H}(\mathbf{curl};\Omega)} \\
&\leq \mathcal{C}h^s\|\mathbf{E}\|_{\mathbf{H}^s(\mathbf{curl};\Omega)} + \overline{C}\gamma(h) \\
&\leq Ch^s\|\mathbf{E}\|_{\mathbf{H}^s(\mathbf{curl};\Omega)}
\end{aligned}$$

as $\gamma(h) = h^s\|\mathbf{E}\|_{\mathbf{H}^s(\mathbf{curl};\Omega)}$ and $C := \mathcal{C} + \overline{C}$. Finally, since $\epsilon \in W^{1,\infty}(\Omega)$, we by (20) get that

$$\begin{aligned}
\|\nabla\cdot(\epsilon\mathbf{E} - \epsilon\mathbf{E}_h)\|_{H^{-\nu}(\Omega)} &\leq \|\nabla\cdot(\epsilon\mathbf{E} - \epsilon\tilde{\mathbf{E}}_h)\|_{H^{-\nu}(\Omega)} + \|\nabla\cdot(\epsilon\tilde{\mathbf{E}}_h - \epsilon\mathbf{E}_h)\|_{H^{-\nu}(\Omega)} \\
&\leq \mathcal{C}h^{\nu+\nu'-1}\left(\|\nabla\times\mathbf{E}\|_{\mathbf{L}^2(\Omega)} + \|\rho\|_{H^{\nu'-1}(\Omega)}\right) \\
&\quad + \|\nabla\cdot(\epsilon\tilde{\mathbf{E}}_h - \epsilon\mathbf{E}_h)\|_{H^{-\nu}(\Omega)},
\end{aligned}$$

with, by (21) and (24),

$$\begin{aligned}
\|\nabla\cdot(\epsilon\tilde{\mathbf{E}}_h - \epsilon\mathbf{E}_h)\|_{H^{-\nu}(\Omega)} &\leq \mathcal{C}h^{\nu+\nu'-1}\|\nabla\times(\tilde{\mathbf{E}}_h - \mathbf{E}_h)\|_{\mathbf{L}^2(\Omega)} \\
&\leq \mathcal{C}\overline{C}h^{\nu+\nu'-1}\gamma(h) \\
&\leq Ch^{\nu+\nu'-1}\left(\|\nabla\times\mathbf{E}\|_{\mathbf{L}^2(\Omega)} + \|\rho\|_{H^{\nu'-1}(\Omega)}\right),
\end{aligned}$$

as the mesh size h is small enough such that

$$\mathcal{C}\overline{C}\gamma(h) \leq C\|\nabla\times\mathbf{E}\|_{\mathbf{L}^2(\Omega)} + \|\rho\|_{H^{\nu'-1}(\Omega)},$$

which finishes the proof. □

Acknowledgments

The author would like to thank the referees and the editors for their valuable comments and suggestions which helped to improve our paper.

References

[1] C. Amrouche, C. Bernardi, M. Dauge, and V. Girault, *Vector potentials in three-dimensional non-smooth domains* (English, with English and French summaries), Math. Methods Appl. Sci. **21** (1998), no. 9, 823–864, DOI 10.1002/(SICI)1099-1476(199806)21:9¡823::AID-MMA976¿3.0.CO;2-B. MR1626990

[2] Régine Barthelmé, Patrick Ciarlet Jr., and Eric Sonnendrücker, *Generalized formulations of Maxwell's equations for numerical Vlasov-Maxwell simulations*, Math. Models Methods Appl. Sci. **17** (2007), no. 5, 657–680, DOI 10.1142/S0218202507002066. MR2325835

[3] Susanne C. Brenner and L. Ridgway Scott, *The mathematical theory of finite element methods*, 3rd ed., Texts in Applied Mathematics, vol. 15, Springer, New York, 2008, DOI 10.1007/978-0-387-75934-0. MR2373954

[4] F. Brezzi, *On the existence, uniqueness and approximation of saddle-point problems arising from Lagrangian multipliers* (English, with French summary), Rev. Française Automat. Informat. Recherche Opérationnelle Sér. Rouge **8** (1974), no. R-2, 129–151. MR365287

[5] Patrick Ciarlet Jr. and Simon Labrunie, *Numerical analysis of the generalized Maxwell equations (with an elliptic correction) for charged particle simulations*, Math. Models Methods Appl. Sci. **19** (2009), no. 11, 1959–1994, DOI 10.1142/S0218202509004017. MR2588956

[6] Patrick Ciarlet, Haijun Wu, and Jun Zou, *Edge element methods for Maxwell's equations with strong convergence for Gauss' laws*, SIAM J. Numer. Anal. **52** (2014), no. 2, 779–807, DOI 10.1137/120899856. MR3188392

[7] Martin Costabel, *A remark on the regularity of solutions of Maxwell's equations on Lipschitz domains*, Math. Methods Appl. Sci. **12** (1990), no. 4, 365–368, DOI 10.1002/mma.1670120406. MR1048563

[8] Gabriel N. Gatica, *A simple introduction to the mixed finite element method*, SpringerBriefs in Mathematics, Springer, Cham, 2014. Theory and applications, DOI 10.1007/978-3-319-03695-3. MR3157367

[9] R. Hiptmair, *Finite elements in computational electromagnetism*, Acta Numer. **11** (2002), 237–339, DOI 10.1017/S0962492902000041. MR2009375

[10] Urve Kangro and Roy Nicolaides, *Divergence boundary conditions for vector Helmholtz equations with divergence constraints*, M2AN Math. Model. Numer. Anal. **33** (1999), no. 3, 479–492, DOI 10.1051/m2an:1999148. MR1713234

[11] A.B. Langdon, *On enforcing Gauss' law in electromagnetic particle-in-cell codes*, Comput. Phys. Comm., 70(1992), 447–450.

[12] Peter Monk, *Finite element methods for Maxwell's equations*, Numerical Mathematics and Scientific Computation, Oxford University Press, New York, 2003, DOI 10.1093/acprof:oso/9780198508885.001.0001. MR2059447

[13] J.-C. Nédélec, *Mixed finite elements in* $\mathbf{R}^3$, Numer. Math. **35** (1980), no. 3, 315–341, DOI 10.1007/BF01396415. MR592160

[14] Q. Tran, H. Antil and H. Díaz, *Optimal control of parameterized stationary Maxwell's system: reduced basis, convergence analysis, and a posteriori error estimations*, Mathematical Control and Related Fields (2022), doi:10.3934/mcrf.2022003.

[15] Irwin Yousept, *Optimal control of Maxwell's equations with regularized state constraints*, Comput. Optim. Appl. **52** (2012), no. 2, 559–581, DOI 10.1007/s10589-011-9422-2. MR2925786

[16] Irwin Yousept and Jun Zou, *Edge element method for optimal control of stationary Maxwell system with Gauss law*, SIAM J. Numer. Anal. **55** (2017), no. 6, 2787–2810, DOI 10.1137/17M1117021. MR3723330

Department of Radiology, School of Medicine, University of California – Davis, Sacramento, California 95817, USA

Email address: ntqtran@ucdavis.edu

Contemporary Mathematics
Volume **784**, 2023
https://doi.org/10.1090/conm/784/15757

Quasi-reversibility method and neural network machine learning for forecasting of stock option prices

Mikhail V. Klibanov, Kirill V. Golubnichiy, and Andrey V. Nikitin

Abstract. This paper presents a novel way to predict options price for one day in advance, utilizing the method of Quasi-Reversibility for solving the Black-Scholes equation. The Black-Scholes equation is solved forwards in time, which is an ill-posed problem. Thus, Tikhonov regularization via the Quasi-Reversibility Method is applied. This procedure allows to forecast stock option prices for one trading day ahead of the current one. To enhance these results, the Neural Network Machine Learning is applied on the second stage. Real market data are used. Results of Quasi-Reversibility Method and Machine Learning method are compared in terms of accuracy, precision and recall.

1. Introduction

This paper discusses a new empirical mathematical model for generating more accurate option trading strategy using initial and boundary conditions for the underlying stock. The idea was initially proposed in [**5**]. The basis for this idea is the Black-Scholes equation. In mathematical finance, the Black-Scholes equation is a parabolic partial differential equation that determines the dynamics of the price of European options [**7**].

The time at a given time t will occur is τ,

$$\tau = T - t. \tag{1.1}$$

$f(s)$ be the payoff function of that option at the maturity time $t = T$ and s is the stock price. Let's assume that the risk-free interest rate equals zero. The function $u(s, \tau)$ is the price of that option and the variable τ is the one defined in (1.1). Let's assume that this function $u(s, \tau)$ satisfies the Black-Scholes equation with the volatility coefficient σ [**11**, Chapter 7, Theorem 7.7]:

$$\begin{aligned} \frac{\partial u(s,\tau)}{\partial \tau} &= \frac{\sigma^2}{2} s^2 \frac{\partial^2 u(s,\tau)}{\partial s^2}, \\ u(s,0) &= f(s), \end{aligned} \tag{1.2}$$

The payoff function is $f(s) = \max(s - K, 0)$, where K is the strike price [**11**] and $s > 0$.

The option price function is defined by the Black-Scholes formula:

$$u(s.\tau) = s\Phi(\theta_+(s,\tau)) - e^{-r\tau} K\Phi(\theta_-(s,\tau)), \tag{1.3}$$

Key words and phrases. The Black-Scholes equation, Ill-posed problem, regularization method, parabolic equation with the reversed time, Machine Learning, neural network.

Based on the Itô formula, we have:

$$du = \left(-\frac{\partial u(s,T-t)}{\partial \tau} + \frac{\sigma^2}{2} s^2 \frac{\partial^2 u(s,T-t)}{\partial s^2}\right)dt + \sigma s \frac{\partial u(s,T-t)}{\partial s} dW. \tag{1.4}$$

If equation (1.2) is solved forwards in time to forecast prices of stock options is an ill-posed inverse problem. For this reason, we used the Method of Quasi-Reversibility (QRM) that is a version of the Tikhonov regularization method. Uniqueness, stability and convergence theorems for this method were formulated in [**5**] and [**3**], also, see [**10**] for proofs.

We have four major questions that we raise in this paper:

(1) What is the forecast interval of the options prices?
(2) What are the boundary and initial conditions on the interval for the Black-Scholes equation?
(3) What are the values of the volatility coefficient in the future?
(4) How to solve the Black-Scholes equation forwards in time t?

The first three questions are addressed in our new mathematical model. We use the regularization method of [**4**] to address the fourth question. Theorems about stability and convergence of this method are formulated. These theorems were proven in [**4**] for a general parabolic equation of the second order where the main key of this method is based on the method of Carleman estimates.

Let the function $f \in L_2(0,\pi)$ and let $T = const. > 0$. To demonstrate that our problem is ill-posed, we consider the example of based on the problem for the heat equation with the reversed time

$$u_t + u_{xx} = 0, (x,t) \in (0,\pi) \times (0,T), \tag{1.5}$$

with Dirichlet boundary conditions

$$u(0,t) = 0, u(\pi,t) = 0, \tag{1.6}$$

and the initial condition

$$u(x,0) = f(x). \tag{1.7}$$

The unique solution of this problem is:

$$u(x,t) = \sum_{n=1}^{\infty} f_n \sin(nx) e^{n^2 t}.$$

Consider

$$u_N(x,t) = \sum_{n=1}^{N} f_n \sin(nx) e^{n^2 t}.$$

Then

$$\|u_N(x,T)\|^2_{L_2(0,\pi)} = \sum_{n=1}^{N} f_n^2 e^{2n^2 T} \approx f_N^2 e^{2N^2 T} \to \infty$$

as $N, T \to \infty$. Hence, the problem (1.5)-(1.7) is severely unstable.

We conclude therefore that to obtain a more or less accurate solution of the Black-Scholes equation forwards in time, we need to solve it on a short time interval $(0,T)$. To get better accuracy, the regularization method works only for a short time interval.

Section 3 presents our mathematical model, the method of Quasi Reversibility as well as the trading strategy. The Quasi Reversibility Method is based on the minimization of a Tikhonov-like functional $J_\beta(u)$. We do this using conjugate gradient method. The minimization process was performed by Hyak Next Generation Supercomputer of the research computing club of University of Washington. The code was parallelized in order to maximize the performance on supercomputer clusters.

The historical data for stock options was collected from the Bloomberg terminal [**2**] of University of Washington. From this data, we obtained about 177,000 minimizers.

Due to ill-posedness of the problem the solution is very sensitive to the noise in the initial data (stock and option prices for the three days preceding the day of forecast). Given results of the Quasi Reversibility Method, we apply on the second stage Machine Learning to reduce the probability of non-profitable trades caused by wrong option price prognosis because of the noise in input data.

Section 4 is dedicated to application of binary classification and regression Neural Network Machine Learning.

Sections 5 and 6 present our results and the summary.

Python with the SciPy and Torch modules were used for implementation of the method of Quasi-Reversibility and Neural Network Machine Learning (binary classification and regression).

2. The new mathematical model and the method of Quasi-Reversibility

Let's denote s as the stock price, t as the time and $\sigma(t)$ as the volatility of the option. The historical implied volatility listed on the market data of [**2**] is used in our particular case. We assume that $\sigma = \sigma(t)$ to avoid other historical data for the volatility. Let's call $u_b(t)$ and $u_a(t)$ the bid and ask prices of the options at the moment of time t and $s_b(t)$ and $s_a(t)$ the bid and ask prices of the stock at the moment of time t. It is also known that

$$u_b(t) < u_a(t) \tag{2.1}$$

and

$$s_b(t) < s_a(t) \tag{2.2}$$

Let's introduce

$$f_s(t) = \frac{s_a(t)}{s_b(t)} - 1 \tag{2.3}$$

and

$$f_u(t) = \frac{u_a(t)}{u_b(t)} - 1 \tag{2.4}$$

Based on real market data we have observed that usually

$$0 \leq f_s(t) \leq 0.003 \tag{2.5}$$

and

$$0 \leq f_u(t) \leq 0.27 \tag{2.6}$$

The idea is to approximate the Black-Scholes equation solutions

$$Lu = u_t + \frac{\sigma^2(t)}{2} s^2 u_{ss} = 0, (s,t) \in (s_b(0), s_a(0)) \times (0, 2\tau) = X_{2\tau}, \tag{2.7}$$

with Dirichlet boundary conditions

$$u\left(s_b,t\right)=u_b\left(t\right),u\left(s_a,t\right)=u_a\left(t\right),\ t\in\left[0,2\tau\right], \tag{2.8}$$

and the initial condition

$$u\left(s,0\right)=f\left(s\right),\quad s\in\left[s_b\left(0\right),s_a\left(0\right)\right]. \tag{2.9}$$

Where L is the partial differential operator of the Black-Scholes equation. Based on Bloomerg terminal we used with End of Day Underlying Price Last, End of Day Underlying Price Bid, End of Day Underlying Price Ask, t is time, $\sigma\left(t\right)$ is the volatility of the stock option. It was used Implied Volatility Using Last Trade Price (IVOL).

$u\left(s,t\right)$ is the price of the stock option. End of Day Option Price Last, End of Day Option Price Bid and End of Day Option Price Ask are the notation that we applied in our algorithm.

Problem 1. *Find the function* $u\in H^2\left(X_{2\tau}\right)$ *satisfying conditions* (2.7)-(2.9).

This problem considers as ill-posed since we solve equation (2.7) forwards in time.

Remarks 3.1: *We increase here the required smoothness of the solution from* $H^{2,1}\left(X_{2\tau}\right)$ *to* $H^2\left(X_{2\tau}\right)$.

Our algorithm based on solving the inverse problem for the Black-Scholes with reversed time equation has five steps:

Step 1 (Dimensionless variables).

We require to make our equation dimensionless. $s_b<s_a$. Let's denote $s_b=s_b(0)$, $s_a=s_a(0)$. Dimensionless variables were applied x,t' such that

$$x=\frac{s-s_b}{s_a-s_b} \tag{2.10}$$

$$t'=\frac{t}{255} \tag{2.11}$$

and now we can say that s is x and t is t'.

According to these substitutions, the equation becomes

$$Ru=u_t+\sigma^2\left(t\right)A(x)u_{xx}, \tag{2.12}$$

where

$$A(x)=\frac{255}{2}\frac{\left[x(s_a-s_b)+s_b\right)]^2}{(s_a-s_b)^2} \tag{2.13}$$

$$X_{2\tau}=\left\{(x,t)\in(0,1)\times(0,2\tau)\right\}. \tag{2.14}$$

$$u(x,0)=g(x),x\in(0,1) \tag{2.15}$$

$$u(0,t)=u_b(t),u(1,t)=u_a(t). \tag{2.16}$$

And the operator L in (2.7) is the operator R

Step 2 (Interpolation and extrapolation).

Our goal is to forecast option price from 'today' to 'tomorrow' and 'the day after tomorrow'. We do have 255 trading days annually. For this reason, let's introduce $\tau>0$ as our unit of time for which we want to make our prediction the option price. Because we predict option prices having the information of these prices, as well as of other parameters for 'today', 'yesterday' and 'the day before yesterday', we consider τ is one trading day. 'One day' $\tau = 1/255$. 'Today' $t = 0$. 'Tomorrow' $t = \tau$. 'The day after tomorrow' $t = 2\tau$. The variable $s-$ is for interval, i.e $s \in [s_b(0), s_a(0)]$. We applied the idea associated with interpolation discrete values of functions $u_b(t)$, $u_a(t)$, and $\sigma(t)$ between these three points (the day before yesterday, yesterday and today) and then extrapolation functions $u_b(t)$, $u_a(t)$ between three points (today, tomorrow and the day after tomorrow). Where $t = -2\tau$ is "the day before yesterday", $t = -\tau$ is "yesterday" and $t = 0$ is "today". We used quadratic polynomials for both approximation and extrapolation of values of functions. Thus, these three functions $u_b(t)$, $u_a(t)$, and $\sigma(t)$ was obtained for a small future time interval, i.e $(0, 2\tau)$. ([**5**]). Where $u_b(t)$, $u_a(t)$ were applied for boundary conditions and $\sigma(t)$ is coefficient function for our problem. The initial condition was set as $u(x,0) = g(x) = x(u_a(0) - u_b(0))x + u_b(0)$. This function is the result of approximation by linear function due to the fact that the interval between bid and ask prices is relatively small. The domain was $X_{2\tau} = \{(x,t) : x \in (0,1), t \in (0, 2\tau)\}$.

Step 3 (Statement of the Problem).

Problem 2. *Assume that functions*

$$u_b(t), u_a(t) \in H^2[0, 2\tau], \sigma(t) \in C^1[0, 2\tau]. \tag{2.17}$$

Find the solution $u \in H^2(X_{2\tau})$ of the following initial boundary value problem:

$$Ru = 0 \text{ in } X_{2\tau}, \tag{2.18}$$

$$u(0,t) = u_b(t), u(1,t) = u_a(t), t \in (0, 2\tau), \tag{2.19}$$

$$u(x,0) = g(x), x \in (0,1), \tag{2.20}$$

where the partial differential operator R is defined in (2.12), *the function $A(x)$ is defined in* (2.13), *the initial condition $g(x)$ is defined in* (2.15), *and the domain $X_{2\tau}$ is defined in* (2.14).

THEOREM 2.1. *The following problem* (2.12)-(2.15) *has one solution $u \in H^{2,1}(X_{2\tau})$.*

The proof of this theorem is [**3**].

Step 4 (Numerical method of solving the problem. Regularization).

Due to the ill-posedness of the problem, we can not say about existence of the solution. Thus, it was applied the regularization method:

Let's consider function $F(x,t) = x(u_a(t) - u_b(t)) + u_b(t)$, $(x,t) \in X_{2\tau}$. This function $F \in H^2(X_{2\tau})$. It follows from (2.15) and (2.16) that

$$F(x,0) = g(x), \tag{2.21}$$

$$F(0,t) = u_b(t), F(1,t) = u_a(t). \tag{2.22}$$

We used an unbounded differential operator $R : H^{2,1}(X_{2\tau}) \to L^2(X_{2\tau})$, where $H^{2,1}(X_{2\tau})$ is a dense linear set in the space $L^2(X_{2\tau})$. Where

$$Ru = u_t + \sigma^2(t) A(x) u_{xx} \tag{2.23}$$

Let's introduce Tikhonov-like functional as:

$$J_\beta(u) = \int_{X_{2\tau}} (Ru)^2 \, dsdt + \beta \|u\|^2_{H^2(X_{2\tau})}, \tag{2.24}$$

where $\beta \in (0,1)$ is the parameter of regularization. To solve the problem, we minimized the functional $J_\beta(u)$ on the set

$$V = \left\{u \in H^2(X_{2\tau}) : u(0,t) = u_b(t), u(1,t) = u_a(t), u(x,0) = g(x)\right\}. \tag{2.25}$$

Step 5 (Minimization Problem).

Minimization Problem 1. $J_\beta : H^2(X_{2\tau}) \to \mathbb{R}$ *is the regularization Tikhonov functional.*

We have used the converting of our partial derivatives from (2.24) into finite differences. A finite difference grid was applied to cover the domain $X_{2\tau}$. The minimization process was to differentiate our functional $J_\beta(u)$ with respect to the values of the function $u(x,t)$ at each grid points via conjugate gradient method. The point $u = 0$ was used for the starting point. Based on computational study with simulated data we have realized that the optimal value of the regularization parameter would be $\beta = 0.01$.

Minimization Problem 1 is a QRM for Problem 2. This is an version of the QRM for problem (2.18)-(2.20). In section 4 we discuss the theory of this specific version of the QRM. In particular, Theorem 4.2 of section 4 presents uniqueness of the solution $u \in H^2(X_{2\tau})$ of Problem 2 and implies an estimate of the stability of this solution with respect to the noise in the data. Theorem 4.3 of section 4 shows existence and uniqueness of the minimizer $u_\beta \in H^{2,1}(X_{2\tau})$ of the functional $J_\beta(u)$ on the set V defined in (2.25). We call such a minimizer "regularized solution" [**9**]. Theorem 4.4 estimates convergence rate of regularized solutions to the exact solution of Problem 2 with the noiseless data. Such estimates depend on the noise level in the data. All proof of these theorems are presented in [**10**].

3. Analysis

This section is devoted to convergence analysis for Problem 2 of subsection 3.2. This problem is the initial boundary value problem for parabolic equation (2.18) with the reversed time. The QRM and convergence analysis for this problem for a more general parabolic operator in $\mathbb{R}^n$ with arbitrary variable coefficients was proposed in [**4**]. Then theorems were presented in [**5**]. However a stability estimate was not a part of [**5**], such an estimate was proven in [**4**]. The same is true for the convergence theorems of QRM in [**4, 5**]. The smallness assumption was lifted in [**3**] via a new Carleman estimate. Results of [**3**] for a 1-D case were significantly modified in this section. Our computations below on a small time interval $(0, 2\tau) = (0, 0.00784)$ (see [**4, 5**], [**8**, Theorem 1 of section 2 in Chapter 4] might result in the requirement of even a smaller length of that interval.

3.1. Problem statement. Let's consider a number $T > 0$ and introduce Q_T as:

$$Q_T = \{(x,t) \in (0,1) \times (0,T)\}.$$

Consider two numbers $b_0, b_1 > 0$ and $b_0 < b_1$. Let the function $b(x,t) \in C^1\left(\overline{Q}_T\right)$ satisfies:

$$\|b\|_{C^1(\overline{Q}_T)} \le b_1, \; b(x,t) \ge b_0 \text{ in } Q_T. \tag{3.1}$$

We also have functions $\psi_0(t), \psi_1(t) \in H^2(0,T)$. In the above case of subsection 3.2,

$$T = 2\tau, b(x,t) = \sigma^2(t)\, A(x), \psi_0(t) = u_b(t), \psi_1(t) = u_a(t).$$

We now formulate Problem 3, which is a slight generalization of Problem 2.

Problem 3. *Find a solution $v \in H^2(Q_T)$ of the following (IBVP):*

$$Nv = v_t + b(x,t)\, v_{xx} = 0 \text{ in } Q_T, \tag{3.2}$$

$$v(0,t) = \psi_0(t), v(1,t) = \psi_1(t),\ t \in (0,T), \tag{3.3}$$

$$v(x,0) = z(x) = \psi_0(0)(1-x) + \psi_1(0)\, x,\ x \in (0,1). \tag{3.4}$$

Remark 4.1. *Because Problem 2 is less general than Problem 3, then this analysis of converegence for Problem 3 also works for Problem 2.*

We use the linear function for $v(x,0)$ in (3.4) is to simplify the initial condition in (2.20). Now problem 3 is an IBVP for the parabolic equation (3.2) with the reversed time. For this reason, the problem can be considered as ill-posed. Assume that the boundary with a noise of the level $\nu > 0$ in (3.3) are in place. Here ν is a sufficiently small number, i.e.

$$\|\psi_0 - \psi_0^*\|_{H^1(0,T)} < \nu, \|\psi_1 - \psi_1^*\|_{H^1(0,T)} < \nu, \tag{3.5}$$

where functions $\psi_0^*, \psi_1^* \in H^2(0,T)$ are "ideal" noiseless data. we assume that there exists an exact solution $v^* \in H^2(Q_T)$ of problem (3.2)-(3.4) with these noiseless data (based on on the theory of Ill-Posed problems). Below we present estimates how this noise affects the accuracy of the solution of Problem 3 and also discuss the convergence rate of numerical solutions obtained by QRM to the exact one as $\nu \to 0$.

Let's introduce the version of functional (2.24):

$$I_\beta(v) = \int_{Q_T} (Nv)^2\, dxdt + \beta \|v\|^2_{H^2(Q_T)}. \tag{3.6}$$

We also have the set $W \subset H^2(Q_T)$,

$$W = \left\{v \in H^2(Q_T) : v(0,t) = \psi_0(t), v(1,t) = \psi_1(t), v(x,0) = z(x)\right\}. \tag{3.7}$$

The solution of Problem 3 is approximate solution by solving the following problem:

Minimization Problem 2. *Minimize the functional $I_\beta(v)$ on the set W given in* (3.7).

Minimization Problem 2 is QRM for Problem 3.

3.2. Theorems. This subsection presents four theorems for Problem 3. All proofs might be found in [**10**]. First, let's introduce the Carleman Weight Function $\phi_\alpha(t)$ with $\alpha > 2$ for the operator $\partial_t + b(x,t)\,\partial_x^2$ as:

$$\phi_\alpha(t) = e^{(T+1-t)^\alpha},\ t \in (0,T). \tag{3.8}$$

As a result, the function $\phi_\alpha(t)$ is decreasing on $[0,T]$, $\phi'_\alpha(t) < 0$,

$$\max_{[0,T]} \phi_\alpha(t) = \psi_\alpha(0) = e^{(T+1)^\alpha},\ \min_{[0,T]} \phi_\alpha(t) = \phi_\alpha(T) = e. \tag{3.9}$$

Denote

$$H_0^2(Q_T) = \left\{u \in H^2(Q_T) : u(0,t) = u(1,t) = 0\right\}. \tag{3.10}$$

$$H_{0,0}^2(Q_T) = \left\{u \in H_0^2(Q_T) : u(x,0) = 0\right\}. \tag{3.11}$$

Theorem 4.1 (Carleman estimate). *Let the coefficient $b(x,t)$ of the operator N satisfies conditions* (3.1). *Then there exist a sufficiently large number $\alpha_0 = \alpha_0(T, b_0, b_1) > 2$ and a constant $C = C(T, b_0, b_1) > 0$, both depending only on listed parameters, such that the following Carleman estimate holds for the operator N :*

$$\int_{Q_T} (Nu)^2 \phi_\alpha^2 dxdt \geq C\sqrt{\alpha} \int_{Q_T} u_x^2 \psi_\alpha^2 dxdt + C\alpha^2 \int_{Q_T} u^2 \phi_\alpha^2 dxdt$$

$$-C\sqrt{\alpha} \|u\|_{H^2(Q_T)}^2 - C\lambda (T+1)^\alpha e^{2(T+1)^\alpha} \|u(x,0)\|_{L_2(0,1)}^2, \tag{3.12}$$

$$\forall \alpha \geq \alpha_0, \forall u \in H_0^2(Q_T).$$

Carleman estimate (3.12) is the MAIN TOOL to proofs of Theorems 4.2, 4.4.

Theorem 4.2 (Hölder stability estimate for Problem 3 and uniqueness). *Let the coefficient $b(x,t)$ of the operator N satisfies conditions* (3.1). *Let's assume that the functions $v \in H^2(Q_T)$ and $v^* \in H^2(Q_T)$ are solutions of Problem 3 with the vectors of data $(\psi_0(t), \psi_1(t))$ and $(\psi_0^*(t), \psi_1^*(t))$ respectively, where $\psi_0, \psi_1, \psi_0^*, \psi_1^* \in H^2(0,T)$. Assume also that error estimates* (3.5) *of the boundary data is in place. Choose an arbitrary number $\epsilon \in (0,T)$. Denote*

$$\lambda = \lambda(T,\epsilon) = \frac{\ln(T+1-\epsilon)}{\ln(T+1)} \in (0,1). \tag{3.13}$$

Then there exists a sufficiently small number $\nu_0 = \nu_0(T_1, b_0, b_1) \in (0,1)$ and a constant $C_1 = C_1(T, b_0, b_1, \epsilon) > 0$, both depending only on listed parameters, such that the following stability estimate holds for all $\nu \in (0, \nu_0)$:

$$\|v_x - v_x^*\|_{L_2(Q_{T-\epsilon})} + \|v - v^*\|_{L_2(Q_{T-\epsilon})} \leq \tag{3.14}$$

$$\leq C_1 \left(1 + \|v - v^*\|_{H^2(Q_T)}\right) \exp\left[-\left(\ln \nu^{-1/2}\right)^\lambda\right].$$

Below $C = C(T, b_0, b_1) > 0$ and $C_1 = C_1(T, a_0, b_1) > 0$ denote different constants depending only on listed parameters.

Corollary 4.1 (uniqueness). *Let the coefficient $b(x,t)$ of the operator N satisfies conditions* (3.1). *Then Problem 3 has at most one solution (uniqueness).*

Proof. If $\nu = 0$, then (3.14) implies that $v(x,t) = v^*(x,t)$ in $Q_{T-\epsilon}$. Since $\epsilon \in (0,T)$ is an arbitrary number, then $v(x,t) \equiv v^*(x,t)$ in Q_T. □

Theorem 4.3 (existence and uniqueness of the minimizer). *Let functions $\psi_0(t), \psi_1(t) \in H^2(0,T)$. Let W be the set defined in (3.7). Then there exists unique minimizer $v_{\min} \in W$ of functional* (3.6) *and*

$$\|v_{\min}\|_{H^2(Q_T)} \leq \frac{C}{\sqrt{\beta}} \left(\|\psi_0\|_{H^2(0,T)} + \|\psi_1\|_{H^2(0,T)}\right). \tag{3.15}$$

In the theory of Ill-Posed Problems, this minimizer $v_{\min}$ is called "regularized solution" of Problem 3 [**9**]. According to the theory of Ill-Posed problems, it is important to establish convergence rate of regularized solutions to the exact one v^*. In doing so, one should always choose a dependence of the regularization parameter β on the noise level ν, i.e. $\beta = \beta(\nu) \in (0,1)$ [**9**].

Theorem 4.4 (convergence rate of regularized solutions). *Let $v^* \in H^2(Q_T)$ be the solution of Problem 3 with the noiseless data $(\psi_0^*(t), \psi_1^*(t))$. Let functions $\psi_0, \psi_1, \psi_0^*, \psi_1^* \in H^2(0,T)$. Let $v_{\min} \in W$ be the unique minimizer of functional*

(3.6) on the set W. Assume that error estimates (3.5) hold. Choose an arbitrary number $\epsilon \in (0,T)$. Let $\lambda = \lambda(T,\epsilon) \in (0,1)$ be the number defined in (3.13) and let

$$\beta = \beta(\nu) = \nu^2, \tag{3.16}$$

Then there exists a sufficiently small number $\nu_0 = \nu_0(T, b_0, b_1) \in (0,1)$ depending only on listed parameters such that the following convergence rate of regularized solutions $v_{\min}$ holds for all $\nu \in (0,\nu_0)$:

$$\|\partial_x v_{\min} - \partial_x v^*\|_{L_2(Q_{T-\epsilon})} + \|v_{\min} - v^*\|_{L_2(Q_{T-\epsilon})} \tag{3.17}$$

$$\leq C_1 \left(1 + \|v^*\|_{H^2(Q_T)} + \|\psi_0^*\|_{H^2(0,T)} + \|\psi_1^*\|_{H^2(0,T)}\right) \exp\left[-\left(\ln \nu^{-1/2}\right)^{\lambda}\right].$$

3.3. Trading Strategy: We use minimizers obtained from the method of Quasi-Reversibility to build a strategy for trading options. Let's define

$$REAL(0) = \frac{u_a(0) + u_b(0)}{2} \tag{3.18}$$

$$REAL(\tau) = \frac{u_a(\tau) + u_b(\tau)}{2} \tag{3.19}$$

$$EST(\tau) = u_\beta(1/2, k\tau) \tag{3.20}$$

or if it was not applied dimensionless

$$EST(\tau) = u_\beta(\frac{s_a + s_b}{2}, k\tau) \tag{3.21}$$

where $k = 1$
Here $EST(\tau)$ means minimizer.
Let's buy an option if the following holds

$$EST(\tau) \geq REAL(0) \tag{3.22}$$

The predicted outcome of option trade is Positive if

$$EST(\tau) \geq REAL(0) \tag{3.23}$$

Definition 1.
It is True Positive if

$$EST(\tau) \geq REAL(0) \tag{3.24}$$

and

$$REAL(\tau) \geq REAL(0) \tag{3.25}$$

Definition 2.
It is True Negative if

$$EST(\tau) < REAL(0) \tag{3.26}$$

and

$$REAL(\tau) < REAL(0) \tag{3.27}$$

Definition 3.
It is False Positive if

$$EST(\tau) \geq REAL(0) \tag{3.28}$$

and

$$REAL(\tau) < REAL(0) \tag{3.29}$$

Definition 4.

It is False Negative if

$$EST(\tau) < REAL(0) \tag{3.30}$$

and

$$REAL(\tau) \geq REAL(0) \tag{3.31}$$

The accuracy of trading strategy is defined as

$$Accuracy = \frac{TP + TN}{\sum options} \tag{3.32}$$

where TP is a summation of True Positive and TN is a summation of True Negative and $\sum options$ is a summation of options in data set.

The precision of trading strategy is defined as

$$Precision = \frac{TP}{TP + FP} \tag{3.33}$$

where FP is a summation of False Positive.

The recall of trading strategy is defined as

$$Recall = \frac{TP}{TP + FN} \tag{3.34}$$

where FN is a summation of False Negative.

The average relative error of trading strategy is defined as

$$Error = \frac{1}{N} \sum \left| \frac{EST(\tau) - REAL(\tau)}{REAL(\tau)} \right| \tag{3.35}$$

4. Application of Neural Network Machine Learning

The Black-Scholes equation gives fair value of options in perfect market. However, real options prices contain some level of noise. We try to filter mispredictions (i.e. where minimizers result in False Positive or False Negative) caused by input noise using Machine Learning to improve accuracy, precision and recall of the trading strategy. We built a neural network with 13 element input vector and 3 fully connected hidden layers. (See Fig 1). Input vector consists of minimizers (for $t = \tau, 2\tau$) obtained from the method of Quasi-Reversibility, stock ask and bid price (for $t = 0$), option ask and bid price and volatility (for $t = -2\tau, -\tau, 0$).

FIGURE 1

All vectors and labels are split into three parts: training, validation and test sets. The training set is used for weight learning. Validation set is used for tuning of the neural network hyper-parameters. Test set is for generating the outcomes of trading strategy.

We collected historical option and stock prices along with implied volatility on companies consisting of Russel 2000 index [**6**].

Table 1.

Set	Dates	Number of options
Training	2016/09/14-2018/05/31	132,912
Validation	2018/06/01-2018/06/29	13,401
Test	2018/07/02-2018/08/17	23,549

We compared the profitability of the trading strategy based on the original minimizer set with the profitability of the output of Machine Learning.

4.1. Machine Learning Input Vector Normalization.

$$\mu = \frac{u_a(0) + u_a(-\tau) + u_a(-2\tau) + u_b(0) + u_b(-\tau) + u_b(-2\tau)}{6} \tag{4.1}$$

$$op_n = \frac{u(t) - \mu}{\sigma} \tag{4.2}$$

$$s_n = \frac{(s - st) - \mu}{\sigma} \tag{4.3}$$

where op_n is a normalized option price, s_n is a normalized stock price normalization, s is the stock, st is the strike and σ is the standard deviation.

4.2. Binary classification. Supervised Machine Learning has been applied to the neural network for the Cross Entropy Loss function with regularization:

$$L(\theta) = \frac{1}{m}\sum_{i=1}^{m}[-y^{(i)}\log(h_\theta(x^{(i)}) - (1 - y^{(i)})\log(1 - h_\theta(x^{(i)}))] + \frac{\lambda}{2m}\sum_{j=1}^{n}\theta_j^2 \tag{4.4}$$

Where θ are weights which are optimized by minimizing the loss function using the method of gradient descent. λ is a parameter of regularization. $x^{(i)}$ is our normalized 13 - dimensional vectors. h_θ is output of the neural network. m is the number of vectors in the training set. $y^{(i)}$ is our labels (the ground truth). The trading strategy is defined by

$$H_c = \begin{cases} 1, & \text{if } h_\theta > c \\ 0, & \text{otherwise.} \end{cases} \tag{4.5}$$

where c is the threshold obtained by maximizing accuracy on validation set. The labels are set to 1 for profitable trades and 0 otherwise.

4.3. Regression model. Similarly, instead of using binary classification, we can use the same ML architecture to predict the option price for tomorrow (τ). We have the same input features as the classification neural network. Regression learning uses mean squared error as the loss function:

$$L(\bar{h}_\theta, \bar{y}) = \frac{1}{m} \sum_n (\bar{h}_\theta - \bar{y}_n)^2 \tag{4.6}$$

Where m is the size of the data set, $\bar{h}_\theta$ is the predicted value and $\bar{y}$ is the real value ($Real(\tau)$).

5. Results

The following graph shows the accuracy of the results on validation set. We use it to determine the optimal value of hyper-parameter c (the threshold value of binary classification).

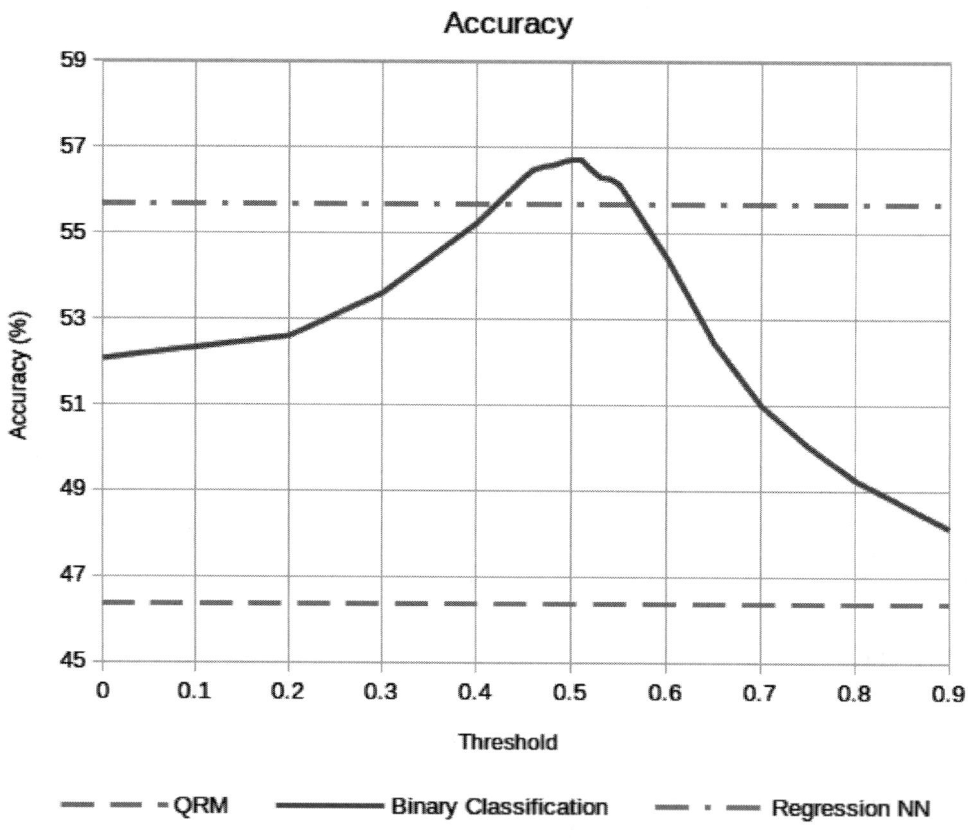

FIGURE 2. Accuracy. Threshold dependency

Observation 1.

The accuracy was improved by both Machine Learning methods compared to the method of Quasi-Reversibility. Based on this graph we set $c = 0.5$.

The next graph presents Recall and Precision diagram built on validation data set.

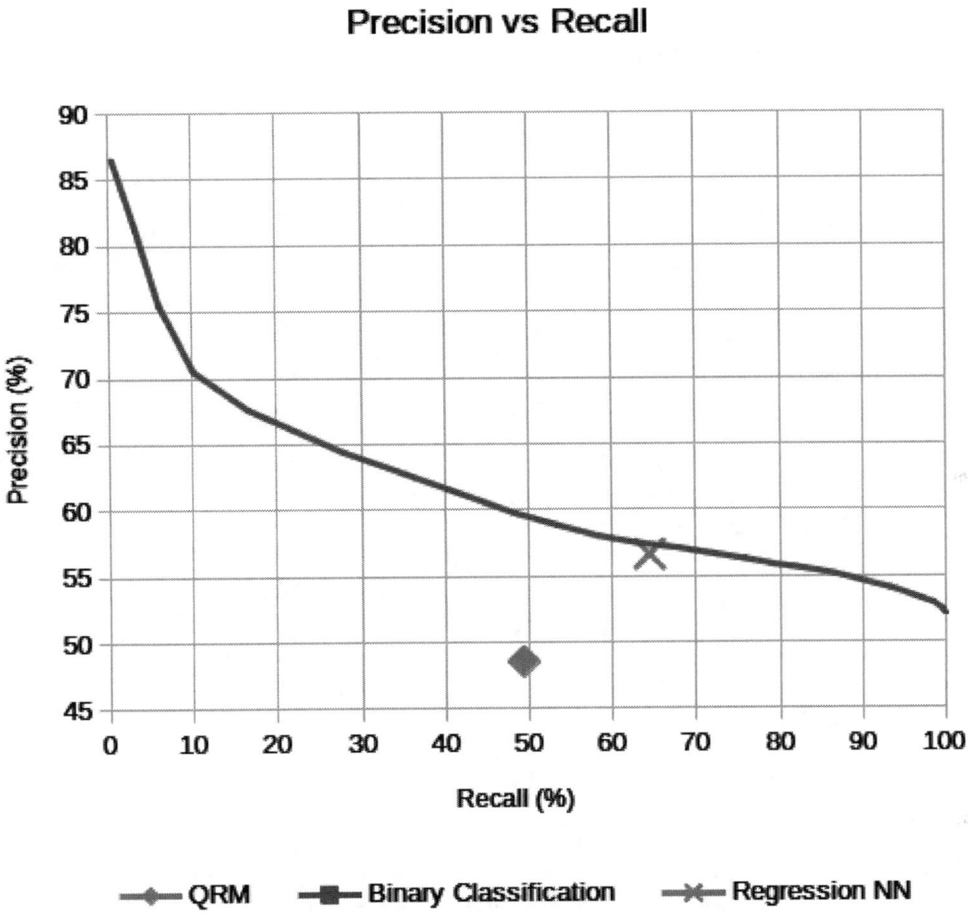

FIGURE 3. Precision vs Recall. Threshold dependency. Here **X** (Cross) indicates the position of Recall and Precision produced by Regression Neural Network and ◊ (Diamond) indicates the position of Recall and Precision produced by QRM.

Observation 2.

Binary Classification and Regression produced similar results that improved both precision and recall compared to the method of Quasi-Reversibility.

Further we divided our test data into bins where (horizontal axis, see Figure 4) each bin is determined by $\frac{s-st}{s}$ with step size 0.1. for each bin we calculated precision (see Figure 4).

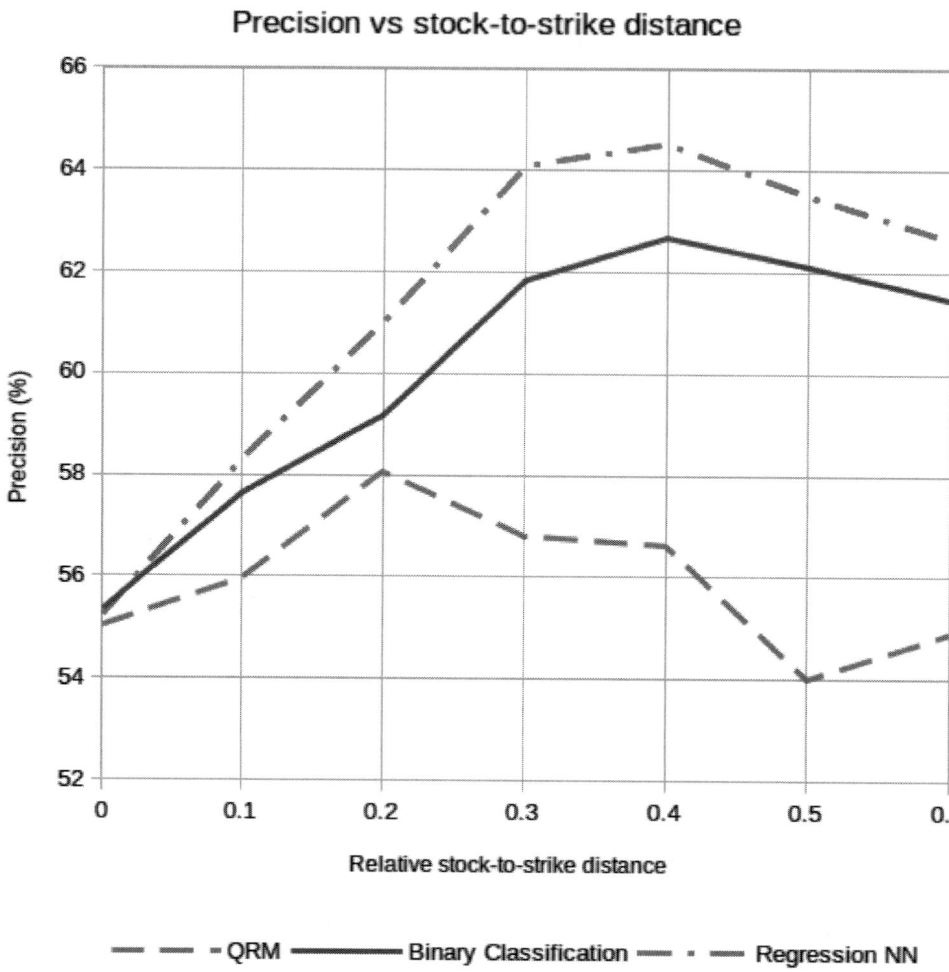

FIGURE 4. Binary Classification, Regression NN and the method of Quasi-Reversibility.

Observation 3.

To our surprise, when stock price was close to the strike price Machine Learning and the method of Quasi-Reversibility give similar precision (bin 0). With stock price diverging from the strike price Machine Learning produced better precision.

The following tables summarize the accuracy, precision and recall for all methods on test data.

Table 2. Final results on Test Data.

Method	Accuracy	Precision	Recall	Error
QRM	49.77%	55.77%	52.43%	12 %
Binary Classification	56.36%	59.56%	70.22%	NA
Regression NN	55.42%	60.32%	61.29%	NA

Table 3. Percentages of options with profits/losses for three different methods.

Method	Profitable options	Options with loss
QRM	55.77%	44.23 %
Binary Classification	59.56%	40.44%
Regression NN	60.32%	39.68%

6. Summary

To predict prices of stock options, we used two empirical mathematical models for Black-Scholes equation. The results achieved by solving the equation forwards in time (as an ill-posed problem) and applying Supervised Machine Learning (Binary Classification and Regression Neural Network, and using these methods with the real market data, show that this methodology produce promising results, potential applications within real-world trading and investment strategies.

The comparison of our methods resulted in the following two conclusions:

(1) The predictions of the method of Quasi-Reversibility ended up being profitable for 55.77% of the options. Compare this to a 59.56% profitability rate for the Binary Classification method, and a 60.32% profitability rate for Regression Neural Network, used on the same data set and with the same trading strategy.
(2) As shown by figures in section 5, option price forecasting using Machine Learning gives us significant accuracy and profit improvements over the method of Quasi-Reversibility. However, when stock price is close to the strike price both models give similar results.

The authors hypothesize that options traders can generate significant profits using trading strategies reliant on predictions generated with these methods.

References

[1] Anatoly B. Bakushinskii, Michael V. Klibanov, and Nikolaj A. Koshev, *Carleman weight functions for a globally convergent numerical method for ill-posed Cauchy problems for some quasilinear PDEs*, Nonlinear Anal. Real World Appl. **34** (2017), 201–224, DOI 10.1016/j.nonrwa.2016.08.008. MR3567958

[2] `https://bloomberg.com`.

[3] Michael V. Klibanov and Anatoly G. Yagola, *Convergent numerical methods for parabolic equations with reversed time via a new Carleman estimate*, Inverse Problems **35** (2019), no. 11, 115012, 30, DOI 10.1088/1361-6420/ab2777. MR4027423

[4] Michael V. Klibanov, *Carleman estimates for the regularization of ill-posed Cauchy problems*, Appl. Numer. Math. **94** (2015), 46–74, DOI 10.1016/j.apnum.2015.02.003. MR3342583

[5] Michael V. Klibanov, Andrey V. Kuzhuget, and Kirill V. Golubnichiy, *An ill-posed problem for the Black-Scholes equation for a profitable forecast of prices of stock options on real market data*, Inverse Problems **32** (2016), no. 1, 015010, 16, DOI 10.1088/0266-5611/32/1/015010. MR3465302

[6] `https://money.cnn.com/data/markets/russell`.

[7] Steven E. Shreve, *Stochastic calculus for finance. II*, Springer Finance, Springer-Verlag, New York, 2004. Continuous-time models. MR2057928

[8] M. M. Lavrent′ev, V. G. Romanov, and S. P. Shishat·skiĭ, *Ill-posed problems of mathematical physics and analysis*, Translations of Mathematical Monographs, vol. 64, American Mathematical Society, Providence, RI, 1986. Translated from the Russian by J. R. Schulenberger; Translation edited by Lev J. Leifman. MR847715

[9] A. N. Tikhonov, A. V. Goncharsky, V. V. Stepanov, and A. G. Yagola, *Numerical methods for the solution of ill-posed problems*, Mathematics and its Applications, vol. 328, Kluwer Academic Publishers Group, Dordrecht, 1995. Translated from the 1990 Russian original by R. A. M. Hoksbergen and revised by the authors, DOI 10.1007/978-94-015-8480-7. MR1350538

[10] M.V. Klibanov, A. A. Shananin, K. V. Golubnichiy and S. M. Kravchenko, *Forecasting Stock Options Prices via the Solution of an Ill-Posed Problem for the Black-Scholes Equation*, arXiv preprint `arXiv:2202.07174`.

[11] T. Bjork, *Arbitrage Theory in Continuous Time*, Oxford University Press, 1999.

Department of Mathematics and Statistics, University of North Carolina at Charlotte, Charlotte, North Carolina
Email address: `mklibanv@uncc.edu`

Department of Mathematics, University of Washington, Seattle, Washington
Email address: `kgolubni@math.washington.edu`

Amazon Web Services, Seattle, Washington 98121
Email address: `andrey.nikitin@hotmail.com`

Contemporary Mathematics
Volume **784**, 2023
https://doi.org/10.1090/conm/784/15758

Numerical reconstruction for 3D nonlinear SAR imaging via a version of the convexification method

Vo Anh Khoa, Michael Victor Klibanov, William Grayson Powell, and Loc Hoang Nguyen

ABSTRACT. This work extends the applicability of our recent convexification-based algorithm for constructing images of the dielectric constant of buried or occluded target. We are orientated towards the detection of explosive-like targets such as antipersonnel land mines and improvised explosive devices in the non-invasive inspections of buildings. In our previous work, the method is posed in the perspective that we use multiple source locations running along a line of source to get a 2D image of the dielectric function. Mathematically, we solve a 1D coefficient inverse problem for a hyperbolic equation for each source location. Different from any conventional Born approximation-based technique for synthetic-aperture radar, this method does not need any linearization. In this paper, we attempt to verify the method using several 3D numerical tests with simulated data. We revisit the global convergence of the gradient descent method of our computational approach.

1. Introduction

Synthetic-aperture radar (SAR) imaging is a commonly used technique in reconstructing images of surfaces of planets, and in detecting antipersonnel land mines and improvised explosive devices; cf. e.g. [**1, 2, 9, 10**] for some essential backgrounds concerning SAR imaging. In SAR imaging, the migration method is applied quite often; cf. e.g. [**6, 7**]. However, this approach is a kind of small perturbation methods that require an *a priori* knowledge of the distribution of the dielectric constant of the target. Another conventional SAR imaging is based on the linearization via the Born approximation [**9**]. Cf. [**8, 9, 45**], this approximation with its variants is very popular since it provides accurate images of shapes and locations of targets. With an indispensable assumption that the dielectric constant of the target is close to the unity, this linearization process is, however, problematic

2020 *Mathematics Subject Classification.* Primary 78A46, 65L70, 65C20.

Key words and phrases. SAR imaging, coefficient inverse problem, convexification, global convergence, simulated data, delay-and-sum procedure.

This work was supported by US Army Research Laboratory and US Army Research Office grant W911NF-19-1-0044. V. A. Khoa was supported by the Faculty Research Awards Program (FRAP) at Florida A&M University, under the project #007633. L. H. Nguyen was partially supported by National Science Foundation grant DMS-2208159, by funds provided by the Faculty Research Grant program at UNC Charlotte Fund No. 111272, and by the CLAS small grant provided by the College of Liberal Arts & Sciences, UNC Charlotte.

if one wants to produce accurate values of the dielectric constant; see e.g. [**12**]. Since dielectric constants characterize constituent materials of targets, then it is obviously important to accurately compute them.

To address this problem, a new convexification based nonlinear SAR imaging, CONSAR, was proposed in [**12**] for the first time. Testing of CONSAR for both computationally simulated and experimentally collected data has shown that dielectric constants of targets are indeed accurately computed [**12, 21**]. However, these publications are only about obtaining 2D SAR images on the so-called "slant range" plane. In this case both the point source and the radiating antenna move along an interval of a single straight line. Thus, unlike [**12, 21**], the current paper is about testing the CONSAR technique for 3D SAR images. By our experience, collection of experimental SAR data along just one line is time consuming [**21**]. Therefore, we consider here the case when the source and the detector concurrently move along only three (3) parallel lines. In principle, one should have many such lines of course to get accurate solutions. Hence, because of an obvious lack of the information content in the input data, it is unrealistic to expect that resulting SAR images would be accurate in terms of shapes of targets. This is why sizes of computed images of our targets are quite different from the true ones. Nevertheless, we accurately image dielectric constants and locations of targets. Only computationally simulated data are considered here.

In this paper, our target application is the non-invasive inspections of buried targets when one tries to image the buried or occluded target using backscattering data collected by e.g., flying devices or roving vehicles (cf. [**1, 11**]). Together with the identification of shape and location of the target (cf. e.g. [**16, 46–48**]), having knowledge of the dielectric constant of it plays a vital role in the development of future classification algorithms that distinguish between explosives and clutter in military applications. Thus, it is anticipated that the knowledge of dielectric constants of targets should decrease the false alarm rate.

In fact, the conventional time dependent SAR data when the source and the detector move concurrently along an interval of a straight line, depend on two variables. On the other hand, we want to image 3D objects. Therefore, the data for the corresponding Coefficient Inverse Problem (CIP) are underdetermined. This is why we work here with the above mentioned more informative case of three parallel lines. Although these data are also underdetermined, we still hope to reconstruct some approximations of 3D images of targets. It is worth mentioning that solving underdetermined or non overdetermined CIPs is omnipresent in applications that we have mentioned above. This is one of the main challenges that researchers encounter in practice is finding a good approximation of the target's dielectric constant. This is also a long standing research project that the corresponding author and his research team have been working on for almost a decade, see, e.g. [**13–19**] for mathematical models and numerical methods to identify the material of the buried targets.

Solving for the target's dielectric constant in this scenario is a CIP for a hyperbolic PDE. It is well-known that CIPs are both severely ill-posed and nonlinear thus, requiring a careful designation of an appropriate numerical method. Very recently, we have proposed in [**12, 21**] a new convexification-based algorithm to construct a 2D image of the target's dielectric constant using SAR data. In fact, the approach of [**12**, **21**] means a nonlinear SAR imaging, as opposed to the conventional linear one. We point out that the computational approach of [**12**, **21**] to the SAR data is a heuristic one. Nevertheless, it was shown in [**12**] that this approach works well numerically for computationally simulated data. It is more encouraging that it was demonstrated in [**12**, **21**] that this approach works well for experimentally collected data. The same approach to the SAR data is used in this paper. Using multiple source locations running along an interval of a straight line, our computational approach is to solve a number of 1D CIPs.

To solve numerically each of our CIPs, we transform first the hyperbolic equation into a nonlocal nonlinear boundary value problem (BVP), based upon the original idea of [**22**]. An approximate solution of this BVP is then found by minimizing a suitable weighted Tikhonov-like cost functional, which is proved to be globally strictly convex. Its strict convexity is based on the presence of the Carleman Weight Function (CWF) in it. The CWF is the function which is used as the weight function in the Carleman estimate for the principal part of the differential operator involved in that BVP. Ultimately, the 2D image which we construct for the case of each of our three lines can be understood as the collection of 1D cross-sections of the so-called slant-range plane established by the "moving" source/antenna positions. To construct the 3D image, we combine those 2D ones, see sections 3 and 5 for details.

Conventional numerical methods for CIPs are based on minimizations of various least squares cost functionals, including the Tikhonov regularization functional, see, e.g. [**3**–**5**]. It is well known, however, that such a functional is typically non convex and has, therefore, multiple local minima and ravines, which, in turn plague the minimization procedure. The convexification method was initially proposed in [**23**, **24**] with the goal to avoid that phenomenon of multiple local minima and ravines of conventional least squares cost functionals for CIPs. While publications [**23**, **24**] are purely theoretical ones, the paper [**25**] has generated many follow up numerical studies of the convexification of this research group, see, e.g. the above cited publications of this research group and references cited therein. We also refer to the recently published book [**26**].

The convexification uses a significantly modified idea of the Bukhgeim–Klibanov method [**27**] of 1981. However, in [**27**] and many follow up publications, only proofs of global uniqueness and stability theorems for CIPs were studied, see, e.g. [**28**–**32**] and references cited therein. On the other hand, the convexification is going in the numerical direction: it is designed for constructions of *globally convergent* numerical methods for CIPs. This is unlike the conventional locally convergent numerical methods for CIPs, which are based on minimizations of those functionals.

As stated above, depending on the differential operator involved in the CIP, the convexification method is based on an appropriate Carleman Weight Function to construct a weighted Tikhonov-like cost functional. The Carleman Weight Function plays two roles. First, it convexifies the conventional Tikhonov functional. Second, it helps to control nonlinearities involved in the underlying CIP.

Definition 1. *We call a numerical method for a CIP globally convergent if there is a theorem which claims that this method reaches at least one point in a sufficiently small neighborhood of the true solution of that CIP without any advanced knowledge of a small neighborhood of this solution.*

The convexification is a globally convergent numerical method because one can prove the convergence of the scheme towards the true solution starting from any point located in a given bounded set of a suitable Hilbert space and the diameter of this set can be arbitrary large.

As to the moving sources, we remark that in the publication [**18**] a version of the convexification method was developed for the first time both analytically and numerically for a 3D inverse medium problem for the Helmholtz equation with the backscattering data generated by point sources running along a straight line and at a fixed frequency. Then this result was confirmed on experimental data in [**19, 20**]. In [**18–20**], we were working with the data in the frequency domain, whereas the SAR data are the time-dependent ones. In addition, in the case of the SAR data, both source and detector move concurrently along a straight line. On the other hand, in [**18–20**] the source moves along a part of a straight line, whereas the detector moves independently along a part of a plane. We also refer to works of Novikov and his group [**33, 34**] for a different approach to tackle CIPs with the moving source. For some other works related to inverse problems with a moving point source, we refer to [**49**] for reconstructing the trajectory of the moving point source with an application to gesture recognition.

We also add that our recent publications [**12, 21**] about CONSAR were focusing only on the 2-D reconstructions. Therefore, the main difference between these papers and the current one is that here we obtain a 3-D image of the target dielectric constant, assuming that the data are collected from multiple lines of sources.

Our paper is organized as follows. In sections 2, 3 and 4, we revisit our setting for the CIP imaging and our convexification-based method proposed in [**21**]. In section 5, we verify the 3-D numerical performance of our method using computationally simulated data. Finally, we present concluding remarks in section 6.

2. Statement of the forward problem

Prior to the statement of the forward problem, we discuss the setting of SAR imaging. We focus only on the stripmap SAR processing; see e.g. [**35**] for details of this setup. With the non-invasive inspection of buildings and land mines detection being in mind, we prepare to have a radiating antenna and a receiver in the SAR device. As the antenna radiates pulses of the time-resolved component of the electric wave field, the receiver collects the backscattering signal. Our SAR device moves along a straight line. In this work, we assume that it moves it along multiple parallel lines, which, we hope, may create a 3-D image of the target. It is worth mentioning that in our inverse setting in section 3 below, we assume that the antenna and the receiver coincide, and form a point, which moves along certain parallel lines. This assumption is a typical one in SAR imaging; cf. [**9**]. Indeed, in practice the antenna and the receiver are rather close to each other, and the distance between them is much less than the distance between the line over which they concurrently move and the targets of interest. However, when generating the SAR data in computational simulations, we solve a forward problem, in which we assume that the antenna is a disk.

We denote below $\mathbf{x} = (x, y, z) \in \mathbb{R}^3$. For $z^0 > 0$, we consider the source location $\mathbf{x}^0 = (x^0, y_j^0, z^0) \in L_j^{\text{src}}$ for $j = \overline{1, j_0}$, $j_0 \in \mathbb{N}^*$. By choosing a certain value of j_0 and varying x^0 in the interval $(-L, L)$ for $L > 0$, we define a finite set of lines L_j^{src} of sources/receivers as follows:

$$L_j^{\rm src} := \left\{ \mathbf{x}^0 = \left(x^0, y_j^0, z^0\right) \in \mathbb{R}^3 : x^0 \in (-L, L), j = \overline{1, j_0} \right\}. \tag{2.1}$$

Therefore, each line of sources $L_j^{\rm src}$ is parallel to the $x-$axis. The antenna and the receiver coincide and are located at the point $\mathbf{x}^0 = \left(x^0, y_j^0, z^0\right) \in L_j^{\rm src}$. The number y_j^0 is the same for all points of $L_j^{\rm src}$ and this number is changing as lines $L_j^{\rm src}$ changes. The number z^0 is the same for all lines (2.1). Assuming that the parameter $x^0 \in (-L, L)$ in (2.1), we actually assume that the antenna/receiver point characterized by x^0 moves along each line $L_j^{\rm src}$, $j = \overline{1, j_0}$.

Let the number $R > 0$. Our domain of interest is

$$\Omega = (-R, R)^3 \subset \mathbb{R}^3.$$

This domain consists of two parts. The lower part $\Omega \cap \{z < 0\}$ mimics the sandbox containing the unknown buried object. Meanwhile, the upper part $\Omega \cap \{z > 0\}$ is assumed to be the air part, where we move our SAR device. For simplicity, we assume that the interface between the air and sand parts is located at $\{z = 0\}$. Let $\varepsilon_r(\mathbf{x}) \in C^1\left(\mathbb{R}^3\right)$ be a function that represents the spatially distributed dielectric constant. We assume that

$$\begin{cases} \varepsilon_r(\mathbf{x}) \geq 4, \text{ for } \{z < 0\} \cap \Omega, \\ \varepsilon_r(\mathbf{x}) = 1, \text{ for } \{z > 0\} \cap \Omega. \end{cases} \tag{2.2}$$

Physically, assumption (2.2) is reasonable. Cf. e.g. [**19**], we know that the dielectric constant of the air/vacuum is identically one. Meanwhile, for the other materials (for example, the dry sand which follows the non-invasive inspections of our interest), the dielectric constant is larger than the unity. As mentioned particularly in [**19, 20**] for identifying experimental fake landmines, the dielectric constant of the dry sand is 4, and the materials of landmines are commonly possessing very large dielectric constants. At least, $\varepsilon_r = 6$ is taken into account when a wood-based landmine is used to pose difficulties for detection services. Henceforth, our assumption on ε_r inside of the sandbox is reasonable.

Let $T > 0$ be the observation time. Practically, T is found by doubling the distance from the line of sources to the furthest point of the sandbox. For every source location $\mathbf{x}^0 \in L_j^{\rm src}$ (see (2.1)), we consider the following forward problem:

$$\begin{cases} \nu^{-2}(\mathbf{x})\, u_{tt} = \Delta u + F\left(\mathbf{x}, \mathbf{x}^0, t\right), \\ u\left(\mathbf{x}, \mathbf{x}^0, 0\right) = u_t\left(\mathbf{x}, \mathbf{x}^0, 0\right), \\ \left[\partial_n u\left(\mathbf{x}, \mathbf{x}^0, t\right) + \nu^{-1}(\mathbf{x})\, u\left(\mathbf{x}, \mathbf{x}^0, t\right)\right]\big|_{\partial\Omega} = 0, \\ \mathbf{x}^0 \in L_j^{\rm src}, \mathbf{x} \in \Omega, t \in (0, T), j = \overline{1, j_0}, \end{cases} \tag{2.3}$$

where $n = n(\mathbf{x}), \mathbf{x} \in \partial\Omega$ is the unit outward looking normal vector on $\partial\Omega$. In (2.3), $u(\mathbf{x}, \mathbf{x}^0, t)$ is the amplitude of a time-resolved component of the electric wave field. The function $\nu(\mathbf{x}) = c_0/\sqrt{\varepsilon_r(\mathbf{x})}$ represents the speed of light in the medium with c_0 is the speed of light in the vacuum. Here, the source function F is defined as

$$F\left(\mathbf{x}, \mathbf{x}^0, t\right) = \begin{cases} \mathrm{Re}\left(\chi_D\left(\mathbf{x}, \mathbf{x}^0\right) e^{-\mathrm{i}\omega_0 t} e^{-\mathrm{i}\alpha_0 (t - \tau_0/2)^2}\right) & \text{if } t \in (0, \tau_0], \mathbf{x}^0 \in L_j^{\rm src}, \\ 0 & \text{if } t > \tau_0, \end{cases} \tag{2.4}$$

where $\mathrm{i} = \sqrt{-1}$. This source function models well the linear modulated pulse, where ω_0 is the carrier frequency, α_0 is the chirp rate [**9**], and τ_0 is its duration.

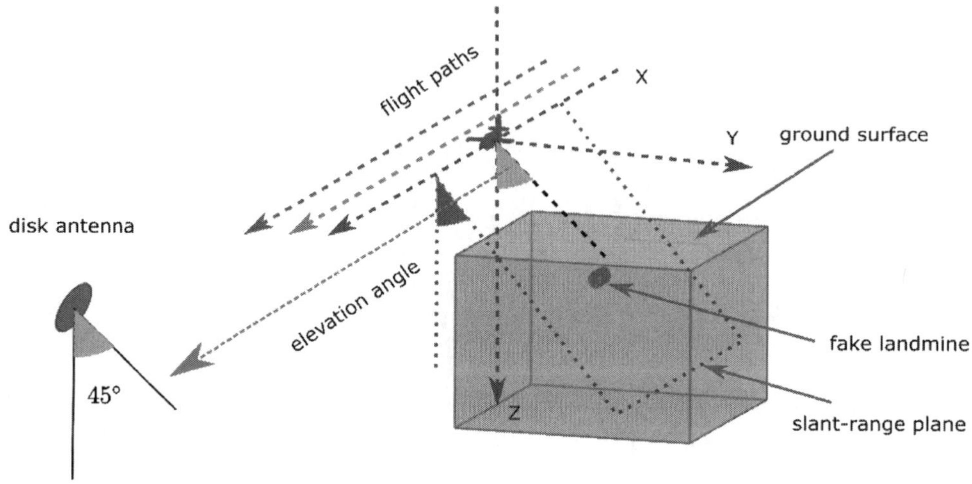

FIGURE 1. A schematic diagram of our SAR imaging setting. A flying device moves along multiple parallel straight lines in x-direction to collect the backscattering data. On this device, we assume that the transmitter and the detector have the same position. The transmitter is a disk-shaped antenna with an elevation angle of $\pi/4$ (yellow part). For our purpose of mimicking the landmine detection, we assume the unknown object is buried in a dry sand region. The notion of our convexification-based method is solving 1D CIPs for every source location. Under a certain elevation angle (red part), solutions of these 1-D CIPs allow us to approximately reconstruct a 3-D image.

The function χ_D in (2.4) represents the disk-shaped antenna of our interest with its radius $D > 0$, thickness $\widetilde{D} > 0$ and centered at a point $\mathbf{x} \in L_j^{\rm src}$. We rotate the disk via a coordinate transformation $\mathbf{x} \mapsto \mathbf{x}' = (x', y', z')$ defined by

$$\begin{aligned} x' &= x - x^0, \\ y' &= \cos(\theta)(y - y_j^0) - \sin(\theta)(z - z^0), \\ z' &= \sin(\theta)(y - y_j^0) + \cos(\theta)(z - z^0), \end{aligned} \tag{2.5}$$

where θ is the elevation angle. Hereby, the function χ_D is given by

$$\chi_D(\mathbf{x}, \mathbf{x}^0) = \begin{cases} 1 & \text{if } \sqrt{x'^2 + y'^2} < D \text{ and } |z'| \leq \widetilde{D}, \\ 0 & \text{otherwise.} \end{cases} \tag{2.6}$$

We consider the so-called absorbing boundary conditions in (2.3). Mathematically, the solution to the acoustic wave equation is sought on the whole space. As introduced above, the computational domain Ω is, however, finite and the boundary condition should be equipped with the underlying PDE. Moreover, at this point, numerical simulations show the non-physical back reflections of the waves incident at the boundaries. Therefore, the absorbing boundary conditions are used to minimize these spurious reflections; cf. e.g. [**36**] and references cited therein.

3. Statement of the inverse problem

Our application being in mind is to identify the material of the buried object. Therefore, our inverse problem aims to compute the spatial distribution of the dielectric constant $\varepsilon_r(\mathbf{x})$ involved in the coefficient $\nu(\mathbf{x})$ in the PDE of (2.3).

3.1. Statement of the problem. SAR Inverse Problem. *Suppose that functions $G_j(\mathbf{x}^0, t) = u(\mathbf{x}^0, \mathbf{x}^0, t)$ for $t \in (0, T)$, $\mathbf{x}^0 \in L_j^{\text{scr}}$ and $j = \overline{1, j_0}$ are given as our SAR data. Find a function $\varepsilon_r(\mathbf{x}) \in C^1(\mathbb{R}^3)$ satisfying conditions in* (2.2).

Cf. [**12, 21**], our strategy is to image the unknown function on the so-called slant-range plane, denoted by P. We define this plane in the following manner. For each antenna/source location $\mathbf{x}_j^{0,m} = (x_m^0, y_j^0, z^0) \in L_j^{\text{scr}}$, we define the central line of the antenna $CRL(\mathbf{x}_j^{0,m})$ that passes through the point $\mathbf{x}_j^{0,m}$, $m = 1, \ldots, N$, where N is the total number of locations of the source on the line L_j^{scr}. Each line $CRL(\mathbf{x}_j^{0,m})$ is orthogonal to the x-axis and has a certain angle with the plane $\{y = y_j^0\}$. Recall that the line L_j^{scr} is parallel to the $x-$axis, see (2.1). Denoted by θ, that angle is called the elevation angle presumably determined by the propagating direction of the radiated energy of the antenna, see (2.4)-(2.6). Henceforth, for each line L_j^{scr}, the slant-range plane P_j is defined as the plane passing through the central lines of antennas and the line of sources moving along L_j^{scr}. Hence, multiple lines of sources give multiple slant-range planes. As a by-product, the solution obtained on this plane is called the slant-range image $\widetilde{\varepsilon}_{r,j}(\mathbf{x})$ of the unknown dielectric constant $\varepsilon_r(\mathbf{x})$ on the slant-range plane P_j. To obtain an approximate image of the function $\varepsilon_r(\mathbf{x})$, we then combine slant-range images $\varepsilon_{r,j}(\mathbf{x})$ on planes P_j, $j = \overline{1, j_0}$. For clarity, we depict an illustration of our SAR imaging setting in Figure 1.

To obtain the image in the slant range plane P_j, we solve many 1-D Coefficient Inverse Problems (CIPs) along lines of antennas $CRL(\mathbf{x}_j^{0,m}), m = 1, \ldots, N$. Let $\widetilde{\varepsilon}_{r,j,m}(\mathbf{x})$ be the solution of this CIP along the line $CRL(\mathbf{x}_j^{0,m})$. Then the resulting 2-D function $\widetilde{\varepsilon}_{r,j}(\mathbf{x})$ is obtained by averaging of these functions over $m = 1, \ldots, N$, see details in section 5.

3.2. 1D Coefficient Inverse Problems. We now present our computational approach to calculate the slant-range dielectric constant $\widetilde{\varepsilon}_{r,j}(\mathbf{x})$ on each slant range plane P_j. For every $j = \overline{1, j_0}$, we consider the $m-$th source location $\mathbf{x}_j^{0,m}$ for $m = \overline{1, N}$. Observe that when introducing a pseudo variable $x \in \mathbb{R}$ and considering a smooth function $c(x)$ for our unknown function $\widetilde{\varepsilon}_r(\mathbf{x})$ along each central line $CRL(\mathbf{x}_j^{0,m})$ of the antenna passing through the source $\mathbf{x}_j^{0,m}$, we can scale to $x \in (0, 1)$ based upon the length of that central line. This way, we treat $c(x)$ as the solution of a 1-D CIP solved along each central line of the source $\mathbf{x}_j^{0,m}$.

By (2.2), for a number $\overline{c} \geq 4$ we consider the function $c(x) \in C^3(\mathbb{R})$ such that

$$c(x) = \begin{cases} \in [4, \overline{c}] & \text{for } x \in (0, 1), \\ 1 & \text{for } x \leq 0 \text{ and } x \geq 1. \end{cases} \tag{3.1}$$

Hereby, we consider the following forward problem:

$$\begin{cases} \frac{c(x)}{c_0^2} u_{tt} = u_{xx} & \text{for } x \in \mathbb{R}, t \in (0, T), \\ u(x, 0) = 0, \quad u_t(x, 0) = \delta(x) & \text{for } x \in \mathbb{R}. \end{cases} \tag{3.2}$$

In (3.2), the variable t is considered in nanosecond (ns), whilst the unit for x is meter (m). Physically, the speed of light in the vacuum $c_0 = 0.3$ (m/ns). Considering dimensionless variables X and τ,

$$X = x/(0.3\text{m}) \text{ and } \tau = t/\text{ns}, \tag{3.3}$$

we arrive at

$$u_{XX} = (0.3\text{m})^2 u_{xx}, \quad u_{\tau\tau} = (1\text{ns})^2 u_{tt}. \tag{3.4}$$

Thus, we obtain the dimensionless regime of the forward problem (3.2) becomes:

$$\begin{cases} \widetilde{c}(X) u_{\tau\tau} = u_{XX}, \quad \widetilde{c}(X) = c(0.3X), \\ u(X,0) = 0, u_\tau(X,0) = \delta(X). \end{cases} \tag{3.5}$$

The change of variables (3.3) means that $X = 1$ implies $x = 0.3$ (m) in the reality, and $\tau = 1$ indicates $t = 1$ (ns). We now state our coefficient inverse problem.

1-D Coefficient Inverse Problem (1-D CIP). *Assume that the following functions $g_0(\tau)$ and $g_1(\tau)$ are given:*

$$u(0,\tau) = g_0(\tau), \quad u_X(0,\tau) = g_1(\tau), \quad \tau \in (0,T). \tag{3.6}$$

Determine the function $\widetilde{c}(X)$ in (3.5) *such that the corresponding function $c(x) \in C^3(\mathbb{R})$ satisfies conditions* (3.1).

To solve this 1D CIP numerically, we employ a version of [**12, 21**] of the convexification method. First, we change variables as:

$$Y = Y(X) = \int_0^X \sqrt{\widetilde{c}(s)}ds. \tag{3.7}$$

Observe that $dY/dX = \sqrt{\widetilde{c}(X)} \geq 1$. Consider the following functions:

$$w(Y,\tau) = u(X(Y),\tau)\widetilde{c}^{1/4}(X(Y)), \quad Q(Y) = \widetilde{c}^{-1/4}(X(Y)),$$

$$p(Y) = \frac{Q''(Y)}{Q(Y)} - 2\left[\frac{Q'(Y)}{Q(Y)}\right]^2. \tag{3.8}$$

The function $p(Y)$ is smooth and $p(Y) = 0$ for $Y < 0$ and $Y > \sqrt{\overline{c}}$. Moreover, the following PDE holds with initial and boundary conditions holds:

$$w_{\tau\tau} = w_{YY} + p(Y)w \quad \text{for } Y \in \mathbb{R}, \tau \in \left(0, \widetilde{T}\right), \tag{3.9}$$

$$w(Y,0) = 0, \quad w_\tau(Y,0) = \delta(Y) \quad \text{for } Y \in \mathbb{R}, \tag{3.10}$$

$$w(0,\tau) = g_0(\tau), \quad w_Y(0,\tau) = g_1(\tau) \quad \text{for } \tau \in \left(0, \widetilde{T}\right), \tag{3.11}$$

where the number $\widetilde{T} \geq 2\sqrt{\overline{c}}$ depends on T.

REMARK 3.1. Cf. [**12, 21**] in this SAR inception, the smoothness of the unknown dielectric constant $c(x)$ should be $C^3(\mathbb{R})$. This assumption is used when we revisit our theorems below. In this regard, the function $p(Y)$ belongs to $C^1(\mathbb{R})$, see (3.8). The number $\widetilde{T} \geq 2\sqrt{\overline{c}}$ is conditioned because it guarantees the uniqueness of the CIP (3.9)–(3.11), see [**37**, Theorem 2.6 of Chapter 2]. As soon as the function $p(Y)$ is reconstructed, the original function $c(x)$ is reconstructed using (3.7)and (3.8) via the procedure described in [**22**, section 7.2].

From now onward, we follow a novel transformation commenced in [**22**] to obtain a nonlocal nonlinear PDE for (3.9)–(3.11). When doing so, we take an arbitrary $b \geq \sqrt{\overline{c}}$ and then take $\mu \in (0, 2\alpha b)$ for $\alpha \in (0, 1/2)$. We introduce the following 2D regions:

$$\mathcal{K} := \left\{ (Y, \tau) \in \mathbb{R}^2 : Y \in (0, b), \tau \in \left(0, \widetilde{T}\right) \right\}, \tag{3.12}$$

$$\mathcal{K}_{\alpha,\mu,b} := \left\{ (Y, \tau) \in \mathbb{R}^2 : Y + \alpha\tau < 2\alpha b - \mu \text{ for } Y, \tau > 0 \right\}. \tag{3.13}$$

Henceforth, we consider a new function $v(Y, \tau)$,

$$v(Y, \tau) = w(Y, \tau + Y) \in C^3(Y \geq 0, \tau \geq 0). \tag{3.14}$$

Using multi-variable chain rules, we obtain the following PDE for $v(Y, \tau)$ from (3.9):

$$v_{YY} - 2v_{Y\tau} + p(Y)\, v = 0 \quad \text{for } (Y, \tau) \in \mathcal{K}. \tag{3.15}$$

Furthermore, by (3.11), it yields $v(Y, 0) = 1/2$ for $Y \in (0, b)$, which then results in the fact that

$$p(Y) = 4v_{Y\tau}(Y, 0), \quad Y \in (0, b). \tag{3.16}$$

Differentiating both sides of (3.15) with respect to τ and setting

$$V(Y, \tau) = v_\tau(Y, \tau) \in C^2(\overline{\mathcal{K}}), \tag{3.17}$$

we obtain the following overdetermined boundary value problem for a nonlinear hyperbolic equation:

$$V_{YY} - 2V_{Y\tau} + 4V_Y(Y, 0)\, V = 0 \quad \text{for } (Y, \tau) \in \mathcal{K}, \tag{3.18}$$

$$V(0, \tau) = q_0(\tau) := g_0'(\tau) \quad \text{for } \tau \in \left(0, \widetilde{T}\right), \tag{3.19}$$

$$V_Y(0, \tau) = q_1(\tau) := g_0''(\tau) + g_1'(\tau) \quad \text{for } \tau \in \left(0, \widetilde{T}\right), \tag{3.20}$$

$$V_Y(b, \tau) = 0 \quad \text{for } \tau \in \left(0, \widetilde{T}\right). \tag{3.21}$$

Boundary condition (3.21) follows from the absorbing boundary condition, which was proven in [**38**] for our 1-D case.

Remark 3.2. Essentially, our heuristic computational approach allows us to solve the non-overdetermined 3D CIP via the solutions of many overdetermined boundary value problems. The transformations we have presented are necessary because we want to use a suitable Carleman estimate. This estimate is applied to deal with nonlinearities involved in the PDE operator. Frequently, the nonlinearity is terms are not avoidable when working on underdetermined. and non-overdetermined CIPs (cf. e.g. [**13, 17, 18**]).

In our SAR perspective, we observe that after computing $V(Y, \tau)$ for $(Y, \tau) \in \mathcal{K}$ from (3.18)–(3.21), we then obtain

$$p(Y) = 4V_Y(Y, 0), \quad Y \in (0, b), \tag{3.22}$$

by (3.16). Hence, the target unknown coefficient $c(x)$ of the original 1D CIP is sought after getting $Q(Y)$ in (3.8).

4. A version of the convexification method: theorems revisited

Consider the subspace $H_0^k(\mathcal{K}) \subset H^k(\mathcal{K})$ for $k \in \mathbb{N}^*$. In this work, we introduce

$$H_0^2(\mathcal{K}) := \left\{u \in H^2(\mathcal{K}) : u(0,\tau) = u_Y(0,\tau), u_Y(b,\tau) = 0\right\},$$
$$H_0^4(\mathcal{K}) := H^4(\mathcal{K}) \cap H_0^2(\mathcal{K}).$$

As mentioned above, several transformations lead us to an overdetermined BVP. Since this problem is nonlinear, then any conventional optimization-based approach, including the Tikhonov regularization and the least-square method may suffer from the phenomenon of local minima and ravines. This is the main reason which has originally prompted the corresponding author to develop the convexification approach [**23, 24**]. The main ingredient of the method and its variants is using an appropriate Carleman Weight Function to "convexify" the cost functional. The associated high-order regularization terms usually play the role in controlling nonlinear terms involved in the PDE operator. Thereby, one obtains a unique minimizer and then, the global convergence of the gradient-like minimization procedure is attained. Depending on the principal parts of differential operators, one can choose different Carleman Weight Functions, see, e.g. [**26**] for a variety of Carleman estimates for partial differential operators. In this work, we consider the following Carleman Weight Function for the linear operator $\partial_Y^2 - 2\partial_Y\partial_\tau$ (see (3.18)):

$$\psi_\lambda(Y,\tau) = e^{-2\lambda(Y+\alpha\tau)}, \quad \lambda \geq 1, \alpha \in (0,1/2).$$

Denoting

$$\mathcal{S}(V) = V_{YY} - 2V_{Y\tau} + 4V_Y(Y,0)V, \; (Y,\tau) \in \mathcal{K}, \tag{4.1}$$

we come up with the following weighted Tikhonov-like functional $J_{\lambda,\gamma} : H^4(\mathcal{K}) \to \mathbb{R}_+$:

$$J_{\lambda,\gamma}(V) = \int_{\mathcal{K}} [\mathcal{S}(V)]^2 \psi_\lambda dY d\tau + \gamma \|V\|^2_{H^4(\mathcal{K})}, \quad \gamma \in (0,1). \tag{4.2}$$

We choose the high-order regularization term in $H^4(\mathcal{K})$ because of the embedding $H^4(\mathcal{K}) \subset C^2(\overline{\mathcal{K}})$. This C^2 regularity is helpful in controlling our nonlinear term in (3.18); see [**22**, Theorem 2]. Following our convexification framework in, e.g., [**17, 18, 22**], we introduce the set $B(r, q_0, q_1)$,

(4.3)

$$B(r,q_0,q_1) = \left\{u \in H^4(\mathcal{K}) : u(0,\tau) = q_0(\tau), u_Y(0,\tau) = q_1(\tau), u_Y(b,\tau) = 0, \|u\|_{H^4(\mathcal{K})} < r\right\}.$$

in which $r > 0$ is an arbitrary number. However, in our computational practice we use the $H^2(\mathcal{K})$ −norm, see section 5.1.

Minimization Problem. *Minimize the cost functional $J_{\lambda,\gamma}(V)$ defined in (4.2) on the set $\overline{B(r,q_0,q_1)}$.*

We are now in the position to go over our analysis in [**22, 38**] of the convexification-based method for the nonlinear boundary value problem (3.18)–(3.21).

THEOREM 4.1 (Carleman estimate [**38**]). *There exist constants $C = C(\alpha, \mathcal{K}) > 0$ and $\lambda_0 = \lambda_0(\alpha, \mathcal{K}) \geq 1$ such that for all functions $u \in H_0^2(\mathcal{K})$ and all $\lambda \geq \lambda_0$ the* *following Carleman estimate holds true:*

$$\int\limits_{\mathcal{K}} (u_{YY} - 2u_{Y\tau})^2 \psi_\lambda dY d\tau \geq C \int\limits_{\mathcal{K}} \left(\lambda \left(u_Y^2 + u_\tau^2\right) + \lambda^3 u^2\right) \psi_\lambda dY d\tau$$

$$+ C \int\limits_0^b \left(\lambda u_Y^2 + \lambda^3 u^2\right)(Y, 0)\, e^{-2\lambda Y} dY - Ce^{-2\lambda \widetilde{T}} \int\limits_0^b \left(\lambda u_Y^2 + \lambda^3 u^2\right)\left(Y, \widetilde{T}\right) dY.$$

THEOREM 4.2 (Global strict convexity [**22**]). *For any $\lambda, \gamma > 0$ and functions $V \in \overline{B(r, q_0, q_1)}$, the cost functional $J_{\lambda,\gamma}$ defined in (4.2) has the Fréchet derivative $J'_{\lambda,\gamma}(V) \in H_0^4(\mathcal{K})$. Let $\lambda_0 = \lambda_0(\alpha, \mathcal{K}) \geq 1$ be the constant of Theorem 4.1 and let $\widetilde{T} \geq 2b$. Then there exists a sufficiently large number $\lambda_1 = \lambda_1(\mathcal{K}, r, \alpha) \geq \lambda_0$ such that for all $\lambda \geq \lambda_1$ and all $\gamma \in [2e^{-\lambda\alpha\widetilde{T}}, 1)$, the cost functional (4.2) is strictly convex on the set $\overline{B(r, q_0, q_1)}$. Moreover, for all $V_1, V_2 \in \overline{B(r, q_0, q_1)}$ the following estimate holds true:*

(4.4)

$$J_{\lambda,\gamma}(V_2) - J_{\lambda,\gamma}(V_1) - J'_{\lambda,\gamma}(V_1)(V_2 - V_1) \geq Ce^{-2\lambda(2\alpha b - \mu)} \|V_2 - V_1\|^2_{H^1(\mathcal{K}_{\alpha,\mu,b})}$$
$$+ Ce^{-2\lambda(2\alpha b - \mu)} \|V_2(y, 0) - V_1(y, 0)\|^2_{H^1(0, 2\alpha b - \mu)} + \frac{\gamma}{2} \|V_2 - V_1\|^2_{H^4(\mathcal{K})},$$

where the constant $C = C(\alpha, \mu, b, r) > 0$ depends only on listed parameters. Furthermore, the functional $J_{\lambda,\gamma}(V)$ has unique minimizer $V_{\min}$ on the set $\overline{B(r, q_0, q_1)}$ and

$$J'_{\lambda,\gamma}(V_{\min})(V - V_{\min}) \geq 0, \ \forall V \in \overline{B(r, q_0, q_1)}.$$

By the conventional regularization theory (cf. [**39**]), the existence of the ideal noiseless data $g_0^*(\tau), g_1^*(\tau)$ is assumed a priori. The existence of the corresponding exact coefficient $p^*(Y)$ and the exact function $w^*(Y, \tau)$ in (3.9)–(3.11) is assumed as well. Moreover, this function $p^*(Y)$ should also satisfy conditions $p^*(Y) = 0$ for $Y < 0$ and $Y > \sqrt{\bar{c}}$ as ones for $p(Y)$. Having the function $w^*(Y, \tau)$, we apply the above transformations (3.14), (3.17)to obtain the corresponding function $V^*(Y, \tau)$. Henceforth, it is natural to assume that functions q_0^* and q_1^* are the noiseless data q_0 and q_1 respectively; see (3.19) and (3.20). Thus, we assume below that the function $V^* \in B(r, q_0^*, q_1^*)$. Hence, it follows from (3.18)–(3.21) and (4.1) that

$$\mathcal{S}(V^*) = 0, \quad (Y, \tau) \in \mathcal{K}, \text{ and } p^*(Y) = 4v_Y^*(Y, 0), \quad Y \in (0, b).$$

Consider a sufficiently small number $\sigma \in (0, 1)$ that characterizes the noise level between the data q_0, q_1 and q_0^*, q_1^*. To obtain the zero boundary conditions at $\{x = 0\}$ in (3.19) and (3.20), consider two functions $F \in B(r, q_0, q_1)$ and $F^* \in B(r, q_0^*, q_1^*)$. We assume that

$$\|F - F^*\|_{H^4(\mathcal{K})} < \sigma. \tag{4.5}$$

Consider functions W, W^*,

$$W^* = V^* - F^*, W = V - V \in B_0(2r) := \left\{u \in H_0^4(\mathcal{K}) : \|u\|_{H^4(\mathcal{K})} < 2r\right\}.$$

Besides, the function $W + F \in B(3r, q_0, q_1)$, $\forall W \in B_0(2r)$. We modify the cost functional $J_{\lambda,\gamma}(V)$ as

$$\widetilde{J}_{\lambda,\gamma}(W) = J_{\lambda,\gamma}(W + F), \ \forall W \in B_0(2r).$$

It follows from Theorem 2 that the functional $\widetilde{J}_{\lambda,\gamma}$ is also globally strict convex on the ball $\overline{B_0(2r)}$ for $\lambda \geq \lambda_2 = \lambda_1(\mathcal{K}, 3r, \alpha) \geq \lambda_1(\mathcal{K}, r, \alpha)$. Also, by Theorem 2, for each value of the parameter $\lambda \geq \lambda_2$, the functional $\widetilde{J}_{\lambda,\gamma}(W)$ has a unique minimizer $W_{\min,\lambda,\gamma}$ on the set $\overline{B_0(2r)}$ and

$$\widetilde{J}'_{\lambda,\gamma}(W_{\min,\lambda,\gamma})(W - W_{\min,\lambda,\gamma}) \geq 0, \ \forall W \in \overline{B_0(2r)}. \tag{4.6}$$

It follows from the proof of Theorem 5 of [**22**] that inequality (4.6) plays an important role in the proof of Theorem 3.

THEOREM 4.3 (Accuracy estimate of the minimizer [**22**]). *Assume that* (4.5) *holds true and let* $\widetilde{T} \geq 4b$. *We choose*

$$\beta = \frac{\alpha\left(\widetilde{T} - 4b\right) + \mu}{2(2\alpha b - \mu)}, \quad \rho = \frac{1}{2}\min\{\beta, 1\}.$$

Let $\lambda_1 = \lambda_1(\mathcal{K}, r, \alpha)$ *by the number of obtained Theorem 4.2 and let* $\lambda_2 = \lambda_1(\mathcal{K}, 3r, \alpha) \geq \lambda_1(\mathcal{K}, r, \alpha)$. *Let* $\sigma_0 \in (0,1)$ *be a sufficiently small number such that* $\ln \sigma_0^{-1/(2(2\alpha b - \mu))} \geq \lambda_2$. *For any* $\sigma \in (0, \sigma_0)$, *we choose*

$$\lambda = \lambda(\sigma) = \ln \sigma^{-1/(2(2\alpha b - \mu))} > \lambda_2, \tag{4.7}$$

$$\gamma = \gamma(\sigma) = 2e^{-\lambda\alpha\widetilde{T}} = 2\sigma^{(\alpha\widetilde{T})/(2(2\alpha b - \mu))}. \tag{4.8}$$

and let the regularization parameter $\gamma \in [2e^{-\lambda\alpha\widetilde{T}}, 1)$. *Let* $V_{\min,\lambda,\gamma}(Y, \tau)$ *be the unique minimizer of the functional* $J_{\lambda,\gamma}(V)$ *on the set* $B(r, q_0, q_1)$. *Let the function* $p_{\min,\lambda,\gamma}(Y)$ *be defined as:* $p_{\min,\lambda,\gamma}(Y) = 4\partial_Y V_{\min,\lambda,\gamma}(Y, 0)$, *see* (3.22). *Then there exists a constant* $C = C(\mathcal{K}, \alpha, \mu, b, r) > 0$ *depending only on listed parameters such that*

$$\|V_{\min,\lambda,\gamma} - V^*\|_{H^1(\mathcal{K}_{\alpha,\mu,b})} \leq C\sigma^\rho, \tag{4.9}$$

$$\|p_{\min,\lambda,\gamma} - p^*\|_{L^2(0,2\alpha b - \mu)} \leq C\sigma^\rho. \tag{4.10}$$

Now, we state a theorem about the global convergence of the gradient descent method of the minimization of the functional $J_{\lambda,\gamma}(V)$. Let $\omega \in (0,1)$ be the step size of this method. Fix an arbitrary number $\vartheta \in (0, 1/3)$. We restrict the starting point, denoted by $V^{(0)}$, to be an arbitrary point in the set $B(\vartheta r, q_0, q_1) \subset B(r, q_0, q_1)$. Then the gradient descent method reads as:

$$V^{(n)} = V^{(n-1)} - \omega J'_{\lambda,\gamma}\left(V^{(n-1)}\right), \quad n = 1, 2, \ldots, \tag{4.11}$$

where $V^{(n)}$ approaches the minimizer $V_{\min,\lambda,\gamma}$ as n is large. This scheme is well-defined because $J'_{\lambda,\gamma}(V^{(n-1)}) \in H_0^4(\mathcal{K})$; see Theorem 4.2. Hence, functions $V^{(n)}$ satisfy the same boundary conditions as in (4.3) for all n.

Below, we assume that our minimizer $V_{\min,\lambda,\gamma}$ belongs to the set $B(\vartheta r, q_0, q_1)$, which is an interior point of the set $B(r, q_0, q_1)$. This enabled us in [**21**] to prove the global convergence of the gradient descent scheme (4.11). This assumption is plausible because by (4.9)the distance between the ideal solution V^* and the minimizer $V_{\min,\lambda,\gamma}$ is small with respect to the noise level $\sigma \in (0,1)$, although only in the $H^1(\mathcal{K}_{\alpha,\mu,b})$ –norm rather than in the stronger norm $H^4(\mathcal{K})$, see (4.3).

THEOREM 4.4 (Global convergence of the gradient descent method [**21**]). *Suppose that the parameters* λ *and* γ *are taken as in Theorem 4.2. Then there exists* *a sufficiently small constant* $w_0 \in (0,1)$ *such that the sequence of the gradient descent method* $\left\{V^{(n)}\right\}_{n=0}^{\infty} \subset B\left(r, q_0, q_1\right)$ *for every* $\omega \in (0, \omega_0)$. *Moreover, for every* $\omega \in (0, \omega_0)$ *there exists a number* $\theta = \theta(\omega) \in (0,1)$ *such that*

$$\left\|V_{\min,\lambda,\gamma} - V^{(n)}\right\|_{H^4(\mathcal{K})} \leq \theta^n \left\|V_{\min,\lambda,\gamma} - V^{(0)}\right\|_{H^4(\mathcal{K})}, \quad n = 1, 2, \ldots$$

Combining Theorems 4.3 and 4.4, we can prove that the functions $V^{(n)}$ converge to the ideal solution V^* as long as the noise level σ in the data tends to zero. In addition, we obtain the convergence of the corresponding sequence $\left\{p^{(n)}\right\}_{n=0}^{\infty}$ towards the ideal function p^*. It works via the expression $p^{(n)}(Y) = 4\partial_Y V_Y^{(n)}(Y, 0)$ in (3.22).

THEOREM 4.5 ([**21**]). *Suppose all conditions of Theorems 4.3 and 4.4 hold. Then the following convergence estimate is valid:*

$$\begin{aligned}\left\|V^* - V^{(n)}\right\|_{H^1(\mathcal{K}_{\alpha,\mu,b})} &+ \left\|p^* - p^{(n)}\right\|_{L^2(0,2\alpha b-\mu)} \\ &\leq C\sigma^{\rho} + C\theta^n \left\|V_{\min,\lambda,\gamma} - V^{(0)}\right\|_{H^4(\mathcal{K})}, \quad n = 1, 2, \ldots,\end{aligned}$$

where the constant $C = C(\mathcal{K}, \alpha, \mu, b, r) > 0$ *depends only on listed parameters.*

It is now worth mentioning that in our convergence results above, r is arbitrary and the starting point $V^{(0)}$ is arbitrary in $B(\vartheta r, q_0, q_1)$ with ϑ being fixed in $(0, 1/3)$. Therefore, our numerical method for solving the underlying nonlinear inverse problem does not require any advanced knowledge of a small neighborhood of the true solution. In this sense, the method is globally convergent, see Definition 1 in Introduction.

REMARK 4.6. Even though the values of λ in our CWF should be large in our theory, our rich computational experience working with the convexification tells us that we can choose λ between 1 and 3. For this We refer the reader to our previous publications on the convexification; see [**17**–**22**, **40**, **41**] some other references cited therein. In the present work, we choose $\lambda = 1.05$, and it works well.

5. Numerical studies

To generate the data for the inverse problem, we have solved the forward problem (2.3)-(2.6) by the standard implicit scheme. Therefore, we describe in this section our numerical solution of only the inverse problem.

It is worth noting that we suppose to have a buried object in the sand region $\{z < 0\} \cap \Omega$ of our computational domain Ω, see (2.2). The dielectric constant of this object is assumed to be different from that of the dry sand. Hence, the sand region is actually heterogeneous. The solution of problem (2.3) includes the signal from the sand, which causes a significant challenge when working on the inversion. In this paper, we follow the heuristic approach commenced in works of this research group with experimentally collected data [**17**, **19**] to get rid from the sand signals. To do so, for each source location, we consider $u_{\text{ref}}(\mathbf{x}, t)$ as the reference data. The reference data are generated for the case when an inclusion is not present in the simulated sandbox. This means that we take $\varepsilon_r(\mathbf{x}) = 4$ for $z \leq 0$ and $\varepsilon_r(\mathbf{x}) = 1$ for $z > 0$. Indeed, the dielectric constant of the dry sand equals 4 (cf. [**19**, **20**]). We remark that we only need to generate these reference data once for any numerical examples. Let $u^*(\mathbf{x},t)$ be the simulated data in the case when an inclusion is present in the sandbox. Then, our target data after filtering out the sand signals are defined by

$$u_{\text{sc}}(\mathbf{x},t) = u^*(\mathbf{x},t) - u_{\text{ref}}(\mathbf{x},t).$$

And this is the function we work with to solve our inverse problem.

5.1. Parameters. In this part, we show numerically how our proposed method works to create 3-D images from SAR data taken on three lines $L_j^{\text{scr}}, j = 1,2,3$. Experimentally, the data collection is very expensive. Therefore, it is pertinent to use only three lines in these numerical studies. This prepares a playground to work with experimental SAR data in the next publication. Different from [**12**] in which the Lippmann–Schwinger equation was used to generate the data, our SAR data are computed by the finite difference solver of section 5.1. Cf. [**12, 21**], we use the standard finite difference approximation to compute the minimizer of $J_{\lambda,\gamma}$ defined in (4.2). In this regard, the discrete operators are formulated to approximate the operator $\mathcal{S}$. Since the integration in $J_{\lambda,\gamma}$ is over the two-dimensional rectangle $\mathcal{K}$, we introduce its equidistant step size as (Δ_Y, Δ_τ) for the grid points (Y,τ) in that domain. Then the discrete solution is sought by the minimization of the corresponding discrete cost functional with respect to the values of the function V at grid points. As in [**12**], we only consider the regularization term in the $H^2(\mathcal{K})-$norm for the numerical inversion; compared with the $H^4(\mathcal{K})-$norm in the theory. This reduces the complexity of computations and thus, saves the elapsed time.

REMARK 5.1. Below we work only with dimensionless variables, also, see (3.3). In this context, 1 in the dimensionless spatial variable is 0.3 (meter) in reality, and 1 in the dimensionless time variable is 1 (nanosecond).

As to our forward problem (2.3)-(2.6), we consider $\Omega = (-2,2)^3$. Essentially, our domain of interest is $(-0.6 \text{ m}, 0.6 \text{ m})^3$ that corresponds to a scanning ground region of $|0.6-(-0.6)|^3/2 = 0.864$ (m^3). For L_j^{src}, we take $j \in \overline{1,j_0}$ with $j_0 = 3$ lines of sources with the increment 0.05 (1.5 cm), which is the distance between 3 lines of source. The central source position of our entire configuration is located at $x^0 = 0, y_2^0 = -1.7, z^0 = 0.8$. With the increment 0.05, our source y-locations are $y_1^0 = -1.75, y_2^0 = -1.7, y_3^0 = -1.65$, respectively. The length of each of these lines equals to 1.4 (0.42 m), i.e. $L = 0.7$, and every source line has 28 source positions, i.e. $M = 28$. Our circular disk antenna is taken with the diameter $D = 0.1$ (3 cm) and its associated parameters are chosen by $\tau_0 = 1$ (ns), $\omega_0 = 6\pi \times 10^8$ (Hz), $\alpha_0 = 8\pi \times 10^6/\tau_0$ and $\widetilde{D} = h_{\mathbf{x}} = |\Omega|/(N_{\mathbf{x}}-1)$ with $N_{\mathbf{x}} = 81$. As a result, the carrier wavelength is $\widetilde{\lambda}_0 = 2\pi c_0/\omega_0 = 1$ (m), and the Fresnel number $\text{Fr} \approx D^2/(\widetilde{\lambda}_0 |z^0|) = 1.25 \times 10^{-3} \ll 1$ guarantees well the far field zone; see Fraunhofer condition in [**44**, section 7.1.3]. The refinement $N_{\mathbf{x}}$ is chosen because $h_{\mathbf{x}} = 0.05$ is compatible with an increment of 1.5 (cm).

For our numerical results, the parameters were chosen as

$$\lambda = 1.05, \alpha = 0.49, \gamma = 10^{-10}$$

and the discretization $(\Delta_Y, \Delta_\tau) = (8 \times 10^{-3}, 8 \times 10^{-3})$. As mentioned above, even though our theory is valid for large λ, we have observed numerically in all our previous publications on the convexification that $\lambda \in [1,3]$ works well; see our previous numerical results in, e.g., [**17–22**] and [**40, 41**]. The elevation angle is

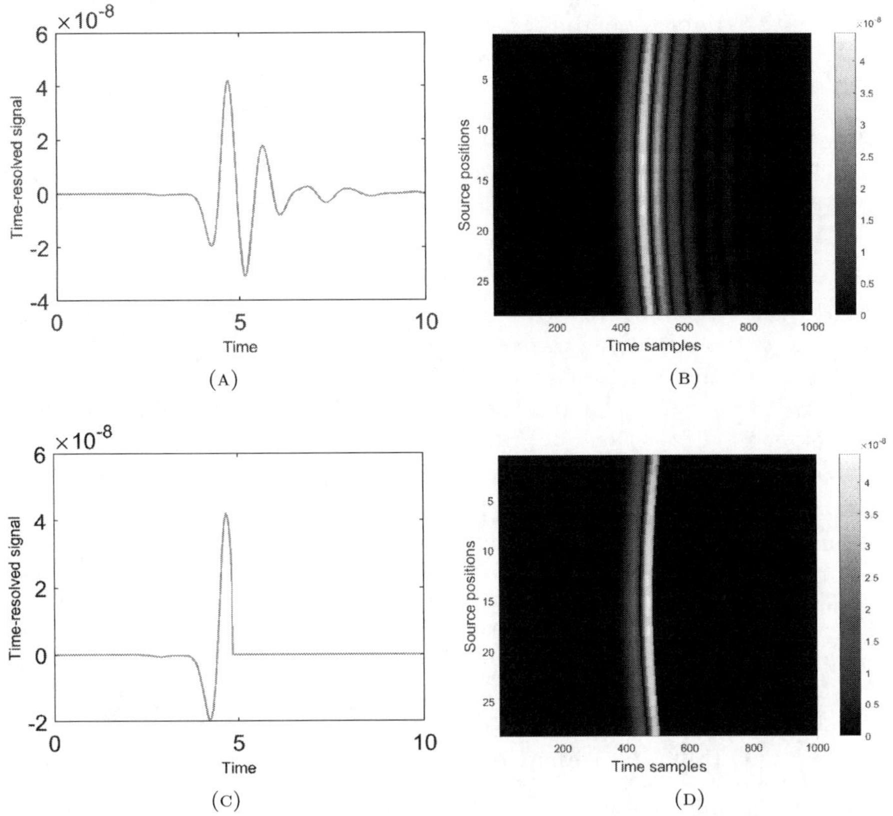

FIGURE 2. A typical sample of the time-dependent computationally simulated data for our Model 1. This is dimensionless time $\tau = t/\mathrm{ns}$, where t is time in nanoseconds (ns), see (3.3). (a) The simulated data at the central source position $x^0 = 0, y_2^0 = -1.7$ before truncation with $T = 10$. (b) The corresponding simulated data with all source positions and 1000 time samples for the time domain. (c) The simulated data after truncation. (d) The corresponding truncated data with all source positions and 1000 time samples.

taken by $\theta = \pi/4$. Similar to our first work on SAR [**12**], we multiply the simulated delay-and-sum data by a calibration factor CF. Here, we choose $CF = 1.75 \times 10^{15}$. Since our method relies on solving 1D problems, which is an approximate model, we then have no choice but to use that calibration factor.

5.2. Reconstruction results. We consider two models for our numerical testing. In both cases the objects are centered at the point $(0, 0, -0.14)$.

- **Model 1**: An ellipsoid with principal semi-major axis and two semi-minor axes, respectively, being 0.2 (6 cm), 0.12 (3.6 cm) and 0.04 (1.2 cm) in x, y and z directions respectively. This ellipsoid has the dielectric constant $\varepsilon_{\text{true}}\,(\text{ellip}) = 15$.

- **Model 2**: A rectangular prism with dimensions $0.18 \times 0.08 \times 0.08$ (L×W×H) corresponding to $5.4 \times 2.4 \times 2.4$ in cm^3. It has the dielectric constant $\varepsilon_{\text{true}}\,(\text{prism}) = 23.8$.

REMARK 5.2. We recall in the context of (3.4) that 1 in the dimensionless spatial variable is 0.3 (m) in reality. Therefore, the top point of the surface of our simulated objects is close to the air/ground interface $\{z = 0\}$ within $|-0.14 + 0.04| \times 30 = 3$ (cm) of the burial depth, which is typical in de-mining operations; cf. e.g. [**42**].

These two examples model well the so-called "appearing" dielectric constant of metallic objects we experimented with in [**43**]. It was established numerically in [**43**] that the range of the appearing dielectric constant of metals is $[10, 30]$. Moreover, this range includes the dielectric constant 23.8 of the experimental watered glass bottle in [**21**]. Both metallic and glass bodied land mines are frequently used in the battlefields.

Suppose that our SAR data are $G(\mathbf{x}_j^{0,m}, t)$ for each $j = \overline{1, j_0}$ and $m = \overline{1, N}$. Recall that $\mathbf{x}_j^{0,m} = \left(x_m^0, y_j^0, z^0\right) \in L_j^{\text{scr}}$ is the source location number m on the line of sources L_j^{scr} number j. Prior of getting the data g_0 and g_1 in (3.6), we perform a data preprocessing procedure to filter out unwanted artifacts observed in the time-resolved signals. First, we eliminate small "peaks" of the signals by setting:

(5.1)
$$\widetilde{G}\left(\mathbf{x}_j^{0,m}, t\right) = \begin{cases} 0 & \text{if } \left|G\left(\mathbf{x}_j^{0,m}, t\right)\right| < 0.1 \max_{[0,T]\times[1,M]} \left|G\left(\mathbf{x}_j^{0,m}, t\right)\right|, \\ G\left(\mathbf{x}_j^{0,m}, t\right) & \text{otherwise.} \end{cases}$$

Second, we use the built-in function `findpeaks` in MATLAB to count the number of the peaks. Then, we deliberately keep only two first peaks of the signals because we believe that those are the most important ones for the object. The resulting signals are denoted by $\widehat{G}(\mathbf{x}_j^{0,m}, t)$. We continue our preprocessing procedure by applying the delay-and-sum technique to $\widehat{G}(\mathbf{x}_j^{0,m}, t)$ as in [**12**]. For brevity, we do not detail internal steps of this data preprocessing procedure, but refer to [**12**, section III-A].

Now we focus on the preprocessing on the delay-and-sum data. In this context, for each line L_j^{scr}, $j = 1, 2, 3$ of source, the data are along the slant range planes which we reconstruct. Let P_j be the slant range plane which corresponds to the line of sources $L_j^{\text{scr}}, j = 1, 2, 3$. In this plane P_j, we, slightly abusing the same notations, consider x_s as the source number and y_s as the variable for the axis orthogonal to the line of source L_j^{scr}. Those y_s are actually the transformation of t using time of arrival. Thereby, we denote the delay-and-sum data by $\text{SAD}_j(x_s, y_s)$. We now introduce a truncation to control the size of the computed object. The "heuristically" truncated delay-and-sum data are given by

(5.2)
$$\text{SAD}_j^{\text{tr}}(x_s, y_s) = \begin{cases} \text{SAD}_j(x_s, y_s), & \text{if } \text{SAD}_j(x_s, y_s) \geq 0.95 \max_{x_s, y_s} |\text{SAD}_j(x_s, y_s)|, \\ 0, & \text{otherwise.} \end{cases}$$

This way, we use only the absolute values as the data g_0 for each 1D inverse problem along the central line of the antenna. We also remark that the argument of [**21**, Remark 4] implies that the Neumann data g_1 equals to the derivative of g_0, i.e. $g_1 = g_0'$. Our data preprocessing is exemplified in Figures 2a–2d.

To illustrate how our forward model works, consider Model 1 as an example. This object is centered at $(0, 0, -0.14)$. The central point of the source line $y_2^0 = -1.7$ is located at $(0, -1.7, 0.8)$. Therefore, the distance between these two points is $\sqrt{(-1.7)^2 + (0.8 + 0.14)^2} \approx 1.94$. Hence, the backscattering time-resolved signal should bump at the point $1.94 \times 2 = 3.88$ of the time domain. This is approximately the point we observe in Figure 2a. More precisely, our approximate signal bumps at the point $\tau = 4.13$, which is 10% difference with $\tau = 3.88$. But given that we have the whole ellipsoid rather than just its central point, this difference is acceptable for our simulations. This confirms how reliable our forward solver is. We can stop looking for the wave after the object's bumps, which is after around the point $\tau = 5$; see again Figure 2a. However, we enlarge the time domain up to the point $\tau = T = 10$ to avoid any possible boundary reflections of the wave propagation. Moreover, we take into account 1000 time samples in Figure 2b to identify well the location of every wave with respect to t.

For each line L_j^{scr}, $j = 1, 2, 3$ of sources, we use the truncation (5.2) to form a rectangular area $(x_s, y_s) \in [s_1^j, s_2^j] \times [l_1^j, l_2^j] \subset P_j$. As mentioned above when considering variables x_s, y_s, this rectangle is used in the slant range plane involving the transformation of τ in y using times of arrival.

REMARK 5.3. The selection of this rectangular region explains the reason why our computed objects in Figures 3b, 3c and Figures 4b, 4c have the rectangle-like shape. We are not focusing our work on the shape of the object since it is very challenging. Rather, we are interested in the accuracy of the computed dielectric constant $\widetilde{\varepsilon}_r$ of our two models. We are also interested in the dimensions of the computed object, and the presence of the above-mentioned rectangular region is helpful in controlling those dimensions.

Given a slant range plane P_j, we solve N 1-D CIPs for N sources $\mathbf{x}_j^{0,m} = (x_m^0, y_j^0, z^0) \in L_j^{\text{scr}}, m = 1, \ldots, N$. Let the corresponding discrete solutions be discrete functions defined in P_j, $\widetilde{\varepsilon}_{1,j}(x_m^0, y_s)$. Note again that slightly abusing notation, the variable y_s here is the grid point in P_j. To reduce the artifacts, we average all the 1D solutions in our rectangular area $[s_1^j, s_2^j] \times [l_1^j, l_2^j] \subset P_j$. To do so, for each j, we define first the semi-discrete function $\widetilde{\varepsilon}_{2,j}(x_m^0, y)$ for $x_m^0 \in [s_1^j, s_2^j], y \in [l_1^j, l_2^j]$ as

$$\widetilde{\varepsilon}_{2,j}\left(x_m^0, y\right) = \begin{cases} 30, & \text{if } \max_{y_s \in [l_1^j, l_2^j]}\left(\left|\widetilde{\varepsilon}_1\left(x_m^0, y_s\right)\right|\right) > 500, \\ \max_{y_s \in [l_1^j, l_2^j]}\left(\left|\widetilde{\varepsilon}_1\left(x_m^0, y_s\right)\right|\right), & \text{otherwise,} \end{cases} \tag{5.3}$$

since our domain of interest of the dielectric constant is [10,30]. Thus, for each source line of sources L_j^{scr}, our computed slant-range solution in P_j, denoted by $\widetilde{\varepsilon}_{r,j}(x_m^0, y_s)$, is computed by

$$\widetilde{\varepsilon}_{r,j}\left(x_m^0, y_s\right) = \begin{cases} \left|s_2 - s_1\right|^{-1} \sum_{x_m^0 \in [s_1^j, s_2^j]} \widetilde{\varepsilon}_{2,j}\left(x_m^0, y\right) & \text{in } \left[s_1^j, s_2^j\right] \times \left[l_1^j, l_2^j\right], \\ 1 & \text{elsewhere.} \end{cases} \tag{5.4}$$

We define our computed slant-range solution in each slant range plane P_j by (5.4) because from the above preprocessing data, we expect our inclusion in only the rectangle area. Therefore, it is relevant that the dielectric constant outside of that area should be unity. Finally, having all computed slant-range solutions $\widetilde{\varepsilon}_{r,1}, \widetilde{\varepsilon}_{r,2}, \widetilde{\varepsilon}_{r,3}$, we assign one value of the final dielectric constant $\widetilde{\varepsilon}_r$ of our object by averaging the maximum values of those solutions. In particular, let $\widetilde{\varepsilon}_{\max,j} = \max_{(x_m^0, y_s) \in P_j} \widetilde{\varepsilon}_{r,j}(x_m^0, y_s)$ and let $\mathrm{SL} \subset \mathbb{R}^3$ be the prism formed by 3 slant-range planes P_j. We then compute

$$\widetilde{\varepsilon}_r(x,y,z) = \begin{cases} \frac{1}{3}\left(\widetilde{\varepsilon}_{\max,1} + \widetilde{\varepsilon}_{\max,2} + \widetilde{\varepsilon}_{\max,3}\right) & \text{in } \Omega \cap \mathrm{SL}, \\ 1 & \text{elsewhere}. \end{cases} \tag{5.5}$$

In addition, we do the linear interpolation with respect to j to help improve the resolution of images.

5.3. Computational results. We now first bring in our results for the case of noiseless data and then for the case of noisy data.

5.3.1. *Noiseless data.* In the case of Model 1, our computed dielectric constant for the ellipsoid $\widetilde{\varepsilon}_r(\text{ellip}) = 15.06$, compared with the true value $\varepsilon_{\text{true}}(\text{ellip}) = 15$. For Model 2, the computed dielectric constant for the rectangular prism is $\widetilde{\varepsilon}_r(\text{prism}) = 22.97$, compared with the true value $\varepsilon_{\text{true}}(\text{prism}) = 23.8$. Therefore, our approximations of the dielectric constants are quite accurate.

We now comment on dimensions of computed images. Recall that by Remark 5.3 our images have only rectangular shapes. One can see in Figures 3a and 3b that the dimensions of the computed ellipsoid of Model 1 are about $0.55 \times 0.04 \times 0.08$ ($\mathrm{L} \times \mathrm{W} \times \mathrm{H}$). On the other hand, dimensions of the true ellipsoid are $0.4 \times 0.24 \times 0.08$ ($\mathrm{L} \times \mathrm{W} \times \mathrm{H}$). This means that the computed object is longer in length, but smaller in width and is the same in height. Similarly, for Model 2, one can see in Figures 4a and 4b that the dimensions of the computed prism are $0.5 \times 0.03 \times 0.03$ ($\mathrm{L} \times \mathrm{W} \times \mathrm{H}$), while dimensions of the true prism are $0.36 \times 0.16 \times 0.16$ ($\mathrm{L} \times \mathrm{W} \times \mathrm{H}$). Hence, the computed prism is longer in length but smaller in width and height than the true prism.

We remark that the `isovalue` for the presentation of the dielectric constants (as well as sizes of imaged targets) in Figures 3b and 4b equals to $\varepsilon_{\text{true}} - 1$ because we want to see how close our approximate dielectric constant to the true one is.

Consider now Figures 3c and 4c in which we have the 2D cross-sections of the computed images by the plane $\{y = 0\}$, the location is not so accurate as we have 10% difference in location from our forward solver. We hope to improve these in our future work.

5.3.2. *Noisy data.* While the above results are for the case of the noiseless data, we now introduce a noise in the data. We add a random additive noise to the truncated simulated data $\widetilde{G}\left(\mathbf{x}_j^{0,m}, t\right)$ (see (5.1)). In this regard, for $j = 1,2,3$ and $m = 1, \ldots, N$, we introduce

$$\widetilde{G}_{\text{noise}}\left(\mathbf{x}_j^{0,m}, t\right) = \widetilde{G}\left(\mathbf{x}_j^{0,m}, t\right) + \sigma \text{rand}\left(\mathbf{x}_j^{0,m}, t\right) \max_{\mathbf{x}_j^{0,m}, t} \left|\widetilde{G}\left(\mathbf{x}_j^{0,m}, t\right)\right|. \tag{5.6}$$

Here, $\sigma \in (0,1)$ represents the noise level and "rand" is a random number uniformly distributed in the interval $(-1,1)$. We use now the noisy data (5.6) for both Models 1 and 2 with $\sigma = 0.05$, which is 5% of noise.

The resulting computed dielectric constant for the ellipsoid is $\widetilde{\varepsilon}_r(\text{ellip}) = 13.52$, compared with the true value $\varepsilon_{\text{true}}(\text{ellip}) = 15$. For Model 2, our computed dielectric constant for the rectangular prism is $\widetilde{\varepsilon}_r(\text{prism}) = 22.08$, compared with the true value $\varepsilon_{\text{true}}(\text{prism}) = 23.8$. In both models images for noisy data are about the same as the ones on Figures 3 and 4 for noiseless data.

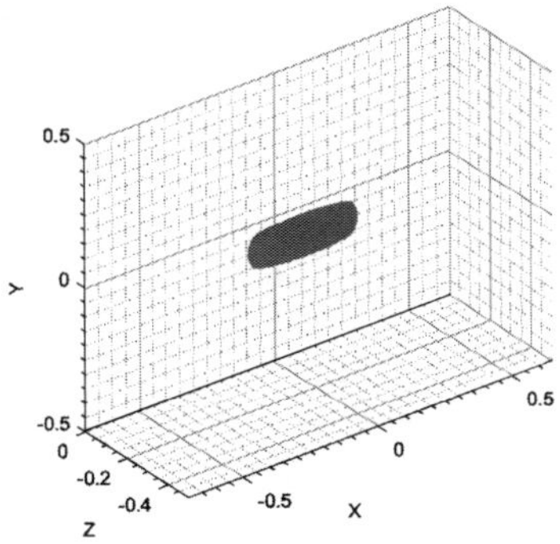

(A) True

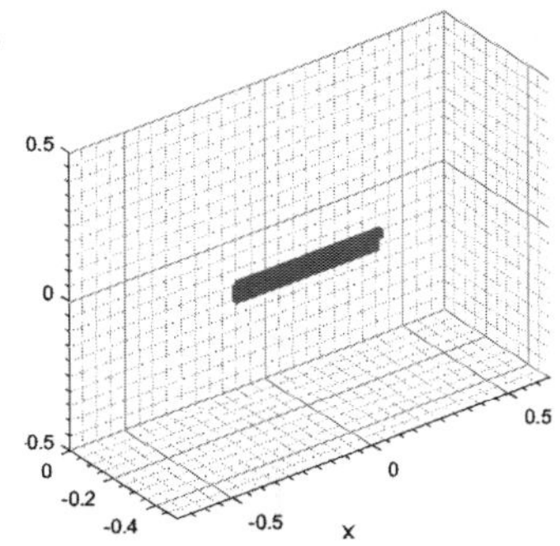

(B) Computed

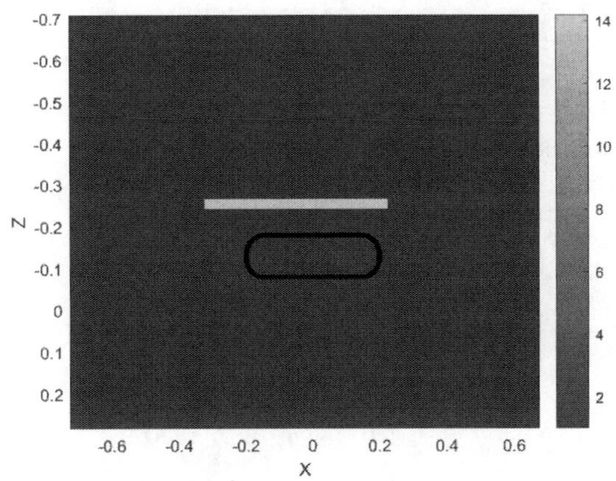

(C) Reconstructed slant-range image

FIGURE 3. Our 3-D reconstruction of the ellipsoid of Model 1 with the true dielectric constant $\varepsilon_{\text{true}}(\text{ellip}) = 15$. This is the case of noiseless data. The ellipsoid is buried in a sandbox with the dielectric constant $\varepsilon(\text{sand}) = 4$. The data are taken in the air with the dielectric constant $\varepsilon(\text{air}) = 1$. (a) True 3D image with dimensions $0.4 \times 0.08 \times 0.12$ (L×W×H). (b) Reconstructed image of the prism with $\widetilde{\varepsilon}_{\text{comp}}(\text{ellip}) = 15.06$. The `isovalue` 14 is used. Sizes of the computed ellipsoid are $0.55 \times 0.04 \times 0.08$ (L×W×H). Hence, the computed object is longer in length, but smaller in width and is the same in height, as compared with the true ellipsoid. (c) The cross-section of the 3-D image by the plane $\{y = 0\}$ of the object superimposed with the true one (solid curve). The image with 5% noise in the data (see (5.6)) is about the same. The computed dielectric constant in the case of noisy data is $\widetilde{\varepsilon}_{\text{comp}}$ (ellip) $= 13.52$.

6. Concluding remarks

In summary, we have numerically tested the convexification-based nonlinear SAR (CONSAR) imaging on identifying images of the dielectric constant of buried

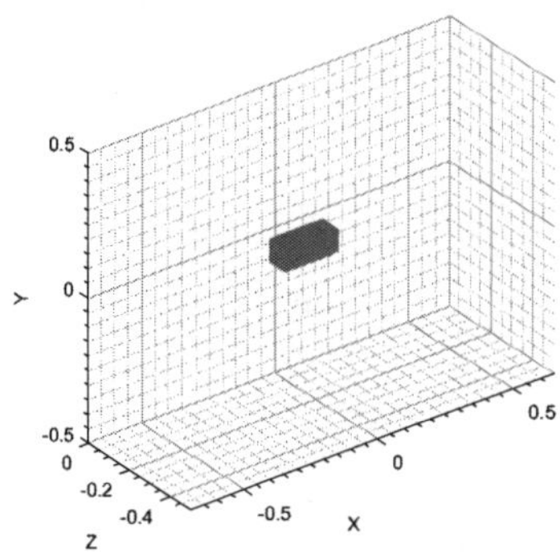

(A) True

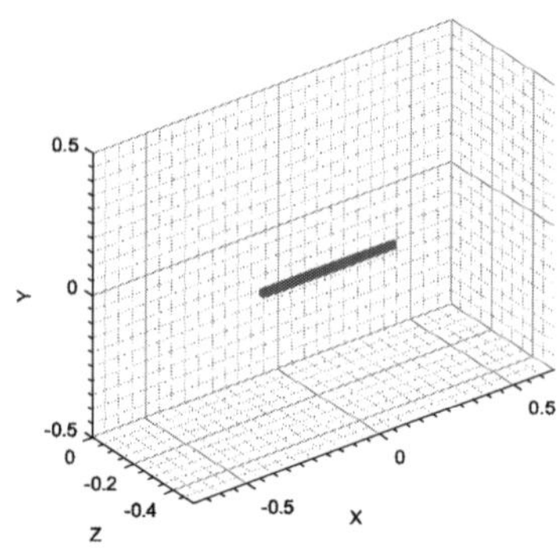

(B) Computed

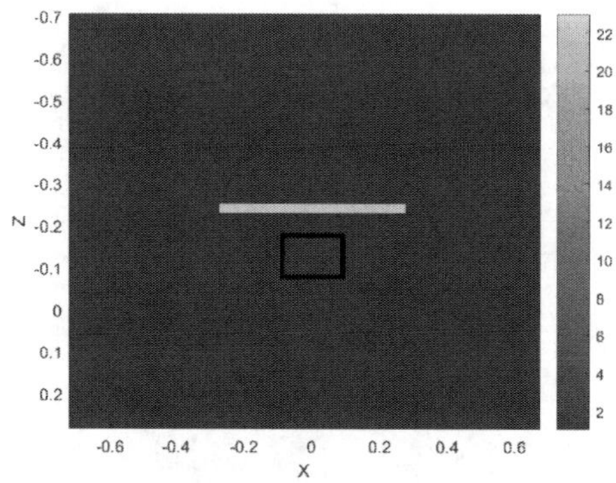

(C) Reconstructed slant-range image

FIGURE 4. Our 3-D reconstruction of the rectangular prism of Model 2 with the true dielectric constant $\varepsilon_{\text{true}}(\text{prism}) = 23.8$. This is the case of noiseless data. The prism is buried in a sandbox with the dielectric constant $\varepsilon(\text{sand}) = 4$. The data are taken in the air with the dielectric constant $\varepsilon(\text{air}) = 1$. (a) True 3D image with dimensions $0.18 \times 0.08 \times 0.08$ (L×W×H). (b) Reconstructed image of the prism with $\widetilde{\varepsilon}_{\text{comp}}(\text{prism}) = 22.97$. The `isovalue` 22.5 is used. Sizes of the computed prism are $0.5 \times 0.03 \times 0.03$ (L×W×H). Hence, the computed prism is longer in length but smaller in width and height than the true prism. (c) The cross-section of the 3-D image by the plane $\{y = 0\}$ of the object superimposed with the true one (solid curve). The image with 5% noise in the data (see (5.6)) is about the same. The computed dielectric constant in the case of noisy data is $\widetilde{\varepsilon}_{\text{comp}}(\text{prism}) = 22.08$.

objects in a three-dimensional setting. We have numerically observed provides accurate values of dielectric constants of targets. However, the computed locations and sizes of tested targets are not as accurate as the convexification usually provides for conventional CIPs, see, e.g. the works of this research group for applications oof the convexification to experimental data [**19, 20**]. We hope to address these questions in follow up publications.

Acknowledgment

Vo Anh Khoa acknowledges Professor Taufiquar Khan (University of North Carolina at Charlotte) for the moral support of his research career.

References

[1] M. Amin, Through-the-wall Radar Imaging, CRC Press, Boca Raton, FL, 2011.

[2] S. A. Carn, Application of synthetic aperture radar (SAR) imagery to volcano mapping in the humid tropics: a case study in East J ava, Indonesia, Bulletin of Volcanology 61 (1-2) (1999) 92–105. doi:10.1007/s004450050265.

[3] G. Chavent, *Nonlinear least squares for inverse problems*, Scientific Computation, Springer, New York, 2009. Theoretical foundations and step-by-step guide for applications. MR2554448

[4] Alexander V. Goncharsky and Sergey Y. Romanov, *Iterative methods for solving coefficient inverse problems of wave tomography in models with attenuation*, Inverse Problems **33** (2017), no. 2, 025003, 24, DOI 10.1088/1361-6420/33/2/025003. MR3626795

[5] Alexander V. Goncharsky and Sergey Y. Romanov, *A method of solving the coefficient inverse problems of wave tomography*, Comput. Math. Appl. **77** (2019), no. 4, 967–980, DOI 10.1016/j.camwa.2018.10.033. MR3913643

[6] F. Soldovieri, R. Solimene, Through-wall imaging via a linear inverse scattering algorithm, IEEE Geoscience and Remote Sensing Letters 4 (4) (2007) 513–517. doi: 10.1109/LGRS.2007.900735.

[7] F. H. C. Tivive, A. Bouzerdoum, M. G. Amin, An SVD-based approach for mitigating wall reflections in through-the-wall radar imaging, in Proceedings of the IEEE National Radar Conference, May 2011, pp. 519–524. doi:10.1109/RADAR.2011.5960591.

[8] Mikhail Gilman and Semyon Tsynkov, *A mathematical model for SAR imaging beyond the first Born approximation*, SIAM J. Imaging Sci. **8** (2015), no. 1, 186–225, DOI 10.1137/140973025. MR3302589

[9] Mikhail Gilman, Erick Smith, and Semyon Tsynkov, *Transionospheric synthetic aperture imaging*, Applied and Numerical Harmonic Analysis, Birkhäuser/Springer, Cham, 2017, DOI 10.1007/978-3-319-52127-5. MR3643084

[10] S. Rotheram, J. T. Macklin, Inverse Methods for O cean Wave Imaging by SAR, in: Inverse Methods in Electromagnetic Imaging, Springer Netherlands, 1985, pp. 907–930. doi:10.1007/978-94-009-5271-3_11.

[11] L. Nguyen, M. Ressler, J. Sichina, Sensing through the wall imaging using the Army Research Lab ultra-wideband synchronous impulse reconstruction (UWB SIRE) radar, in: K. I. Ranney, A. W. Doerry (Eds.), Radar Sensor Technology XII, SPIE, 2008. doi:10.1117/12.776869.

[12] M. V. Klibanov, A. V. Smirnov, V. A. Khoa, A. J. Sullivan, L. H. Nguyen, Through-the-wall nonlinear SAR imaging, IEEE Transactions on Geoscience and Remote Sensing 59 (9) (2021) 7475–7486. doi: 10.1109/TGRS.2021.3055805

[13] Michael V. Klibanov and Nguyen Trung Thành, *Recovering dielectric constants of explosives via a globally strictly convex cost functional*, SIAM J. Appl. Math. **75** (2015), no. 2, 518–537, DOI 10.1137/140981198. MR3323559

[14] Nguyen Trung Thành, Larisa Beilina, Michael V. Klibanov, and Michael A. Fiddy, *Imaging of buried objects from experimental backscattering time-dependent measurements using a globally convergent inverse algorithm*, SIAM J. Imaging Sci. **8** (2015), no. 1, 757–786, DOI 10.1137/140972469. MR3327354

[15] Aleksandr E. Kolesov, Michael V. Klibanov, Loc H. Nguyen, Dinh-Liem Nguyen, and Nguyen T. Thành, *Single measurement experimental data for an inverse medium problem inverted by a multi-frequency globally convergent numerical method*, Appl. Numer. Math. **120** (2017), 176–196, DOI 10.1016/j.apnum.2017.05.007. MR3669730

[16] Dinh-Liem Nguyen, Michael V. Klibanov, Loc H. Nguyen, and Michael A. Fiddy, *Imaging of buried objects from multi-frequency experimental data using a globally convergent inversion method*, J. Inverse Ill-Posed Probl. **26** (2018), no. 4, 501–522, DOI 10.1515/jiip-2017-0047. MR3833435

[17] Michael V. Klibanov, Aleksandr E. Kolesov, and Dinh-Liem Nguyen, *Convexification method for an inverse scattering problem and its performance for experimental backscatter data for*

buried targets, SIAM J. Imaging Sci. **12** (2019), no. 1, 576–603, DOI 10.1137/18M1191658. MR3928344

[18] Vo Anh Khoa, Michael Victor Klibanov, and Loc Hoang Nguyen, *Convexification for a three-dimensional inverse scattering problem with the moving point source*, SIAM J. Imaging Sci. **13** (2020), no. 2, 871–904, DOI 10.1137/19M1303101. MR4097812

[19] Vo Anh Khoa, Grant W. Bidney, Michael V. Klibanov, Loc H. Nguyen, Lam H. Nguyen, Anders J. Sullivan, and Vasily N. Astratov, *Convexification and experimental data for a 3D inverse scattering problem with the moving point source*, Inverse Problems **36** (2020), no. 8, 085007, 34, DOI 10.1088/1361-6420/ab95aa. MR4151408

[20] Vo Anh Khoa, Grant W. Bidney, Michael V. Klibanov, Loc H. Nguyen, Lam H. Nguyen, Anders J. Sullivan, and Vasily N. Astratov, *An inverse problem of a simultaneous reconstruction of the dielectric constant and conductivity from experimental backscattering data*, Inverse Probl. Sci. Eng. **29** (2021), no. 5, 712–735, DOI 10.1080/17415977.2020.1802447. MR4258641

[21] M. V. Klibanov, V. A. Khoa, A. V. Smirnov, L. H. Nguyen, G. W. Bidney, L. H. Nguyen, A. J. Sullivan, and V. N. Astratov, *Convexification inversion method for nonlinear SAR imaging with experimentally collected data*, J. Appl. Ind. Math. **15** (2021), no. 3, 413–436, DOI 10.1134/s1990478921030054. MR4402801

[22] A. V. Smirnov, M. V. Klibanov, A. J. Sullivan, and L. H. Nguyen, *Convexification for an inverse problem for a 1D wave equation with experimental data*, Inverse Problems **36** (2020), no. 9, 095008, 32, DOI 10.1088/1361-6420/abac9a. MR4149877

[23] Michael V. Klibanov and Olga V. Ioussoupova, *Uniform strict convexity of a cost functional for three-dimensional inverse scattering problem*, SIAM J. Math. Anal. **26** (1995), no. 1, 147–179, DOI 10.1137/S0036141093244039. MR1311886

[24] Michael V. Klibanov, *Global convexity in a three-dimensional inverse acoustic problem*, SIAM J. Math. Anal. **28** (1997), no. 6, 1371–1388, DOI 10.1137/S0036141096297364. MR1474219

[25] Anatoly B. Bakushinskii, Michael V. Klibanov, and Nikolaj A. Koshev, *Carleman weight functions for a globally convergent numerical method for ill-posed Cauchy problems for some quasilinear PDEs*, Nonlinear Anal. Real World Appl. **34** (2017), 201–224, DOI 10.1016/j.nonrwa.2016.08.008. MR3567958

[26] Michael V. Klibanov and Jingzhi Li, *Inverse problems and Carleman estimates—global uniqueness, global convergence and experimental data*, Inverse and Ill-posed Problems Series, vol. 63, De Gruyter, Berlin, [2021] ©2021, DOI 10.1515/9783110745481. MR4368863

[27] A. L. Bukhgeĭm and M. V. Klibanov, *Uniqueness in the large of a class of multidimensional inverse problems* (Russian), Dokl. Akad. Nauk SSSR **260** (1981), no. 2, 269–272. MR630135

[28] Larisa Beilina and Michael V. Klibanov, *Relaxation property for the adaptivity for ill-posed problems*, Appl. Anal. **93** (2014), no. 2, 223–253, DOI 10.1080/00036811.2013.768339. MR3169202

[29] Mourad Bellassoued and Masahiro Yamamoto, *Carleman estimates and applications to inverse problems for hyperbolic systems*, Springer Monographs in Mathematics, Springer, Tokyo, 2017, DOI 10.1007/978-4-431-56600-7. MR3729280

[30] M. V. Klibanov and A. Timonov, *Carleman estimates for coefficient inverse problems and numerical applications*, Inverse and Ill-posed Problems Series, VSP, Utrecht, 2004, DOI 10.1515/9783110915549. MR2126149

[31] Michael V. Klibanov, *Carleman estimates for global uniqueness, stability and numerical methods for coefficient inverse problems*, J. Inverse Ill-Posed Probl. **21** (2013), no. 4, 477–560, DOI 10.1515/jip-2012-0072. MR3089754

[32] Masahiro Yamamoto, *Carleman estimates for parabolic equations and applications*, Inverse Problems **25** (2009), no. 12, 123013, 75, DOI 10.1088/0266-5611/25/12/123013. MR3460049

[33] N. V. Alekseenko, V. A. Burov, O. D. Rumyantseva, Solution of the three-dimensional acoustic inverse scattering problem. The modified N ovikov algorithm, Acoustical Physics 54 (3) (2008) 407–419. doi:10.1134/s1063771008030172.

[34] R. G. Novikov, *The $\overline{\partial}$-approach to approximate inverse scattering at fixed energy in three dimensions*, IMRP Int. Math. Res. Pap. **6** (2005), 287–349. MR2202575

[35] G. A. Showman, Stripmap SAR, in: Principles of Modern Radar: Advanced techniques, Institution of Engineering and Technology, pp. 259–335. doi:10.1049/sbra020e_ch7.

[36] R. A. Renaut, *Absorbing boundary conditions, difference operators, and stability*, J. Comput. Phys. **102** (1992), no. 2, 236–251, DOI 10.1016/0021-9991(92)90369-A. MR1187690

[37] V. G. Romanov, Inverse Problems of Mathematical P hysics, Walter de Gruyter GmbH & Co.KG, 2019.

[38] Alexey Smirnov, Michael Klibanov, and Loc Nguyen, *Convexification for a 1D hyperbolic coefficient inverse problem with single measurement data*, Inverse Probl. Imaging **14** (2020), no. 5, 913–938, DOI 10.3934/ipi.2020042. MR4128448

[39] A. N. Tikhonov, Numerical Methods for the S olution of Ill-Posed Problems, Springer Netherlands, Dordrecht, 1995.

[40] Michael V. Klibanov, Aleksandr E. Kolesov, Anders Sullivan, and Lam Nguyen, *A new version of the convexification method for a 1D coefficient inverse problem with experimental data*, Inverse Problems **34** (2018), no. 11, 115014, 29, DOI 10.1088/1361-6420/aadbc6. MR3863309

[41] Michael V. Klibanov and Aleksandr E. Kolesov, *Convexification of a 3-D coefficient inverse scattering problem*, Comput. Math. Appl. **77** (2019), no. 6, 1681–1702, DOI 10.1016/j.camwa.2018.03.016. MR3926836

[42] D. J. Daniels, A review of GPR for landmine detection, Sensing and Imaging: An International Journal 7 (3) (2006) 90–123. doi:10.1007/s11220-006-0024-5.

[43] A. V. Kuzhuget, L. Beilina, M. V. Klibanov, A. Sullivan, L. Nguyen, M. A. Fiddy, Blind backscattering experimental data collected in the field and an approximately globally convergent inverse algorithm, Inverse Problems 28 (9) (2012) 095007. doi:10.1088/0266-5611/28/9/095007.

[44] A. Lipson, S. G. Lipson, H. Lipson, Optical Physics, Cambridge University Press, 2018.

[45] Ai-Guo Wu, Ying Zhang, and Wei-Jun Wu, *Discussion on: "Robust fault detection observer and fault estimation filter design for LTI systems based on GKYP lemma" [MR2731507]*, Eur. J. Control **16** (2010), no. 4, 384–386, DOI 10.1016/S0947-3580(10)70667-1. MR2731508

[46] Jingzhi Li and Hongyu Liu, *Recovering a polyhedral obstacle by a few backscattering measurements*, J. Differential Equations **259** (2015), no. 5, 2101–2120, DOI 10.1016/j.jde.2015.03.030. MR3349431

[47] Jingzhi Li, Peijun Li, Hongyu Liu, and Xiaodong Liu, *Recovering multiscale buried anomalies in a two-layered medium*, Inverse Problems **31** (2015), no. 10, 105006, 28, DOI 10.1088/0266-5611/31/10/105006. MR3405366

[48] Jingzhi Li, Hongyu Liu, and Yuliang Wang, *Recovering an electromagnetic obstacle by a few phaseless backscattering measurements*, Inverse Problems **33** (2017), no. 3, 035011, 20, DOI 10.1088/1361-6420/aa5bf3. MR3626816

[49] Xianchao Wang, Yukun Guo, Jingzhi Li, and Hongyu Liu, *Mathematical design of a novel input/instruction device using a moving acoustic emitter*, Inverse Problems **33** (2017), no. 10, 105009, 19, DOI 10.1088/1361-6420/aa873f. MR3706190

Department of Mathematics, Florida A&M University, Tallahassee, FL 32307
Email address: `anhkhoa.vo@famu.edu`

Department of Mathematics and Statistics, University of North Carolina at Charlotte, Charlotte, NC 28223
Email address: `mklibanv@uncc.edu`

Department of Mathematics, University of Wisconsin Madison, Madison, WI 53706
Email address: `wgpowell@wisc.edu`

Department of Mathematics and Statistics, University of North Carolina at Charlotte, Charlotte, NC 28223
Email address: `loc.nguyen@uncc.edu`

Contemporary Mathematics
Volume **784**, 2023
https://doi.org/10.1090/conm/784/15759

Initial state reconstruction on graphs

Vo Anh Khoa, Mai Thanh Nhat Truong, Imhotep Hogan,
and Roselyn Williams

ABSTRACT. The presence of noise is an intrinsic problem in acquisition processes for digital images. One way to enhance images is to combine the forward and backward diffusion equations. However, the latter problem is well known to be exponentially unstable with respect to any small perturbations on the final data. In this scenario, the final data can be regarded as a blurred image obtained from the forward process, and that image can be pixelated as a network. Therefore, we study in this work a regularization framework for the backward diffusion equation on graphs. Our aim is to construct a spectral graph-based solution based upon a cut-off projection. Stability and convergence results are provided together with some numerical experiments.

1. Statement of the initial state reconstruction on graphs

Regularization of the restoration problem is a denoising technique used in the hope that it can retain the image crucial signal features in the presence of noise (motion artifacts, blurring, distortion) during procurement course. It is worth mentioning that even modern cameras which are able to acquire high resolution images still, nowadays, produce noisy outputs. Noise in those images usually constitutes high frequencies; therefore, one way to denoise the images is to smooth "away" the noise. It is well known that one of the standard filtering processes is Gaussian smoothing; cf. e.g. [**3, 9, 20**]. In particular, let g_0 be a gray scale image as a real-valued mapping; see Figure 1 for an overview of the surface representation of a gray scale image. Let g_0^ε be its noisy image, where $\varepsilon \in (0,1)$ represents the noise level. Suppose that these g_0, g_0^ε belong to $L^2(\mathbb{R}^2)$. Then, the Gaussian smoothing constructs a smoothed version of g_0^ε by the convolution $G_\sigma * g_0^\varepsilon \in C^\infty(\mathbb{R}^2)$,

$$(G_\sigma * g_0^\varepsilon)(x) = \int_{\mathbb{R}^2} \frac{1}{2\pi\sigma^2} e^{-|x-y|^2/(2\sigma^2)} g_0^\varepsilon(y)\, dy, \quad \sigma > 0. \tag{1.1}$$

In terms of Partial Differential Equations (PDEs), the Gaussian convolution is obtained by solving a sequence of the parabolic PDEs forward in time. Indeed, if

2020 *Mathematics Subject Classification.* 65M12, 15A18, 05C50.

Key words and phrases. Initial state reconstruction, Simple graphs, Denoising, Error estimates, Graph Laplacian.

The first author was supported by the Faculty Research Awards Program - FAMU #007633.

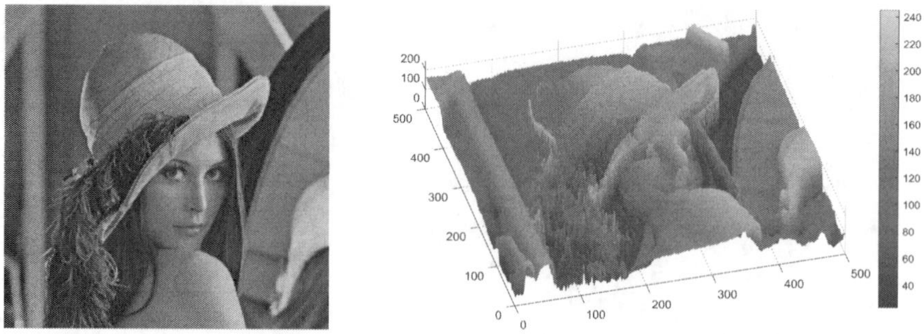

FIGURE 1. Illustration of the surface representation of a gray scale image with size 400×400 (left). An image can be considered as a two-dimensional surface (right) embedded in three-dimensional space. The pixel value ranges from 0 to 255.

FIGURE 2. Illustration of the forward process for a color scale image (left) after $t = 0.001$. The obtained image (right) is very blurry due to the smoothness property of the parabolic equation.

we consider

$$\begin{cases} \partial_t u - \Delta u = 0 & \text{in } \mathbb{R}^2 \times (0,T), \\ u(x,0) = g_0^\varepsilon(x) & \text{in } \mathbb{R}^2, \end{cases} \tag{1.2}$$

then its solution can be expressed as $u(x,t) = \left(G_{\sqrt{2t}} * g_0^\varepsilon\right)(x)$ for $x \in \mathbb{R}^2, t > 0$.

We remark that due to the smoothness property of the parabolic equation, the smoothing process will destroy image characteristics such as lines and edges; see Figure 2. Stemming from the idea in [**7**, **19**], we combine the forward and backward diffusion equations to enhance images. Henceforth, it brings us back to the initial state reconstruction problem in which the noisy final condition as our blurred image is $g_T^\varepsilon = G_{\sqrt{2T}} * g_0^\varepsilon(x)$. Unique continuity of this problem is studied in [**6**].

We would like to mention that image can be pixelated as a network; see Figure 3. Therefore, one can consider the linear diffusion PDE (1.2) on graphs. In this work, we concentrate on the simple graph $G = (V, E)$. It is an ordered pair of sets, where

- $E \subset \{\{x_i, x_j\} \,|x_i, x_j \in V, i \neq j\}$ is a set of undirected edges;
- V is a set of vertices $\{x_i\}_{1 \le i \le n}$ with $n = |V| < \infty$ being the number of vertices.

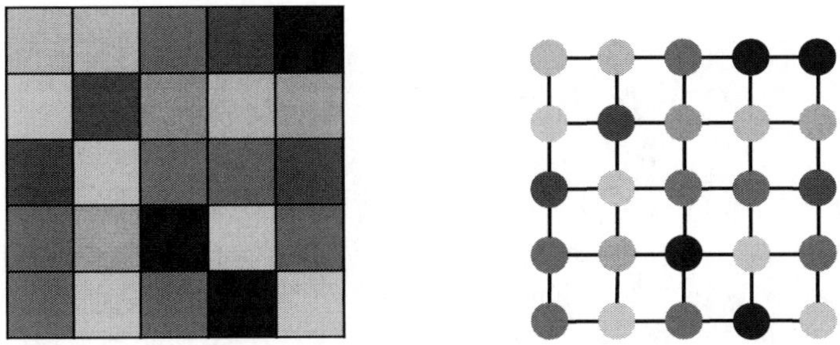

FIGURE 3. Illustration of how part of a gray scale image (left) is pixelated as a network (right). The value of every vertex is based on the pixel value at the corresponding pixel.

Consider $u_i(t) = u(x_i, t)$ as the pixel value at time t and vertex x_i. Note here that the number n should be fixed because of the fixed resolution of the image. Let $\mathbf{A} \in \mathbb{R}^{n\times n}$ be the adjacency matrix, whose the entries A_{ij} are given by

$$A_{ij} = \begin{cases} 1 & \text{if } \{x_i, x_j\} \in E, \\ 0 & \text{otherwise.} \end{cases}$$

From knowledge of the adjacency matrix $\mathbf{A}$, we define the degree $D_i = D(x_i) > 0$ of a vertex x_i in the simple graph G is the number of vertices in G that are adjacent to x_i. As a consequence, it holds true that

$$\sum_{j=1}^{n} A_{ij} = D(x_i) =: D_i \quad \text{for } 1 \le i \le n.$$

By this way, $\mathbf{D} \in \mathbb{R}^{n\times n}$ is the degree matrix which is diagonal with D_i being the degree of the i-th vertex. Henceforth, $\mathbf{D} \in \mathbb{R}^{n\times n}$ is of the following form:

$$\mathbf{D} = \begin{bmatrix} D_1 & 0 & \cdots & 0 & 0 \\ 0 & D_2 & \ddots & & 0 \\ \vdots & \ddots & \ddots & \ddots & \vdots \\ 0 & & \ddots & D_{n-1} & 0 \\ 0 & 0 & \cdots & 0 & D_n \end{bmatrix}.$$

Since we pixelate an image as a square lattice grid (see Figure 3), the distance between two adjacent vertices, as the number of edges in a graph geodesic connecting them, is 1. Therefore, the Laplace operator in the PDE of (1.2) can be approximated using the five-point stencil finite-difference method, resulting in

$$\Delta u_i(t) \approx \sum_{j=1}^{n} A_{ij} (u_j(t) - u_i(t))$$

Observe that

$$\begin{aligned}\sum_{j=1}^{n} A_{ij}\left(u_j(t) - u_i(t)\right) &= \sum_{j=1}^{n} A_{ij} u_j(t) - u_i(t) \sum_{j=1}^{n} A_{ij} \\ &= \sum_{j=1}^{n} A_{ij} u_j(t) - u_i(t) D_i = \sum_{j=1}^{n} A_{ij} u_j(t) - \sum_{j=1}^{n} \delta_{ij} u_j(t) D_j\end{aligned}$$

in which the Kronecker delta has been used. This leads to the so-called Graph Laplacian matrix defined as $\mathbf{L} = \mathbf{D} - \mathbf{A} \in \mathbb{R}^{n\times n}$. In this scenario, the Graph Laplacian matrix is symmetric and its entries are given by

$$L_{ij} = \begin{cases} -1 & \text{if } i \neq j \text{ and } \{x_i, x_j\} \in E, \\ D_i & \text{if } i = j, \\ 0 & \text{otherwise.} \end{cases}$$

From now on, let $U = U(t)$ now be the n-dimensional vector of $u_i(t)$ for $1 \le i \le n$, i.e.

$$U = \begin{pmatrix} u_1 & u_2 & \cdots & u_{n-1} & u_n \end{pmatrix}^{\mathrm{T}}.$$

Suppose that U_T^{ε} is the blurred image g_T^{ε} pixelated on the simple graph under study. Thereby, our initial state reconstruction reads as

$$\begin{cases} \frac{dU}{dt}(t) + \mathbf{L}U(t) = 0 & \text{for } t \in (0,T), \\ U(T) = U_T^{\varepsilon} \in \mathbb{R}^n. \end{cases} \tag{1.3}$$

2. Fourier instability of the initial state reconstruction

In this section, we show the natural instability of the initial state reconstruction problem (1.3). When doing so, we need to introduce the standard inner product

$$\langle A, B\rangle = A^{\mathrm{T}}B = \sum_{i=1}^{n} a_i b_i \quad \text{for } A, B \in \mathbb{R}^n.$$

Hereby, the corresponding ℓ^2-norm $\|A\| = \sqrt{\langle A, A\rangle}$ is given by

$$\|A\|^2 = \sum_{i=1}^{n} a_i^2.$$

Next, we introduce in Definition 1 regarding an eigenvector of a matrix. Then, we state the standard spectral theorem without proof in Theorem 1. These are essential in our way to show the Fourier instability of problem (1.3).

DEFINITION 1. An eigenvector of a matrix $\mathcal{A} \in \mathbb{R}^{n\times n}$ is a (non-zero) vector ϕ such that $\mathcal{A}\phi = \lambda\phi$ for some scalar λ. The value λ is the corresponding eigenvalue, which is a root of the characteristic polynomial of $\mathcal{A}$, $p_{\mathcal{A}} = \det(\lambda I - \mathcal{A})$ where $I \in \mathbb{R}^{n\times n}$ denotes the identity matrix.

THEOREM 1. *Let $\mathbf{M} \in \mathbb{R}^{n\times n}$ be a real and symmetric matrix. Then it has n orthogonal eigenvectors and the eigenvalues are real and non-negative.*

Since the Graph Laplacian matrix $\mathbf{L}$ is real and symmetric, we know that there is an orthonormal basis $\phi_1, \ldots, \phi_n$ of $\mathbb{R}^n$ such that each ϕ_j is an eigenvector of $\mathbf{L}$. Let λ_j be the real eigenvalue corresponding to ϕ_j, i.e. $\mathbf{L}\phi_j = \lambda_j\phi_j$ for $1 \le j \le n$. Moreover, we have $0 \le \lambda_1 \le \lambda_2 \le \ldots \le \lambda_n$, and the sum of these eigenvalues is twice the number of edges of the simple graph under consideration (cf. [**2**]). We find $U(t)$ as a linear combination of the eigenvectors:

$$U(t) = \sum_{i=1}^{n} B_i(t)\,\phi_i. \tag{2.1}$$

It is clear that in this form, $U(t) \in \mathbb{R}^n$ for every t. Now, we multiply both sides of (1.3) (in the sense of the standard product) by $\phi_j \in \mathbb{R}^n$ and thus, arrive at

$$\frac{d}{dt}\langle U(t), \phi_j\rangle + \langle \mathbf{L}U(t), \phi_j\rangle = 0,$$

equivalently,

$$\frac{d}{dt}\langle U(t), \phi_j\rangle + \langle U(t), \mathbf{L}\phi_j\rangle = 0.$$

Since $\mathbf{L}\phi_j = \lambda_j\phi_j$, we obtain the following differential equation:

$$\frac{d}{dt}\langle U(t), \phi_j\rangle + \lambda_j\langle U(t), \phi_j\rangle = 0, \tag{2.2}$$

associated with the following final condition (cf. (1.3)),

$$\langle U(T), \phi_j\rangle = \langle U_T^\varepsilon, \phi_j\rangle. \tag{2.3}$$

Solving (2.2)-(2.3) as a final-value differential problem, we obtain

$$\langle U(t), \phi_j\rangle = e^{\lambda_j(T-t)}\langle U_T^\varepsilon, \phi_j\rangle. \tag{2.4}$$

Then plugging the spectral form of $U(t)$ (cf. (2.1)) in (2.4) and using the orthogonality of ϕ_i yield

$$e^{\lambda_j(T-t)}\langle U_T^\varepsilon, \phi_j\rangle = \langle U(t), \phi_j\rangle = \left\langle \sum_{i=1}^{n} B_i(t)\,\phi_i, \phi_j \right\rangle = \sum_{i=1}^{n} B_i(t)\langle \phi_i, \phi_j\rangle = B_j(t).$$

Henceforth, our solution $U(t)$ to the initial state reconstruction problem (1.3) can be computed as

$$U(t) = \sum_{i=1}^{n} e^{\lambda_i(T-t)}\langle U_T^\varepsilon, \phi_i\rangle\,\phi_i. \tag{2.5}$$

Hence, the initial information of an pixelated noisy image, denoted by $U_0^\varepsilon = U^\varepsilon(0)$ is formulated by

$$U_0^\varepsilon = \sum_{i=1}^{n} e^{\lambda_i T}\langle U_T^\varepsilon, \phi_i\rangle\,\phi_i. \tag{2.6}$$

Observe in (2.5) that the exponential kernel $e^{\lambda_i(T-t)}$ is the main natural factor of the Fourier instability. In this sense, the solution is exponentially unstable when the number of vertices, which is n, becomes larger and larger for a better resolution of image. If we keep refining the mesh of discretization or, in other words, increasing the number of vertices, the reconstruction process will be worse regardless of any numerical method applied.

Our aim in this work is to study a stable approximate solution of the discrete model on graphs (1.3). This new research will formulate a new numerical approach in image denoising among well known approaches such as Perona–Malik [**16**], p-Laplacian [**15**], Mumford–Shah [**8**] and deep learning [**17**].

3. A spectral graph-based approximate solution

For each noise level ε, we consider the set of admissible eigenvalues $\Theta(\varepsilon) = \{i \in \mathbb{N}^* : \lambda_i \le M_\varepsilon\}$. Here, $M_\varepsilon = M(\varepsilon) > 0$ is the so-called regularization parameter which will be chosen later. It should be that $|\Theta(\varepsilon)| \le |V|$ as the resolution of image is fixed. If there exists ε_0 such that $|\Theta(\varepsilon_0)| > |V|$, for any $\varepsilon \ge \varepsilon_0$ we set $M_\varepsilon = \lambda_n$, where we recall that λ_n is the largest eigenvalue of the Graph Laplacian matrix. We thereby construct the following cut-off projection in which the number of vertices n is selected appropriately in terms of ε.

$$\mathbf{P}^\varepsilon U(t) = \sum_{i\in\Theta(\varepsilon)} e^{\lambda_i(T-t)} \langle U(T), \phi_i\rangle \phi_i. \tag{3.1}$$

REMARK 3.1. The underlying cut-off projection is one of the conventional Fourier-based approaches that is "computable" in the regularization theory; cf. [**12, 18**]. We remark that the existing literature on regularization of terminal-value problems is huge. For brevity, we mention here some PDE-based approaches (e.g. quasi-reversibility method [**11, 13**], quasi-boundary value method [**4**], and references cited therein) and some minimization-based approaches (see in [**5**] regarding methods of total variation, sparse representation and others, and convexification in [**10**]).

THEOREM 2. *Consider $U(T) \in \mathbb{R}^n$. Then, it holds true that*

$$\|\mathbf{P}^\varepsilon U(t)\| \le e^{M_\varepsilon(T-t)} \|U(T)\|,$$

PROOF. The proof can be done directly by taking into account the formulation of the cut-off projection in (3.1). Indeed, by the orthogonality of the eigenvectors, we have

$$\begin{aligned}\|\mathbf{P}^\varepsilon U(t)\|^2 &= \langle \mathbf{P}^\varepsilon U(t), \mathbf{P}^\varepsilon U(t)\rangle \\ &= \sum_{i\in\Theta(\varepsilon)}\sum_{j\in\Theta(\varepsilon)} e^{\lambda_i(T-t)}e^{\lambda_j(T-t)} \langle U(T),\phi_i\rangle \langle U(T),\phi_j\rangle \langle \phi_i,\phi_j\rangle \\ &= \sum_{i\in\Theta(\varepsilon)} e^{2\lambda_i(T-t)} |\langle U(T),\phi_i\rangle|^2 .\end{aligned}$$

Thus, it yields that

$$\|\mathbf{P}^\varepsilon U(t)\|^2 \le e^{2M_\varepsilon(T-t)} \|U(T)\|^2,$$

which completes the proof of the theorem. □

Theorem 2 shows that the cut-off projection solution is bounded at each noise level. This result also proves the stability of the approximate solution when the input, $U(T)$, is perturbed by noise. Let $U_0 \in \mathbb{R}^n$ be the pixel values of the ideal image g_0, and let $U_0^\varepsilon \in \mathbb{R}^n$ be the pixel values of the noisy image g_0^ε. We assume that

$$\|U_0 - U_0^\varepsilon\| \le \varepsilon.$$

Henceforth, we obtain

$$\|U_T - U_T^\varepsilon\| \le \|U_0 - U_0^\varepsilon\| \le \varepsilon, \tag{3.2}$$

where U_T and U_T^ε are, respectively, the terminal pixel values of the ideal and noisy images after the forward solver (1.2).

THEOREM 3. *Let $U_T, U_T^\varepsilon \in \mathbb{R}^n$ be the terminal data of the ideal solution $U(t)$ and the noisy solution $U^\varepsilon(t)$, respectively. These data satisfy (3.2). Then, the following stability estimate holds true:*

$$\left\| \mathbf{P}^\varepsilon U(t) - \mathbf{P}^\varepsilon U^\varepsilon(t) \right\| \le e^{M_\varepsilon (T-t)} \varepsilon. \tag{3.3}$$

PROOF. The proof is straightforward because

$$\begin{aligned} \left\| \mathbf{P}^\varepsilon U(t) - \mathbf{P}^\varepsilon U^\varepsilon(t) \right\|^2 &= \sum_{i \in \Theta(\varepsilon)} e^{2\lambda_i (T-t)} \left| \langle U_T - U_T^\varepsilon, \phi_i \rangle \right|^2 \\ &\le e^{2M_\varepsilon (T-t)} \varepsilon^2. \end{aligned}$$

□

Now, we are in a position to state the following theorem for the convergence rate of the cut-off projection.

THEOREM 4. *Let $\mathbf{P}^\varepsilon U^\varepsilon(t) \in \mathbb{R}^n$ be the stable approximate solution of $U(t)$ associated with the terminal data U_T^ε. Then, the following uniform-in-time estimate holds:*

$$\left\| U(t) - \mathbf{P}^\varepsilon U^\varepsilon(t) \right\| \le M_\varepsilon^{-1} \left\| \frac{d}{dt} U(t) \right\| + e^{M_\varepsilon (T-t)} \varepsilon. \tag{3.4}$$

Furthermore, by choosing

$$M_\varepsilon = \frac{1}{T} \ln\left(\varepsilon^{-\gamma}\right), \quad \gamma \in (0,1),$$

we obtain

$$\left\| U(t) - \mathbf{P}^\varepsilon U^\varepsilon(t) \right\| \le \frac{T}{\ln\left(\varepsilon^{-\gamma}\right)} \left\| \frac{d}{dt} U(t) \right\| + \varepsilon^{1-\gamma\left(1-\frac{t}{T}\right)}. \tag{3.5}$$

PROOF. In view of the facts that $|\Theta(\varepsilon)| \le |V|$ and

$$U(t) = \sum_{i=1}^{n} e^{\lambda_i (T-t)} \langle U_T, \phi_i \rangle \phi_i, \tag{3.6}$$

we have

$$U(t) - \mathbf{P}^\varepsilon U(t) = \sum_{i \notin \Theta(\varepsilon)} e^{\lambda_i (T-t)} \langle U_T, \phi_i \rangle \phi_i \tag{3.7}$$

It follows from (3.6) that

$$\frac{d}{dt} U(t) = \sum_{i=1}^{n} -\lambda_i e^{\lambda_i (T-t)} \langle U_T, \phi_i \rangle \phi_i.$$

Combining this with (3.7), we estimate that

$$\begin{aligned} \left\| U(t) - \mathbf{P}^\varepsilon U(t) \right\|^2 &= \sum_{i \notin \Theta(\varepsilon)} \lambda_i^{-2} \lambda_i^2 e^{2\lambda_i (T-t)} \left| \langle U_T, \phi_i \rangle \right|^2. \\ &\le M_\varepsilon^{-2} \left\| \frac{d}{dt} U(t) \right\|^2. \end{aligned}$$

It is equivalent to

$$\left\| U(t) - \mathbf{P}^\varepsilon U(t) \right\| \le M_\varepsilon^{-1} \left\| \frac{d}{dt} U(t) \right\|. \tag{3.8}$$

Henceforth, by combining (3.8), (3.3), and using the triangle inequality, we obtain

$$\begin{aligned}\left\|U(t)-\mathbf{P}^{\varepsilon}U^{\varepsilon}(t)\right\| &\le \left\|U(t)-\mathbf{P}^{\varepsilon}U(t)\right\|+\left\|\mathbf{P}^{\varepsilon}U(t)-\mathbf{P}^{\varepsilon}U^{\varepsilon}(t)\right\| \\ &\le M_{\varepsilon}^{-1}\left\|\frac{d}{dt}U(t)\right\|+e^{M_{\varepsilon}(T-t)}\varepsilon.\end{aligned}$$

This completes the proof of (3.4). Proof of (3.5) also follows. □

REMARK 3.2. In the proof of Theorem 4, we obtain a logarithmic rate of convergence under the assumption that $|\Theta(\varepsilon)|\le|V|$. When ε is so small such that $|\Theta(\varepsilon)|>|V|$, as mentioned above, we set $M_{\varepsilon}=\lambda_n$. This means that $U(t)$ coincides with $\mathbf{P}^{\varepsilon}U(t)$, i.e. $U(t)-\mathbf{P}^{\varepsilon}U(t)=0$. In this scenario, we obtain a Lipschitz rate of convergence. In addition, taking $t=0$ in (3.5) gives us that

$$\left\|U_0-\mathbf{P}^{\varepsilon}U_0^{\varepsilon}\right\|\le\frac{T}{\ln\left(\varepsilon^{-\gamma}\right)}\left\|\frac{d}{dt}U_0\right\|+\varepsilon^{1-\gamma} \tag{3.9}$$

in which we assume the existence of $\frac{d}{dt}U_0\in\mathbb{R}^n$.

For $A\in\mathbb{R}^{n\times n}$, we define the function $A\mapsto\|A\|$ by

$$\|A\|=\sup_{V\in\mathbb{R}^n,\|V\|=1}\|AV\|.$$

This subordinate matrix norm is induced by the vector norm $\|\cdot\|$ and without any confusion, we use the same notations here. Using this matrix norm, the norm $\left\|\frac{d}{dt}U(t)\right\|$ for $t>0$ in Theorem 4 can be estimated by

$$\begin{aligned}\left\|\frac{d}{dt}U(t)\right\| &\le \|\mathbf{L}\|\left\|U(t)\right\|=\lambda_n\left\|U(t)\right\| \\ &\le \left\|U(t)\right\|\max\left\{D(x_i)+D(x_j)\,|x_i,x_j\in V\right\}\le 8\left\|U(t)\right\|\end{aligned}$$

in which we have used the Anderson–Morley upper bound of the largest Laplacian eigenvalue on simple graphs; cf. [**1**]. Consequently, for $t\in(0,T)$, it is natural to obtain from (3.5) that

$$\left\|U(t)-\mathbf{P}^{\varepsilon}U^{\varepsilon}(t)\right\|\le\frac{8T}{\ln\left(\varepsilon^{-\gamma}\right)}\left\|U(t)\right\|+\varepsilon^{1-\gamma\left(1-\frac{t}{T}\right)}.$$

Last but not least, if we impose the compatibility condition on U_0 such that it satisfies the PDE in (1.3), then the convergence result (1.3) holds true without the above-mentioned source condition $\frac{d}{dt}U_0\in\mathbb{R}^n$.

4. Numerical examples

In this section, we provide a series of numerical examples to demonstrate the proposed reconstruction algorithm in denoising corrupted gray images. We use *Set12* – a widely-used test dataset for benchmarking denoising algorithms; cf. [**14**]. The dataset consists of 12 gray images of sizes 256×256 or 512×512; however, we resize the images to the size of 128×128 due to the limitation of computer hardware in creating larger graphs. With this resized dataset, we experiment two types of corrupted gray images (noisy images): normal additive noise, noisy dataset = dataset + εrand (for $|\text{rand}|<0.5$), and the commonly used additive white Gaussian noise (AWGN) with deviation $\sigma_{noise}=\overline{10,50}$. Numerical experiments with the AWGN show the robustness of the projection method, even though currently, the stochasticity of such noise has not been studied in the model description as well

TABLE 1. Quantitative comparison in terms of PSNR for denoising performance between approximate solutions (2.6) and cut-off solutions (3.1) for normal additive noise.

Image name	Approximation (2.6)	Cut-off (3.1)
Camera man	19.25	20.64
House	21.70	22.90
Peppers	19.62	20.54
Starfish	19.46	20.52
Monarch	18.40	19.24
Airplane	19.22	20.39
Parrot	17.69	18.73
Lena	20.26	21.34
Barbara	20.00	20.94
Boat	20.23	21.60
Man	20.56	21.70
Couple	20.51	22.01
Avg.	19.74	20.88

as the theoretical findings. Various values of the deviation of AWGN, σ_{noise}, are taken into account to see the performance of the scheme when σ_{noise} is large.

From the noisy images, the corresponding blurred images U_T^ε are obtained via the forward process in (1.2). For simplicity, we apply in this process the standard forward Euler method with the Courant number of 0.03 for the conditional stability and with $T = 0.5$ that corresponds to the deviation $\sigma = 1$ in the Gaussian blurring process. It is worth noting that since $u(x,t) = \left(G_{\sqrt{2t}} * g_0^\varepsilon\right)(x)$ in (1.2), the time variable t is in connection with the spatial width $\sigma = \sqrt{2t}$ in (1.1). In other words, the larger the final time observation, the flatter the solution behaves. Therefore, the smoothing parameter σ in (1.1) requires to stop the diffusion process at $T = \sigma^2/2$.

Since we work with 128×128 images, the number of vertices is $n = 16384$ in our numerical results. Moreover, we fix $\varepsilon = 0.1$ and $\gamma = 0.5$ that gives $M_\varepsilon \approx 2.30$. For objective assessment, we use peak signal-to-noise ratio (PSNR) as an image quality metric in evaluating denoising performance, which is represented in the logarithmic decibel (dB) scale. In principle, it is computed as

$$\text{PSNR}(U_0, \mathbf{P}U_0^\varepsilon) = 10 \log_{10} \left(\frac{255^2}{\frac{1}{n}\sum_{i=1}^n (u_{0i} - \mathbf{P}u_{0i}^\varepsilon)^2} \right). \tag{4.1}$$

To estimate the PSNR of our images, we compare those to the original ones that are resized from the *Set12* dataset. This comparison is operated by the built-in MATLAB function `psnr`. Last but not least, the algorithm is implemented using MATLAB 2016 and executed using a computer equipped with Intel® Core™ i5-10210U @ 1.60GHz and 8 GB of RAM.

Comments. Figures 4 and 5 show the denoising results of the cut-off solution $\mathbf{P}U_0^\varepsilon$ in (3.1) for all images in *Set12* degraded by normal additive noise. The denoised results exhibit better visual quality than that of noisy images. Especially in regions with rich textures, the noise is reduced while edge features are preserved without significant degradation. Table 1 shows the objective evaluations for denoising results of all images in *Set12* from approximate solution (2.6) and cut-off

(A) *Camera man* image

(B) *House* image

(C) *Peppers* image

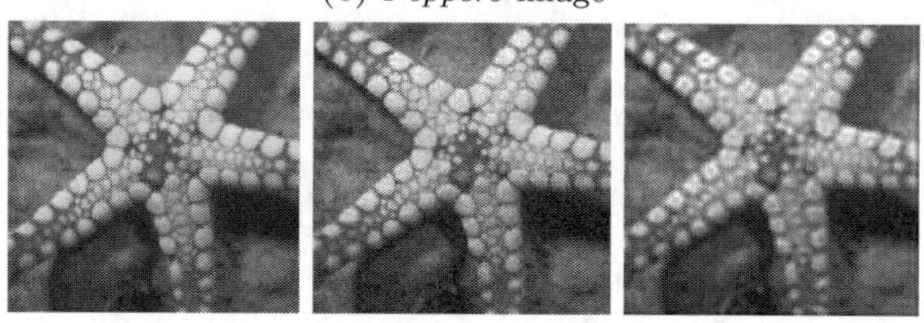

(D) *Starfish* image

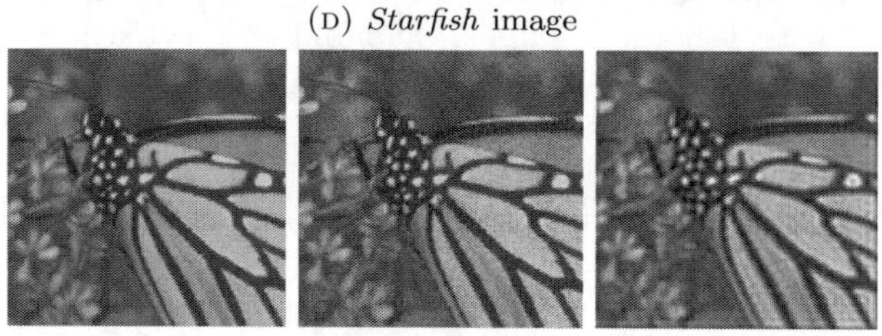

(E) *Monarch* image

(F) *Airplane* image

FIGURE 4. Visual comparison of denoising results on standard test images in the *Set12* dataset with normal noise. Left column: original images; middle column: noisy images; right column: denoised images.

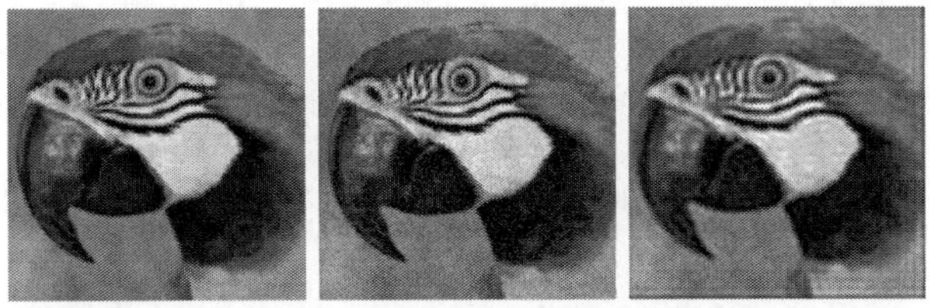

(A) *Parrot* image

(B) *Lena* image

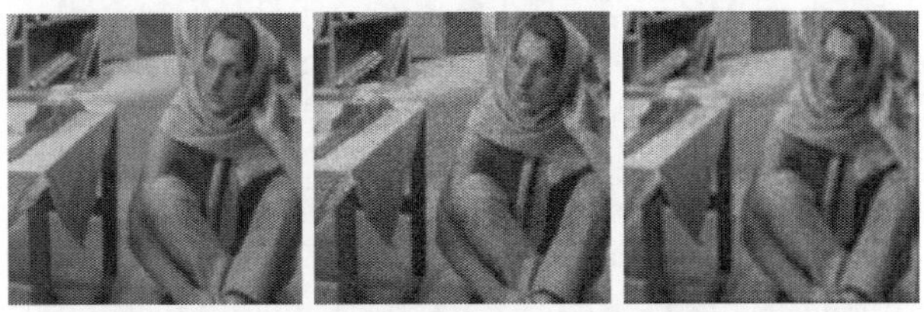

(C) *Barbara* image

(D) *Boat* image

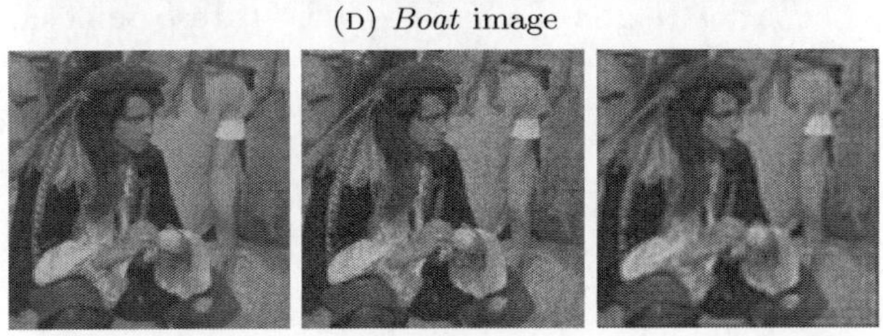

(E) *Man* image

(F) *Couple* image

FIGURE 5. Visual comparison of denoising results on standard test images in the *Set12* dataset with normal noise. Left column: original images; middle column: noisy images; right column: denoised images.

solution $\mathbf{P}U_0^\varepsilon$ in (3.1) in terms of PSNR. As can be seen from the table, the cut-off solution $\mathbf{P}U_0^\varepsilon$ outperforms the approximate solution (2.6), as the PSNR scores of cut-off solutions $\mathbf{P}U_0^\varepsilon$ are higher than those of approximate solutions from 0.84dB (for *Monarch* image) to 1.50dB (for *Couple* image). We detail that 0.84dB for the *Monarch* image is calculated directly from the PSNR difference between the cut-off (19.24dB) and approximate (18.40dB) solutions; see Table 2. The average of PSNR increase is 1.14dB, i.e. 20.88dB of cut-off solutions $\mathbf{P}U_0^\varepsilon$ compared with 19.74dB of approximate solutions (2.6), which is a significant improvement.

To further prove the robustness of the proposed algorithm, we perform the denoising on the images in *Set12* which are degraded by AWGN. Figures 6 and 7 show the denoising results where $\sigma_{noise} = 20$. As the proposed algorithm is not specifically designed for AWGN, the results are not as detailed as those in normal additive noise case. However, the denoised results still show better perceptual quality. The fine details are well preserved while the noise are effectively suppressed. Table 2 shows the objective comparison between the results of approximate solution (2.6) and cut-off solution $\mathbf{P}U_0^\varepsilon$ in (3.1) in terms of PSNR. Similarly to the normal additive noise case, the cut-off solution outperforms the approximate solution. The PSNR scores of cut-off solutions are 2.17dB higher than those of approximation solutions on average. Tables 3 and 4 shows the objective assessments of denoising images degraded by AWGN with various values of σ_{noise}. As σ_{noise} increases, the images are more severely damaged by the increased amount of noise. Consequently, the denoising performance is negatively affected when σ_{noise} is large. However, in all cases, the cut-off solution outperforms the approximate solution by a large margin. When varying $\gamma \in (0,1)$, we also experiment in Table 5 that $\gamma = 0.25$ shows the best quality denoising results of all images degraded by AWGN.

Last but not least, we provide Table 6 to show the time efficiency of the cut-off projection under study. As can readily be expected from the fact that $|\Theta(\varepsilon)| < |V|$, in average, the running time for the cut-off solution $\mathbf{P}U_0^\varepsilon$ is 4.5 times less than that of the approximation (2.6). Therefore, it is potential to adapt the cut-off projection to more complex scenarios (e.g. color images with three-dimensional matrices) in upcoming research projects.

(A) *Camera man* image

(B) *House* image

(C) *Peppers* image

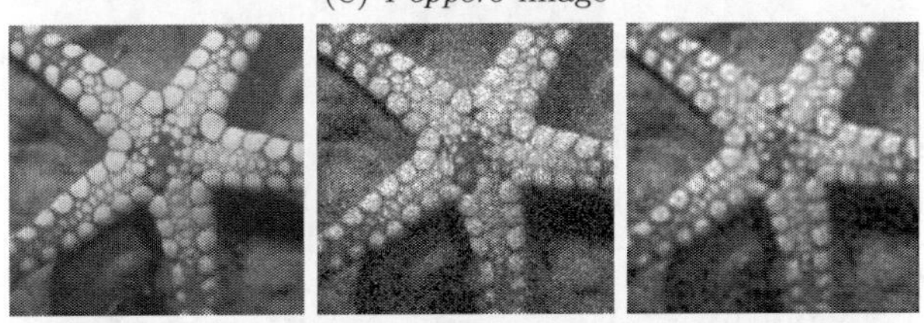

(D) *Starfish* image

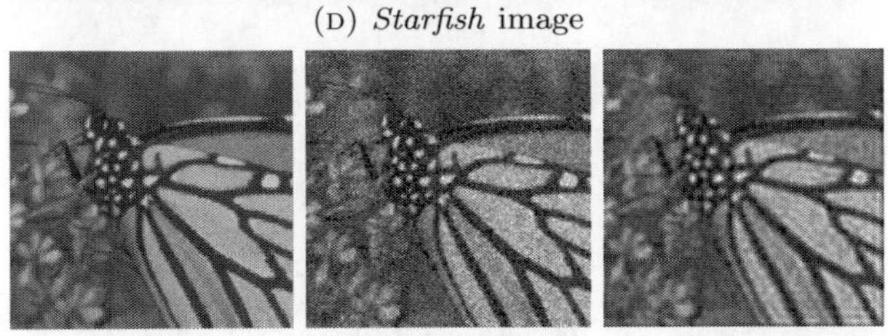

(E) *Monarch* image

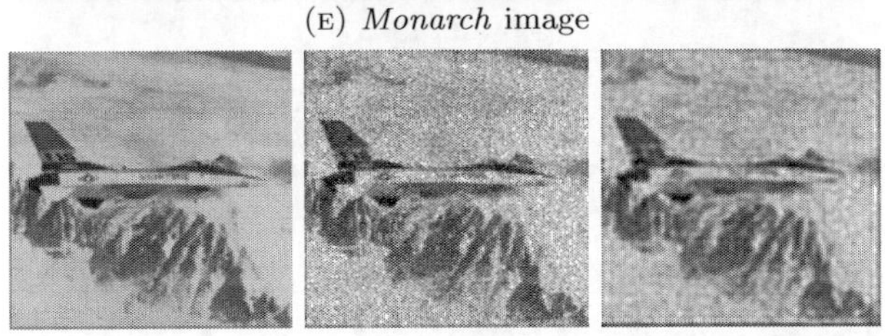

(F) *Airplane* image

FIGURE 6. Visual comparison of denoising results on standard test images in the *Set12* dataset with AWGN. Left column: original images; middle column: noisy images; right column: denoised images.

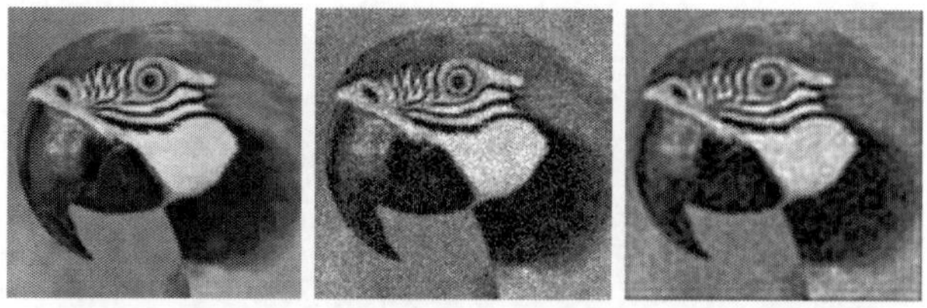

(A) *Parrot* image

(B) *Lena* image

(C) *Barbara* image

(D) *Boat* image

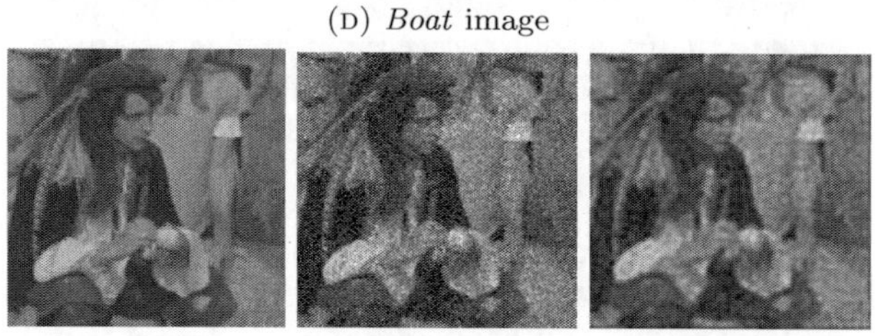

(E) *Man* image

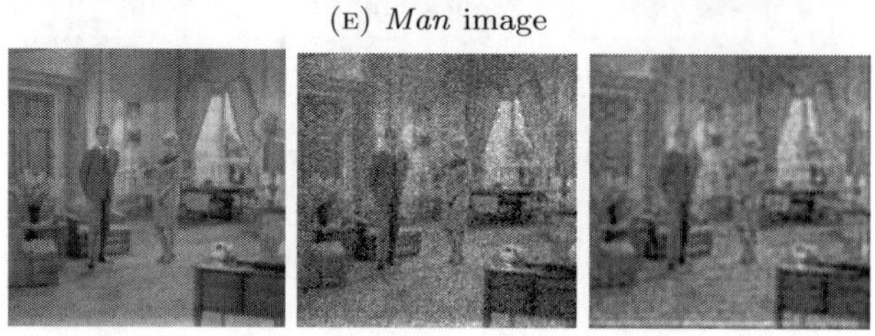

(F) *Couple* image

FIGURE 7. Visual comparison of denoising results on standard test images in the *Set12* dataset with AWGN. Left column: original images; middle column: noisy images; right column: denoised images.

TABLE 2. Quantitative comparison in terms of PSNR for denoising performance between approximate solutions (2.6) and cut-off solutions (3.1) for AWGN.

Image name	Approximation (2.6)	Cut-off (3.1)
Camera man	17.89	20.15
House	19.40	22.00
Peppers	18.10	20.06
Starfish	18.05	20.08
Monarch	17.20	18.86
Airplane	17.76	19.83
Parrot	16.66	18.39
Lena	18.42	20.66
Barbara	18.34	20.37
Boat	18.50	20.94
Man	18.75	21.09
Couple	18.64	21.28
Avg.	18.14	20.31

TABLE 3. Quantitative results in terms of PSNR for denoising performance of cut-off solutions (3.1) for AWGN in various values of σ_{noise}.

σ_{noise} value	10	15	20	25	30	35	40	45	50
Camera man	20.58	20.37	20.18	19.87	19.48	19.04	18.58	18.21	17.98
House	22.88	22.51	22.03	21.54	21.06	20.61	20.01	19.47	18.88
Peppers	20.44	20.29	20.07	19.68	19.41	18.83	18.58	18.10	17.67
Starfish	20.43	20.28	19.96	19.71	19.36	18.92	18.46	18.18	17.80
Monarch	19.19	19.03	18.84	18.61	18.20	17.98	17.67	17.34	17.11
Airplane	20.38	20.15	19.90	19.57	19.20	19.04	18.46	17.97	17.75
Parrot	18.70	18.56	18.38	18.23	17.97	17.65	17.31	16.94	16.48
Lena	21.24	20.97	20.84	20.26	19.98	19.44	18.98	18.48	18.20
Barbara	20.84	20.59	20.30	20.07	19.68	19.26	18.86	18.27	17.99
Boat	21.54	21.18	20.93	20.49	20.18	19.54	19.20	18.70	18.45
Man	21.68	21.53	21.07	20.71	20.13	19.64	19.27	18.73	18.34
Couple	21.99	21.76	21.30	20.87	20.47	19.87	19.48	19.08	18.43
Avg.	20.82	20.60	20.32	19.97	19.59	19.15	18.74	18.29	17.92

TABLE 4. Quantitative results in terms of PSNR for denoising performance of approximate solutions (2.6) for AWGN in various values of σ_{noise}.

σ_{noise} value	10	15	20	25	30	35	40	45	50
Camera man	19.10	18.54	17.83	17.12	16.36	15.56	14.75	14.15	13.44
House	21.44	20.45	19.44	18.38	17.40	16.44	15.50	14.76	13.96
Peppers	19.40	18.81	18.10	17.27	16.51	15.63	14.91	14.20	13.50
Starfish	19.27	18.69	17.91	17.22	16.40	15.61	14.82	14.21	13.56
Monarch	18.27	17.77	17.17	16.49	15.80	15.07	14.46	13.82	13.24
Airplane	19.09	18.48	17.81	17.05	16.28	15.63	14.79	14.03	13.46
Parrot	17.59	17.18	16.66	16.10	15.49	14.84	14.20	13.44	12.85
Lena	20.02	19.27	18.61	17.54	16.83	15.85	15.06	14.29	13.67
Barbara	19.76	19.07	18.28	17.52	16.66	15.88	15.09	14.23	13.66
Boat	20.06	19.31	18.51	17.61	16.77	15.88	15.14	14.34	13.72
Man	20.38	19.69	18.72	17.87	16.85	15.96	15.25	14.36	13.73
Couple	20.32	19.55	18.65	17.80	16.88	15.97	15.18	14.55	13.66
Avg.	19.56	18.90	18.14	17.33	16.52	15.69	14.93	14.20	13.54

TABLE 5. Quantitative results in terms of PSNR for denoising performance of cut-off solutions (3.1) for AWGN in various values of γ.

γ value	0.01	0.25	0.5	0.75	0.99
Camera man	18.02	21.02	20.12	19.13	18.39
House	19.82	23.12	22.07	21.13	20.20
Peppers	17.07	20.78	20.00	19.28	18.64
Starfish	18.07	20.74	20.04	19.25	18.56
Monarch	15.48	19.46	18.85	18.25	17.66
Airplane	18.06	20.78	19.88	19.09	18.38
Parrot	16.34	19.20	18.48	17.70	17.16
Lena	17.88	21.57	20.76	19.89	19.20
Barbara	17.50	21.18	20.35	19.55	18.97
Boat	19.16	21.90	20.96	19.96	19.20
Man	18.60	21.84	21.05	20.20	19.52
Couple	20.05	22.38	21.28	20.36	19.40
Avg.	18.00	21.17	20.32	19.48	18.77

TABLE 6. Comparison of running time in seconds between approximate solutions (2.6) and cut-off solutions (3.1).

Image name	Approximation (2.6)	Cut-off (3.1)
Camera man	1.23	0.28
House	1.15	0.27
Peppers	1.15	0.24
Starfish	1.13	0.26
Monarch	1.14	0.25
Airplane	1.14	0.25
Parrot	1.13	0.26
Lena	1.13	0.26
Barbara	1.16	0.27
Boat	1.16	0.26
Man	1.26	0.28
Couple	1.20	0.27
Avg.	1.17	0.26

Acknowledgment. Vo Anh Khoa would like to thank Professor Paul Sacks (Iowa State University, USA) for great support of his research career.

References

[1] William N. Anderson Jr. and Thomas D. Morley, *Eigenvalues of the Laplacian of a graph*, Linear and Multilinear Algebra **18** (1985), no. 2, 141–145, DOI 10.1080/03081088508817681. MR817657

[2] Andries E. Brouwer and Willem H. Haemers, *Spectra of graphs*, Universitext, Springer, New York, 2012, DOI 10.1007/978-1-4614-1939-6. MR2882891

[3] A. Buades, B. Coll, and J. M. Morel, *Image denoising methods. A new nonlocal principle*, SIAM Rev. **52** (2010), no. 1, 113–147, DOI 10.1137/090773908. Reprint of "A review of image denoising algorithms, with a new one" [MR2162865]. MR2608636

[4] M. Denche and K. Bessila, *A modified quasi-boundary value method for ill-posed problems*, J. Math. Anal. Appl. **301** (2005), no. 2, 419–426, DOI 10.1016/j.jmaa.2004.08.001. MR2105682

[5] L. Fan, F. Zhang, H. Fan, and C. Zhang. Brief review of image denoising techniques. *Visual Computing for Industry, Biomedicine, and Art*, 2(1), 2019.

[6] A. Friedman. *Partial Differential Equations of Parabolic Type*. Dover Pubn Inc, April 2008.

[7] G. Gilboa, N. Sochen, and Y. Y. Zeevi. Forward-and-backward diffusion processes for adaptive image enhancement and denoising. *IEEE Transactions on Image Processing*, 11(7):689–703, 2002.

[8] Miyoun Jung, Xavier Bresson, Tony F. Chan, and Luminita A. Vese, *Nonlocal Mumford-Shah regularizers for color image restoration*, IEEE Trans. Image Process. **20** (2011), no. 6, 1583–1598, DOI 10.1109/TIP.2010.2092433. MR2828604

[9] Pl. Kannappan and P. K. Sahoo, *Rotation invariant separable functions are Gaussian*, SIAM J. Math. Anal. **23** (1992), no. 5, 1342–1351, DOI 10.1137/0523076. MR1177794

[10] Michael V. Klibanov and Anatoly G. Yagola, *Convergent numerical methods for parabolic equations with reversed time via a new Carleman estimate*, Inverse Problems **35** (2019), no. 11, 115012, 30, DOI 10.1088/1361-6420/ab2777. MR4027423

[11] R. Lattès and J.-L. Lions, *Méthode de quasi-réversibilité et applications* (French), Travaux et Recherches Mathématiques, No. 15, Dunod, Paris, 1967. MR0232549

[12] Phan Thanh Nam, Dang Duc Trong, and Nguyen Huy Tuan, *The truncation method for a two-dimensional nonhomogeneous backward heat problem*, Appl. Math. Comput. **216** (2010), no. 12, 3423–3432, DOI 10.1016/j.amc.2010.03.038. MR2661698

[13] Huy Tuan Nguyen, Vo Anh Khoa, and Van Au Vo, *Analysis of a quasi-reversibility method for a terminal value quasi-linear parabolic problem with measurements*, SIAM J. Math. Anal. **51** (2019), no. 1, 60–85, DOI 10.1137/18M1174064. MR3895336

[14] S. Roth and M.J. Black. Fields of experts: a framework for learning image priors. In *Proceedings of the IEEE/CVF Conference on Computer Vision and Pattern Recognition (CVPR)*, volume 2, pages 860–867, 2005.

[15] Kehan Shi, *Coupling local and nonlocal diffusion equations for image denoising*, Nonlinear Anal. Real World Appl. **62** (2021), Paper No. 103362, 17, DOI 10.1016/j.nonrwa.2021.103362. MR4265731

[16] Kehan Shi, *Image denoising by nonlinear nonlocal diffusion equations*, J. Comput. Appl. Math. **395** (2021), Paper No. 113605, 11, DOI 10.1016/j.cam.2021.113605. MR4248452

[17] Ruo-Yu Sun, *Optimization for deep learning: an overview*, J. Oper. Res. Soc. China **8** (2020), no. 2, 249–294, DOI 10.1007/s40305-020-00309-6. MR4114365

[18] Nguyen Huy Tuan, Vo Anh Khoa, Mai Thanh Nhat Truong, Tran The Hung, and Mach Nguyet Minh, *Application of the cut-off projection to solve a backward heat conduction problem in a two-slab composite system*, Inverse Probl. Sci. Eng. **27** (2019), no. 4, 460–483, DOI 10.1080/17415977.2018.1470623. MR3904578

[19] M. Welk, G. Gilboa, and J. Weickert. Theoretical Foundations for Discrete Forward-and-Backward Diffusion Filtering. pages 527–538. Springer Berlin Heidelberg, 2009.

[20] A. M. Wink and J. B. J. M. Roerdink. Denoising functional MR images: A comparison of wavelet denoising and Gaussian smoothing. *IEEE Transactions on Medical Imaging*, 23(3):374–387, 2004.

Department of Mathematics, Florida A&M University, Tallahassee, FL 32307
Email address: anhkhoa.vo@famu.edu

Department of Multimedia Engineering, Dongguk University, South Korea
Email address: mtntruong@mme.dongguk.edu

Department of Mathematics, Florida A&M University, Tallahassee, FL 32307
Email address: imhotep1.hogan@famu.edu

Department of Mathematics, Florida A&M University, Tallahassee, FL 32307
Email address: roselyn.williams@famu.edu

Contemporary Mathematics
Volume **784**, 2023
https://doi.org/10.1090/conm/784/15760

Active control of scalar Helmholtz fields in the presence of known impenetrable obstacles

Lander Besabe and Daniel Onofrei

Abstract. In this paper, we consider the question of actively manipulating scalar Helmholtz fields radiated by a given source that is supported on a compact domain. We claim that the field radiated by the source approximates given scalar fields in prescribed exterior regions while maintaining desired far field patterns in prescribed directions in the presence of exterior known impenetrable obstacles. For simplicity of the exposition, we consider a simplified geometry with only one obstacle, one region of control, and a finite number of far field directions and present a theoretical argument for our claim stated above. Afterwards, we also show how it can be elementarily extended to the general case. Further, we construct a numerical scheme to compute these boundary inputs using the method of moments, the addition theorem, Tikhonov regularization, and Laplace spherical functions.

1. Introduction

1.1. General Background. The study of active control problems for fields modeled by the scalar Helmholtz equation is a very active area of research not only due to its intrinsic importance to acoustic wave phenomena but also because in many situations scalar problems can be used as an approximation to electromagnetic phenomena since they often preserve many important aspects of the problem while avoiding excessive mathematical complexities. In fact, for two-dimensional models, vector electromagnetic problems can be reduced to scalar formulations. Thus, there already exists a rich body of literature concerning the problem of active manipulation of vector and scalar fields in desired regions of space. The active (partial) nulling of acoustic fields was first studied in [**34**] (feed-forward control of sound) and in [**49**] (feedback control of sound). The works [**59**], [**25**], [**36**], [**39**], [**38**], [**37**], [**56**], [**23**], [**55**], [**63**], [**22**], [**3**] discuss methods and source optimization strategies for the active control of scalar fields modeled by the Helmholtz equation. Common applications include active noise cancellation ([**48**], [**24**], [**11**], [**32**], [**42**]), sound synthesis and reproduction ([**3**], [**30**], [**50**],[**68**],[**62**], [**17**]), and active control of acoustic scattered fields with application to cloaking and shielding ([**6**],

2020 *Mathematics Subject Classification.* Primary 35R30, 35J05, 45Q05; Secondary 35R25, 45P05, 93C20, 31B10.

Key words and phrases. Active Cloaking, Helmholtz Equation, Integral equations.

[**67**], [**65**],[**66**],[**35**], [**20**], [**52**], [**41**]). A more comprehensive discussion and analysis of some of the methods employed in these applications is also given in [**10**] for scattered field control and in [**21**] and [**1**] for sound field synthesis.

There is an extensive literature on the active control of Helmholtz fields. In [**36**] and [**39**], active scalar Helmholtz controls were constructed through generalized Calderon potentials and boundary projection operators while in [**47**] and [**67**], the authors employed the Green representation theorem to characterize active controls for the scalar Helmholtz equation. Multizone field synthesis and reproduction were studied using wave-domain methods (such as in [**26**], [**27**]) or modal-domain approaches (see [**46**], [**69**]). An algorithm for the reduction of scattering interference to the overall multizone scalar field synthesis effect was presented in [**68**]. In [**51**] and [**52**] the author made use of boundary integral operators to produce a stable unified control strategy in the case of a single active surface source proving the active control of radiated scalar fields in prescribed exterior region of space. These theoretical results were later applied for the 2D and 3D numerical study in [**28**], [**53**], [**17**], [**18**]. Other ideas that are used in the context of Helmholtz field control are discussed and analyzed in [**4**], [**21**], [**58**] and [**70**]. The monographs [**29**], [**31**], [**9**] and [**7**] discuss at length the problem of scalar fields propagation in oceans and layered media.

The problem of far field synthesis requires the construction of necessary boundary inputs on the active sources for the approximation of a given far field pattern [**5**],[**2**] ("the far field synthesis") (see also the monograph of [**16**] where general radiation theory and source synthesis techniques are discussed). The works [**8, 14, 15, 33**] discuss the problem of pattern synthesis with prescribed exterior nulls in specified parts of the near field region. In [**40, 61**] the authors assume physical optics approximations (i.e., small wavelengths relative to the size of the obstacles) and consider obstacles positioned in specified subregions of the near field (see also the review [**12**], for a genetic optimization algorithm based on the far-field to near field mapping for a solution). In our formulation below, the active field control problem is formulated in homogeneous media in the presence of known impenetrable obstacles and it requires the characterization of optimal boundary inputs on the active sources for the approximation of various prescribed scalar fields in given exterior regions and specified far field directions.

Computational and analytical methods for scalar scattering and radiation are presented in many excellent monograph in the literature and we mention here just several major monographs, [**44, 57, 64**].

In this paper, we build up on our previous work and produce a theoretical result concerning the active control of radiated fields showing that we can approximate given scalar fields in prescribed exterior regions and given far field directions in the presence of known impenetrable obstacles. The paper is organized as follows: In the first part of Section 2 we introduce our main question and present the mathematical formulation of the main problem we solve. Then , in the second part of the section we present the argument leading to our main result. In Section 3 we show several numerical results to support our theoretical findings. The last Section presents our conclusions and remarks about future works.

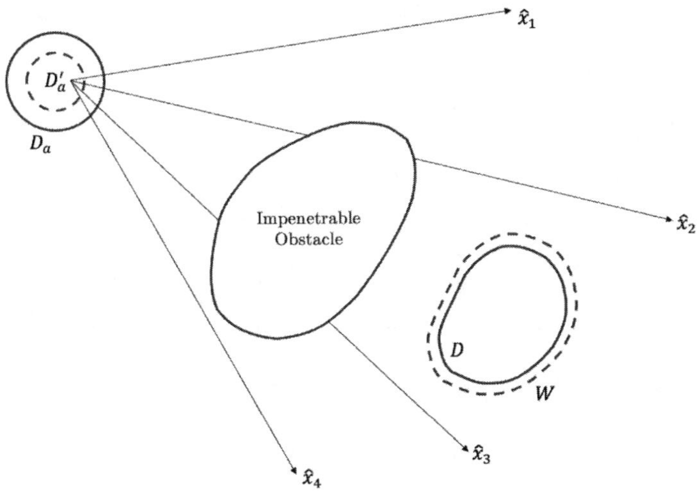

FIGURE 1. Sketch of the problem geometry

2. Statement of the Problem

The main question we plan to address concerns the characterization of physical boundary inputs on a given source support so that the radiated scalar field approximates desired behaviors in prescribed mutually disjoint exterior regions of space and desired far-field patterns in prescribed far-field directions in the presence of known exterior impenetrable obstacles.

For the simplicity of exposition while preserving the generality of the argument (see Remarks 2.2, 2.3, 2.4, 2.5), we will next consider the case where only one sound-soft obstacle is present with only one exterior control region and N far-field directions. See Figure 1 for a sketch of the geometry where only four far-field directions are drawn.

For the mathematical formulation of the problem, we need to introduce several notations. Let D_a be a compact smooth domain representing the physical source support. Consider D a compact control region and O be the open smooth set representing the sound-soft obstacle. Also, let $\hat{\mathbf{x}}_1, \hat{\mathbf{x}}_2, \ldots, \hat{\mathbf{x}}_N$ be unit vectors designating the N given far-field directions. As in one of the previous works, see [**17, 18, 51–54**], we assume a fictitious source domain $D'_a \subset\subset D_a$ and a fictitious control domain $W \supset\supset D$ so that

$$\overline{D_a} \cap \overline{W} = \emptyset, \left(\overline{D_a} \cup \overline{W}\right) \cap \overline{O} = \emptyset,$$

where $\subset\subset$ denotes compact embedding. Without restricting the generality, we also assume that the non-physical fictitious domains D'_a and W are smooth. Mathematically, the problem can be formulated as follows. Consider the incident field $u_{inc}(\mathbf{x}) = \int_{\partial D'_a} w(\mathbf{y})\Phi(\mathbf{x},\mathbf{y})\, dS_y$ and assume there is no coupling between the obstacle O and the source D_a, e.g., the obstacle is in the far-field of the source. Let $0 < \mu \ll 1$, u_1 be a Helmholtz potential in W, $f_1, f_2, \ldots, f_N$ be given complex numbers. The problem is to determine the density w such that the total field u

obtained (radiated and scattered), i.e., $u = u_{inc} + u_s$ where u_s solves

$$
\begin{cases} \Delta u_s + k^2 u_s = 0, & \text{in } \mathbb{R}^3 \setminus \overline{O} \\ \langle \frac{\mathbf{x}}{|\mathbf{x}|}, \nabla u_s(\mathbf{x}) \rangle - iku_s(\mathbf{x}) = o\left(\frac{1}{|x|}\right), & \text{uniformly for all } \frac{\mathbf{x}}{|\mathbf{x}|} \text{ as } \mathbf{x} \to \infty \\ u_s = -u_{inc}, & \text{on } \partial O. \end{cases} \tag{2.1}
$$

satisfies

$$
\begin{cases} \|u - u_1\|_{L^2(\partial W)} < \mu \\ |u_\infty(\hat{\mathbf{x}}_i) - f_i| < \mu, & \text{for } i = 1, \dots, N \end{cases} \tag{2.2}
$$

where $u_\infty(\hat{\mathbf{x}}_i)$ denotes the far-field pattern of u in the direction of the unit vector $\hat{\mathbf{x}}_i$, $i = 1, \dots, N$.

Classical results [**13, 43, 45**] imply the existence of a unique solution. For simplicity, assuming k is not a Helmholtz Dirichlet eigenvalues for the interior problem in W, O, and D'_a. We look for u_s in the form of a double-layer potential

$$
u_s(\mathbf{z}) = \int_{\partial O} v(\mathbf{y}) \frac{\partial \Phi}{\partial \nu_{\mathbf{y}}}(\mathbf{z}, \mathbf{y})\, dS_{\mathbf{y}}, \quad \text{for } z \in \mathbb{R}^3 \setminus \partial O, \tag{2.3}
$$

where $\Phi(\mathbf{x}, \mathbf{y}) = \frac{e^{ik|\mathbf{x}-\mathbf{y}|}}{4\pi|\mathbf{x}-\mathbf{y}|}$ is the fundamental solution of the 3D Helmholtz equation. Imposing the boundary condition on ∂O and using the jump of the double-layer potential operator, we get that v satisfies

$$
\left(\mathcal{K}_O + \frac{1}{2} I\right) v = -\mathcal{P}_O w, \quad \text{on } \partial O,
$$

where $\mathcal{K}_O : L^2(\partial O) \to L^2(\partial O)$ and $\mathcal{P}_O : L^2(\partial D'_a) \to L^2(\partial O)$ are given by

$$
\mathcal{K}_O v(\mathbf{x}) = \int_{\partial O} v(\mathbf{x}) \frac{\partial \Phi}{\partial \nu_y}(\mathbf{x}, \mathbf{y})\, dS_{\mathbf{y}}, \quad \mathbf{x} \in \partial O,
$$

and

$$
\mathcal{P}_O w(\mathbf{z}) = \int_{\partial D'_a} w(\mathbf{y}) \Phi(\mathbf{x}, \mathbf{y})\, dS_{\mathbf{y}}, \quad \mathbf{z} \in \partial O,
$$

respectively. Then, since k is not a resonance, we get

$$
v = -\left(\mathcal{K}_O + \frac{1}{2} I\right)^{-1} \mathcal{P}_O w
$$

and

$$
u_s(\mathbf{z}) = -\int_{\partial O} \left(\mathcal{K}_O + \frac{1}{2} I\right)^{-1} \mathcal{P}_O w(\mathbf{y}) \frac{\partial \Phi}{\partial \nu_{\mathbf{y}}}(\mathbf{z}, \mathbf{y})\, dS_{\mathbf{y}}.
$$

Then the total field in W is given by

$$
\begin{aligned} u(\mathbf{z}) &= u_{inc}(\mathbf{z}) + u_s(\mathbf{z}) \\ &= \mathcal{P}_W w(\mathbf{z}) - \mathcal{D}_{O,W} \left(\mathcal{K}_O + \frac{1}{2} I\right)^{-1} \mathcal{P}_O w(\mathbf{z}), \end{aligned}
$$

where $\mathcal{P}_W : L^2(\partial D'_a) \to L^2(\partial W)$ and $\mathcal{D}_{O,W} : L^2(\partial O) \to L^2(\partial W)$ are defined by

$$
\begin{aligned} \mathcal{P}_W w(\mathbf{z}) &= \int_{\partial D'_a} w(\mathbf{y}) \Phi(\mathbf{z}, \mathbf{y})\, dS_{\mathbf{y}}, \quad z \in \partial W, \\ \mathcal{D}_{O,W} g(\mathbf{z}) &= \int_{\partial O} g(\mathbf{y}) \frac{\partial \Phi}{\partial \nu_{\mathbf{y}}}(\mathbf{z}, \mathbf{y})\, dS_{\mathbf{y}}, \quad z \in \partial W, \quad \text{for } g \in L^2(\partial O). \end{aligned} \tag{2.4}
$$

Moreover, the total field in the far-field direction $\hat{\mathbf{x}}_i$ is given by
(2.5)

$$u_\infty(\hat{\mathbf{x}}_i) = \frac{1}{4\pi}\int_{\partial D'_a} w(\mathbf{y})e^{-ik\hat{\mathbf{x}}_i\cdot\mathbf{y}}\,dS_{\mathbf{y}} - \frac{1}{4\pi}\int_{\partial O}\left(\mathcal{K}_O + \frac{1}{2}I\right)^{-1}\mathcal{P}_O w(\mathbf{y})\frac{\partial e^{-ik\hat{\mathbf{x}}_i\cdot\mathbf{y}}}{\partial\nu_{\mathbf{y}}}\,dS_{\mathbf{y}}$$

for $i = 1, \ldots N$. Hence, we can now introduce the total propagator operator $\mathcal{T}$: $L^2(\partial D'_a) \to L^2(\partial W) \times \mathbb{C}^N$ by

$$\text{(2.6)} \quad \mathcal{T}w = \left(\mathcal{P}_W w(\mathbf{z}) - \mathcal{D}_{O,W}\left(\mathcal{K}_O + \frac{1}{2}I\right)^{-1}\mathcal{P}_O w(\mathbf{z}), u_\infty(\hat{\mathbf{x}}_1), \ldots, u_\infty(\hat{\mathbf{x}}_N)\right)$$

where $\mathcal{P}_W$ and $\mathcal{D}_{O,W}$ are defined in (2.4) and u_∞ is defined in (2.5).

THEOREM 2.1. *The operator $\mathcal{T}$ defined above has a dense range.*

PROOF. To show this, since $\mathcal{T}$ is linear and compact, it suffices to show that $\ker \mathcal{T}^* = \{0\}$ where $\ker \mathcal{T}^*$ denotes the kernel of the adjoint operator T^*. The adjoint operator $\mathcal{T}^* : L^2(\partial W) \times \mathbb{C}^N \to L^2(\partial D'_a)$ is given by

$$\langle \mathcal{T}w, \tilde{\psi}\rangle_{L^2(\partial W)\times\mathbb{C}^N} = \langle w, \mathcal{T}^*\tilde{\psi}\rangle_{L^2(\partial D'_a)}$$

where $\tilde{\psi} = (\psi, c_1, \ldots, c_N)$,

$$\langle \psi_1, \psi_2\rangle_{L^2(\partial W)} = \int_{\partial W} \psi_1\overline{\psi_2}$$

and

$$\langle (c_1, c_2, \ldots, c_N), (d_1, d_2, \ldots, d_N)\rangle_{\mathbb{C}^N} = \sum_{i=1}^N c_i\bar{d}_i$$

for every $\psi_1, \psi_2 \in L^2(\partial W)$ and $(c_1, c_2, \ldots, c_N), (d_1, d_2, \ldots, d_N) \in \mathbb{C}^N$. With these, we get

$$\begin{aligned}
\text{(2.7)} \quad \langle Tw, \tilde{\psi}\rangle_{L^2(\partial)\times\mathbb{C}^N} &= \left\langle \mathcal{P}_W w(\mathbf{z}) - \mathcal{D}_{O,W}\left(\mathcal{K}_O + \frac{1}{2}I\right)^{-1}\mathcal{P}_O w, \psi\right\rangle_{L^2(\partial W)} \\
&\quad + \sum_{i=1}^N u_\infty(\hat{x}_i)\cdot\bar{c_i} \\
&= \left\langle w, \mathcal{P}_W^*\psi - \mathcal{P}_O^*\left(\mathcal{K}'_O + \frac{1}{2}I\right)^{-1}\mathcal{D}_{O,W}^*\psi\right\rangle_{L^2(\partial D'_a)} \\
&\quad + \sum_{i=1}^N \frac{1}{4\pi}\left[\int_{\partial D'_a} w(\mathbf{y})e^{-ik\hat{\mathbf{x}}_i\cdot\mathbf{y}}\,dS_{\mathbf{y}}\right. \\
&\quad \left. - \int_{\partial O}\left(\mathcal{K}_O + \frac{1}{2}I\right)^{-1}\mathcal{P}_O w(\mathbf{z})\frac{\partial e^{-ik\hat{\mathbf{x}}_i\cdot\mathbf{z}}}{\partial\nu_{\mathbf{z}}}\,dS_{\mathbf{z}}\right]\cdot\bar{c_i}
\end{aligned}$$

where $\mathcal{P}_W^* : L^2(\partial W) \to L^2(\partial D'_a)$ is the adjoint of $\mathcal{P}_W$ given by

$$\mathcal{P}_W^*\psi(\mathbf{y}) = \int_{\partial W}\psi(\mathbf{z})\overline{\Phi}(\mathbf{z},\mathbf{y})\,dS_{\mathbf{z}}, \quad \mathbf{y} \in \partial D'_a,$$

$\mathcal{P}_O^* : L^2(\partial O) \to L^2(\partial D'_a)$ is the adjoint of $\mathcal{P}_O$ defined by

$$\mathcal{P}_O^*\xi(\mathbf{y}) = \int_{\partial O}\xi(\mathbf{z})\overline{\Phi}(\mathbf{z},\mathbf{y})\,dS_{\mathbf{z}}, \quad \mathbf{y} \in \partial D'_a,$$

$\mathcal{K}'_O : L^2(\partial O) \to L^2(\partial O)$ is the adjoint of $\mathcal{K}_O$ given in [**13**],

$$\mathcal{K}'_O \xi(\mathbf{x}) = \int_{\partial O} \xi(\mathbf{y}) \overline{\frac{\partial \Phi}{\partial \nu_{\mathbf{y}}}}(\mathbf{x}, \mathbf{y})\, dS_{\mathbf{y}}, \quad \mathbf{y} \in \partial O,$$

and lastly, $\mathcal{D}^*_{O,W} : L^2(\partial W) \to L^2(\partial O)$ is the adjoint of $\mathcal{D}_{O,W}$ given by

$$\mathcal{D}^*_{O,W} \psi(\mathbf{y}) = \int_{\partial W} \psi(\mathbf{z}) \overline{\frac{\partial \Phi}{\partial \nu_{\mathbf{y}}}}(\mathbf{z}, \mathbf{y})\, dS_{\mathbf{y}}, \quad \mathbf{z} \in \partial O.$$

So, algebraic manipulation of (2.7) gives

(2.8)

$$\begin{aligned}\langle Tw, \tilde{\psi} \rangle_{L^2(\partial) \times \mathbb{C}^N} = &\left\langle w, \left(\mathcal{P}^*_W - \mathcal{P}^*_O \left(\mathcal{K}'_O + \frac{1}{2} I \right)^{-1} \mathcal{D}^*_{O,W} \right) \psi \right\rangle_{L^2(\partial D'_a)} \\ &+ \int_{\partial D'_a} w(\mathbf{y}) \overline{\sum_{i=1}^N \frac{1}{4\pi} \left(e^{ik\hat{\mathbf{x}}_i \cdot \mathbf{y}} - \mathcal{P}^*_O \left(\mathcal{K}'_O + \frac{1}{2} I \right)^{-1} \frac{\partial e^{ik\hat{\mathbf{x}}_i \cdot \mathbf{z}}}{\partial \nu_{\mathbf{z}}} dS_{\mathbf{z}} \right) \cdot c_i},\end{aligned}$$

where $\frac{\partial e^{ik\hat{\mathbf{x}}_i \cdot \mathbf{z}}}{\partial \nu_{\mathbf{z}}}$ is considered defined for $\mathbf{z} \in \partial O$. Consequently, from (2.8), we obtain

$$\begin{aligned}\mathcal{T}^* \tilde{\psi}(\mathbf{y}) = &\left(\left(\mathcal{P}^*_W - \mathcal{P}^*_O \left(\mathcal{K}'_O + \frac{1}{2} I \right)^{-1} \mathcal{D}^*_{O,W} \right) \psi \right)(\mathbf{y}) \\ &+ \frac{1}{4\pi} \sum_{i=1}^N \left(e^{ik\hat{\mathbf{x}}_i \cdot \mathbf{y}} - \mathcal{P}^*_O \left(\mathcal{K}'_O + \frac{1}{2} I \right)^{-1} \left(\frac{\partial e^{ik\hat{\mathbf{x}}_i \cdot \mathbf{z}}}{\partial \nu_{\mathbf{z}}} \right)(\mathbf{y}) \right) \cdot c_i,\end{aligned} \tag{2.9}$$

for $\mathbf{y} \in \partial D'_a$ Now, we define

$$\begin{aligned}f(\mathbf{x}) = &\int_{\partial W} \psi(\mathbf{y}) \bar{\Phi}(\mathbf{x}, \mathbf{y})\, dS_{\mathbf{y}} - \int_{\partial O} \left(\mathcal{K}'_O + \frac{1}{2} I \right)^{-1} \mathcal{D}^*_{O,W} \psi(\mathbf{z}) \bar{\Phi}(\mathbf{x}, \mathbf{z})\, dS_{\mathbf{z}} \\ &+ \frac{1}{4\pi} \sum_{i=1}^N \left(e^{ik\hat{\mathbf{x}}_i \cdot \mathbf{x}} - \int_{\partial O} \left(\mathcal{K}'_O + \frac{1}{2} I \right)^{-1} \frac{\partial e^{ik\hat{\mathbf{x}}_i \cdot \mathbf{z}}}{\partial \nu_{\mathbf{z}}} \bar{\Phi}(\mathbf{x}, \mathbf{z})\, dS_{\mathbf{z}} \right) \cdot c_i,\end{aligned} \tag{2.10}$$

for $\mathbf{x} \in \overline{D'_a}$. For $\tilde{\psi} = (\psi, c_1, \ldots, c_N) \in \ker \mathcal{T}^*$, from (2.9) and (2.10),

$$f(\mathbf{y}) = \mathcal{T}^* \tilde{\psi}(\mathbf{y}), \quad \mathbf{y} \in \partial D'_a$$

and thus, f solves the homogeneous Dirichlet problem in D'_a. Therefore, since k is not an interior Helmholtz Dirichlet eigenvalue for D'_a, using uniqueness, we have that

$$f = 0 \text{ on } D'_a,$$

and analytic continuation implies

$$f = 0 \text{ in } \mathbb{R}^3 \setminus \left(\overline{W} \cup \overline{O} \right). \tag{2.11}$$

Continuity of the single-layer implies $f = 0$ on $\partial W \cup \partial O$. Since f satisfies the Helmholtz homogeneous equation in W and k is not an interior Helmholtz Dirichlet eigenvalue in W, from (2.11) and uniqueness of the interior Dirichlet problem in W, we get

$$f = 0 \text{ in } \overline{W}. \tag{2.12}$$

Using the jump relation of the normal derivatives of the single-layer operator on ∂W, we obtain

$$\psi = 0 \text{ on } \partial W. \tag{2.13}$$

Using (2.13) in (2.10), we get

$$f(\mathbf{x}) = \frac{1}{4\pi}\sum_{i=1}^{N}\left(e^{ik\hat{\mathbf{x}}_i\cdot\mathbf{x}} - \int_{\partial O}\left(\mathcal{K}_O' + \frac{1}{2}I\right)^{-1}\frac{\partial e^{ik\hat{\mathbf{x}}_i\cdot\mathbf{z}}}{\partial\nu_{\mathbf{z}}}\bar{\Phi}(\mathbf{x},\mathbf{z})\,dS_{\mathbf{z}}\right)\cdot c_i. \tag{2.14}$$

From (2.11) and (2.12), we derive

$$f = 0 \text{ in } \mathbb{R}^3 \setminus \overline{O}. \tag{2.15}$$

Continuity of the single-layer potential on ∂O gives

$$f = 0 \text{ on } \partial O. \tag{2.16}$$

Since f satisfies the homogeneous Helmholtz equation in O and k is not an interior Helmholtz Dirichlet eigenvalue in O, relation (2.16) and uniqueness of the interior Dirichlet problem in O implies

$$f = 0 \text{ in } O. \tag{2.17}$$

Using again the jump relations of the normal derivatives of the single-layer potential on ∂O, we obtain from (2.15) and (2.17),

$$\sum_{i=1}^{N}\left(\mathcal{K}_O' + \frac{1}{2}I\right)^{-1}\frac{\partial e^{ik\hat{\mathbf{x}}_i\cdot\mathbf{z}}}{\partial\nu_{\mathbf{z}}}\cdot c_i = 0 \text{ on } \partial O. \tag{2.18}$$

With (2.18) in (2.14), (2.15), (2.16), (2.17), we have

$$f(\mathbf{y}) = \frac{1}{4\pi}\sum_{i=1}^{N} e^{ik\hat{\mathbf{x}}_i\cdot\mathbf{y}}c_i = 0 \text{ in } \mathbb{R}^3, \quad \text{for all } \mathbf{y}\in\mathbb{R}^3. \tag{2.19}$$

Proceeding as in [**19**], we first fix a unit vector $\mathbf{y}\in\mathbb{R}^3$ and define $\mathbf{y}_p = p\mathbf{y}$ for $p = 0,\dots,N-1$ so that from (2.19), we obtain the Vandermonde system

$$\underbrace{\begin{bmatrix} 1 & 1 & \cdots & 1 \\ z_1 & z_2 & \cdots & z_N \\ \vdots & \vdots & \ddots & \vdots \\ z_1^{N-1} & z_2^{N-1} & \cdots & z_N^{N-1}\end{bmatrix}}_{:=Z}\begin{bmatrix} c_1 \\ c_2 \\ \vdots \\ c_N\end{bmatrix} = \begin{bmatrix} 0 \\ 0 \\ \vdots \\ 0\end{bmatrix} \tag{2.20}$$

where $z_i = e^{ik\hat{\mathbf{x}}_i\cdot\mathbf{y}}$. The unique solution of (2.20) is the trivial solution where $c_i = 0$ for all $i = 1,\dots,N$ whenever the coefficient matrix Z is nonsingular. The determinant of the Vandermonde matrix above is given by

$$det(Z) = \prod_{1\le j\le 1\le N}(z_j - z_i) \tag{2.21}$$

which is zero if and only if at least one of the differences in the product is zero, i.e., there are indices j_1, j_2 such that $z_{j_1} = z_{j_2}$, equivalently this means that

$$(\hat{\mathbf{x}}_{j_1} - \hat{\mathbf{x}}_{j_2})\cdot\mathbf{y} = \frac{2\pi}{k}M \tag{2.22}$$

for some integer M. Hence, we get $|M| = \frac{k}{2\pi}|(\hat{\mathbf{x}}_{j_1} - \hat{\mathbf{x}}_{j_2})\cdot\mathbf{y}| \le \frac{k}{\pi}$ by triangle inequality and Cauchy-Schwarz inequality. This means that choosing $\mathbf{y}$ on the exterior of the hyperplanes defined by $\binom{N}{2}\left(2\lfloor\frac{k}{\pi}\rfloor+1\right)$ equations of the form (2.22), we observe that this gives a set of N $\mathbf{y}$ values, which are the $\mathbf{y}_p$ vectors, that results to (2.19) having the solution $c_i = 0$ for all $i = 1, \dots, N$ and thus, $\mathcal{T}$ has a dense range. □

REMARK 2.2. Note that the physical source input on ∂D_a is obtained from u_{inc} introduced above either as pressure

$$p(\mathbf{x}) = \int_{\partial D_a'} w(\mathbf{y})\Phi(\mathbf{x},\mathbf{y})\,dS_{\mathbf{y}} \tag{2.23}$$

for $\mathbf{x} \in \partial D_a$ or normal velocity

$$v(\mathbf{x}) = \frac{-i}{\rho c k}\frac{\partial}{\partial \nu_{\mathbf{x}}}\int_{\partial D_a'} w(\mathbf{y})\Phi(\mathbf{x},\mathbf{y})\,dS_{\mathbf{y}} \tag{2.24}$$

for $\mathbf{x} \in \partial D_a$ where ρ denotes the density of the surrounding medium and c denotes the speed of propagation of sound in the respective media.

REMARK 2.3. It can be immediately seen that by using a single layer representation of the scattered field, the arguments presented in Theorem 2.1 can be adapted to the case when a sound-hard obstacle is used.

REMARK 2.4. We use a smooth fictitious source $D_a' \subset\subset D_a$ which simplifies the computation (we assume D_a' is a spherical domain) and ensures that, the boundary inputs given in Remark 2.2 will be smooth.

REMARK 2.5. By employing interior stability estimates, it was shown in [**52**] that L^2-control in ∂W is enough to show smooth interior control. Thus, we have that $\|u - u_1\|_{L^2(D)} < \epsilon$ for some $0 < \epsilon \ll 1$.

REMARK 2.6. If coupling between the obstacle is neglected then the theoretical argument presented in Theorem 2.1 extend immediately to the case of more impenetrable obstacles of the same type and also to the case where one has sound-soft and sound-hard obstacles (in this case, the well-posedness is implied by [**60**]).

REMARK 2.7. The arguments of Theorem 2.1 can be extended to all possible values of k which are made by considering a modified Green's function on a modified layer potential approach in the spirit of [**45**] and [**60**].

3. Numerical Implementation

In this section, we demonstrate a numerical scheme to support the theoretical concepts presented in the previous section. We consider problem (2.1), without far-field constraints, which may be summarized by finding a function $w \in L^2(D_a')$ such that

$$\mathcal{T}w = f \tag{3.1}$$

for a prescribed field $f = (f_1, \dots, f_m) \in Y$, where $\mathcal{T}$ is defined in (2.6), with no far-field direction. In this numerical scheme, we will be using N point sources $\mathbf{y}_j$, $j = 1, \dots, N$, which may be viewed as a discretization of the compact source D_a' described in the previous section or in their own as suggested by the following numerical results, a spherical sound-soft obstacle O, and a control region W. Note that we set the center of the obstacle as the center of the reference 3D system in this implementation. We solve (3.1) by discretizing the control region into a mesh of M collocation points $\mathbf{z}_p$, $p = 1, \ldots, M$ and using far-field approximation and spherical harmonics to represent the density w. Hence, we transform (3.1) into a linear system

$$A_d w_d \approx b \tag{3.2}$$

where A_d is the $M \times N$ coefficient matrix of moments, M is the number of control points and N is the number of point sources we consider. We compute w_d as the Tikhonov solution

$$w_d = (\alpha I + A_d^* A_d)^{-1} A_d^* b, \tag{3.3}$$

where $0 < \alpha \ll 1$ is computed using Morozov discrepancy principle, which is a method to find an optimal parameter $\alpha > 0$ such that

$$\|A_d w_d - b\|_{\ell^2} = \delta,$$

for some accuracy criterion $0 < \delta \ll 1$ and A_d^* is the conjugate transpose of A_d.

3.1. Point sources are placed at least 10 wavelengths from the obstacle and are close to the control region. In this simulation, we assume that the obstacle is a sphere of radius R centered at the origin which is in the far field of the point sources. Thus, as developed in [**13**], where using the far-field approximation, i.e., $|\mathbf{y}_j| \gg 1$

$$u_{inc}(\mathbf{x}) = \sum_{j=1}^{N} w_j \frac{e^{ik|\mathbf{x}-\mathbf{y}_j|}}{4\pi|\mathbf{x}-\mathbf{y}_j|} \approx \sum_{j=1}^{N} w_j \underbrace{\frac{e^{ik|\mathbf{y}_j|}}{|\mathbf{y}_j|}}_{:=d_j} e^{-k\hat{\mathbf{y}}_j \cdot \hat{\mathbf{x}} \cdot R}, \quad \text{for } \mathbf{x} = R \cdot \hat{\mathbf{x}} \in \partial O \tag{3.4}$$

where w_j are scalar entries of the (acoustic) density vector representing each point source $\mathbf{y}_j$ and R is the radius of the sound-soft obstacle O. By using (3.4) and the addition theorem for the spherical harmonic representation of plane waves, see [**13**], we obtain

$$u_{inc}(\mathbf{x}) = \sum_{j=1}^{N} w_j \frac{e^{ik|\mathbf{x}-\mathbf{y}_j|}}{4\pi|\mathbf{x}-\mathbf{y}_j|} \approx \sum_{j=1}^{N} 4\pi w_j d_j \sum_{n=0}^{\infty} \sum_{m=-n}^{n} i^n j_n(kR) Y_n^m(\hat{\mathbf{x}}) \overline{Y_n^m(-\hat{\mathbf{y}}_j)} \tag{3.5}$$

where j_n is the regular spherical Bessel function of order n and Y_n^m are the spherical harmonics. On the other hand, it is know that

$$u_s(\mathbf{z}) = \sum_{n=0}^{\infty} \sum_{m=-n}^{n} a_{mn} h_n^{(1)}(k|\mathbf{z}|) Y_n^m(\hat{\mathbf{z}}), \quad \mathbf{z} \in \mathbb{R}^3 \setminus O \tag{3.6}$$

where $h_n^{(1)}$ is the spherical Hankel function of order n. Using (3.4) – (3.6), the fact that $u_{inc}(\mathbf{x}) = -u_s(\mathbf{x})$ on ∂O, and orthogonality of the spherical harmonics, we get that the coefficients a_{mn} are computed by

$$a_{mn} = -4\pi i^n \frac{j_n(kR) \sum_{j=1}^{N} d_j \overline{Y_n^m(-\hat{\mathbf{y}}_j)} w_j}{h_n^{(1)}(kR)}. \tag{3.7}$$

Now, we know that the total field u is given by $u(\mathbf{z}) = u_{inc}(\mathbf{z}) + u_s(\mathbf{z})$, where u_{inc} is the exact expression of the incident field, i,.e, the first equality, given in (3.5) and u_s is given by (3.6), (3.7).

In all the following simulations, we use the wave number $k = 10$ with direction of propagation $\hat{\mathbf{k}} = [1, 0, 0]$, and a spherical obstacle with radius $R = 1$. We also approximate the acoustic field using 30 harmonic orders, for a total of 961 spherical harmonics, and $N = 150$ point sources, and in this section, we set the point sources as points around a sphere of radius 0.7 centered at [4,-4,-4] so that we set the point sources at least 10 wavelengths away from the obstacle and we define the control region as

$$W = \left\{ (r, \theta, \phi) \in \mathbb{R}^3 : r \in [1, 1.5], \theta \in \left[-\frac{\pi}{8}, \frac{\pi}{10} \right], \phi \in \left[\frac{49\pi}{128}, \frac{79\pi}{128} \right] \right\}. \tag{3.8}$$

shifted to $[4.25, -4, -4]$ so that the region is one wavelength away from the sources, and represent it with 580 control points. Figure 2 shows the geometry of the problem (where a 2D sketch of the 3D geometry is rendered).

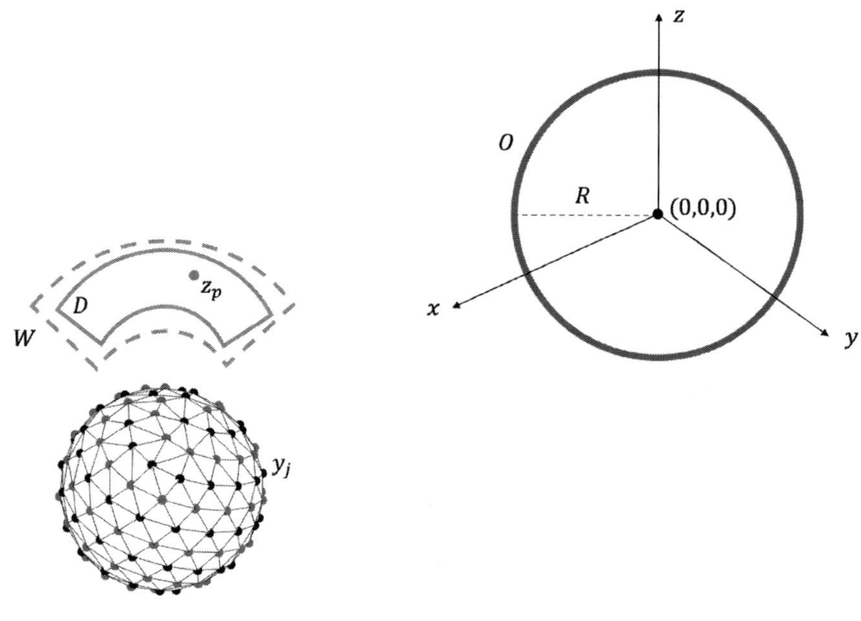

FIGURE 2. Geometry of the numerical scheme

We prescribe a field on W using a plane wave $f(\mathbf{z}) = e^{k \cdot \hat{\mathbf{k}} \cdot \mathbf{z}}$, for $\mathbf{z} \in W$. Since we represent W a set of 580 control points, we let $\mathbf{z} = \mathbf{z}_p$ for $p = 1, \ldots, 580$. Figure 3 shows the real parts of the generated field and the prescribed field on a mesh of W (which is different than the computational mesh of W), and the pointwise relative error on each point $\mathbf{z}_p$. We further note that in this particular simulation, maximum relative error on W is 0.39%. We further reiterate that the figures concerning W are on a slightly shifted mesh of W.

We then indicate the ℓ^2 norm of w_d, which is an indication of the power radiated by the point sources. In this simulation is $\|w_d\|_{\ell^2} = 8.6021$. Figure 4 shows the contribution, in magnitude, of each point source $\mathbf{y}_j$ for the source strength. In Figure 5, we exhibit the interpolated prescribed and generated fields on the $x = 5.40$ cross section of the control region W.

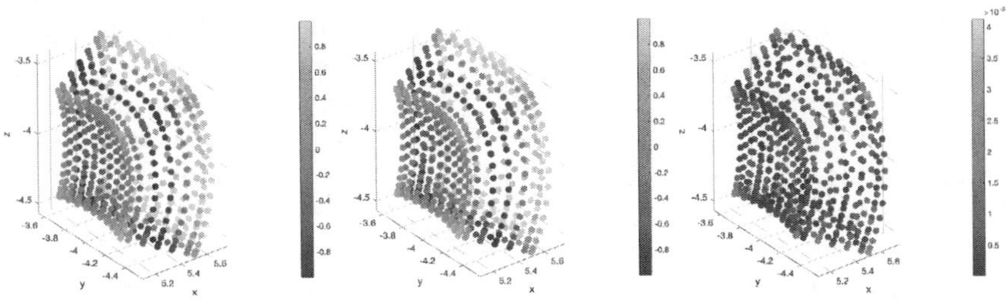

(A) (B) (C)

FIGURE 3. Real parts of the (A) generated and (B) desired fields on a shifted mesh of W, and (C) the pointwise relative error on W

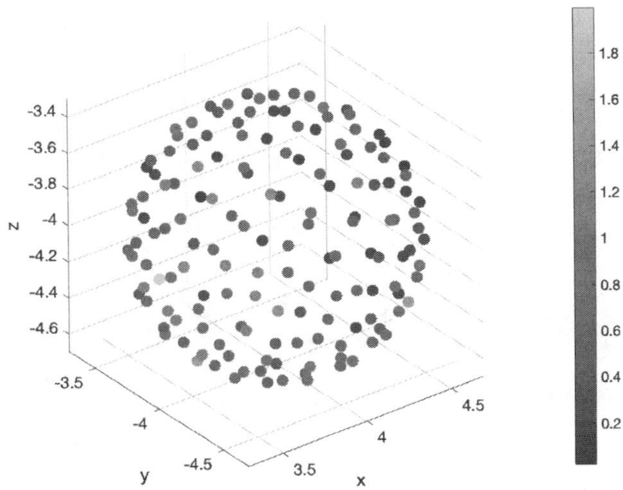

FIGURE 4. Power radiated by each point source $\mathbf{y}_j$ (magnitude)

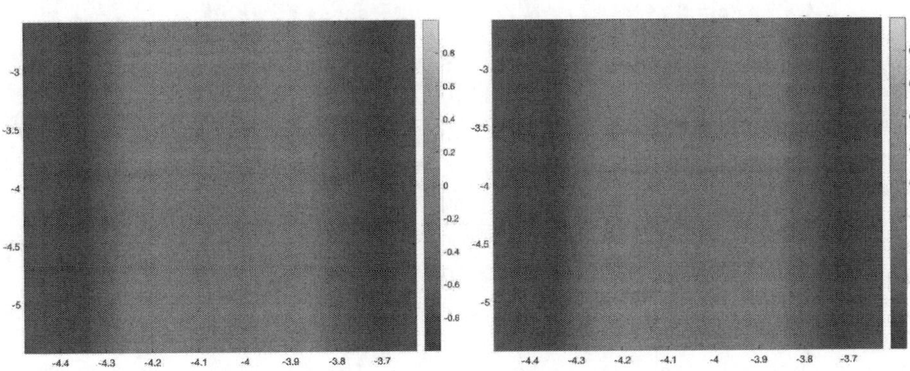

(A) (B)

FIGURE 5. (A) Desired and (B) generated fields on a cross section of W

Next, moving the control region to two wavelengths away from the source by setting its center to $[5, -4, -4]$, the numerical simulation shows that the maximum relative error is on the same order, which is 0.32%. On the other hand, $\|w_d\|_{\ell^2} = 12.5796$, which shows 46% increase from the previous case. However, if we raise the number of point sources to 300, the radiating effort and relative error both remain in the same order as in the first experiment indicating that more sources ease up the total radiating effort. Further, positioning the center of the control region to $[7, -4, -4]$, so that the distance between the point sources and the region is about five wavelengths, we observe that the relative error is kept at the same order, which is 0.29%, and the radiating effort of the point sources rises to $\|w_d\|_{\ell^2} = 27.4611$. Hence, the simulation suggests that the control points may be placed further away from the sources, however, this may increase the power required to be radiated by the sources.

3.2. Point sources are placed at least 10 wavelengths from the obstacle and the control region is behind the obstacle, i.e., Let $\mathbf{y}_j$, $j = 1, \ldots, N$ be the point sources and if we have $|\mathbf{y}_j| \gg |\mathbf{z}|$ for every $j = 1, \ldots, N$, $\mathbf{z} \in W$, we can use the far-field approximation to see that

$$\frac{e^{ik|\mathbf{z}-\mathbf{y}_j|}}{4\pi|\mathbf{z}-\mathbf{y}_j|} \approx \frac{e^{ik|\mathbf{y}_j|}}{|\mathbf{y}_j|} \cdot e^{-ik\mathbf{z}\cdot\hat{\mathbf{y}}_j} \tag{3.9}$$

and with the plane wave representation, Laplace spherical functions, and addition theorem, we get

$$e^{-ik\mathbf{z}\cdot\hat{\mathbf{y}}_j} = 4\pi \sum_{n=0}^{\infty} \sum_{m=-n}^{n} i^n j_n(k|\mathbf{z}|) Y_n^m(\hat{\mathbf{z}}) \overline{Y_n^m(-\hat{\mathbf{y}}_j)}. \tag{3.10}$$

Hence, we see that with these configurations, the incident field may be expressed as

$$u_{inc}(\mathbf{z}) = \sum_{j=1}^{N} 4\pi \frac{e^{ik|\mathbf{z}-\mathbf{y}_j|}}{4\pi|\mathbf{z}-\mathbf{y}_j|} \approx \sum_{j=1}^{N} \frac{e^{ik|\mathbf{y}_j|}}{|\mathbf{y}_j|} \cdot \sum_{n=0}^{\infty} \sum_{m=-n}^{n} i^n j_n(k|\mathbf{z}|) Y_n^m(\hat{\mathbf{z}}) \overline{Y_n^m(-\hat{\mathbf{y}}_j)}, \tag{3.11}$$

and the scattered field u_s is obtained as before given by (3.6) and (3.7). In this section, we place the center of the 150 point sources to $[4, 0, -5]$ with radius of support 0.7 so that the sources are at least 10 wavelengths away from the obstacle and define the control region by

$$W = \left\{ (r, \theta, \phi) \in \mathbb{R}^3 : r \in [1, 1.5], \theta \in \left[\frac{7\pi}{8}, \frac{11\pi}{10}\right], \phi \in \left[\frac{17\pi}{64}, \frac{63\pi}{128}\right] \right\}, \tag{3.12}$$

and move its center to $[-0.75, 0, 0.75]$ so that the region is about one wavelength behind the obstacle, and represent this again using 580 collocation points. This geometry is conveyed in Figure 6 (where a 2D sketch of the 3D geometry is rendered).

In Figure 7, we can see the generated and prescribed field on a shifted mesh of W with the maximum relative error of 0.33% on W while $\|w_d\|_{\ell^2} = 1172.4$. Figure 8 shows the radiating effort contribution of each point source. Meanwhile, in Figure 9, we show the interpolated values for the prescribed and generated acoustic field on the cross section $x = -1.90$ of the control region W. If we take the control region closer to the obstacle, we see that we may be able to keep the pointwise relative error on the same order, however, one needs to increase the radiated effort of the point sources.

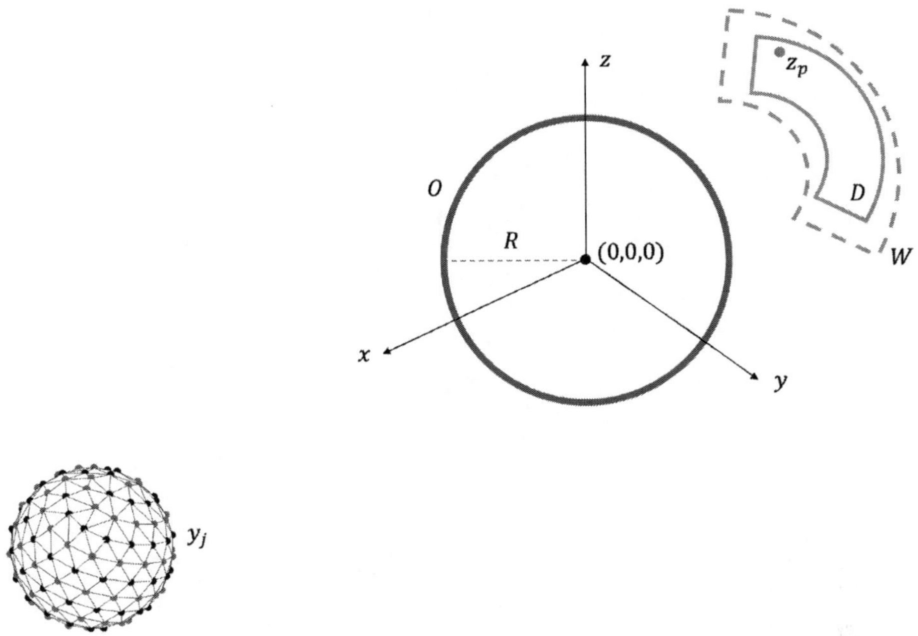

FIGURE 6. Geometry of the numerical scheme

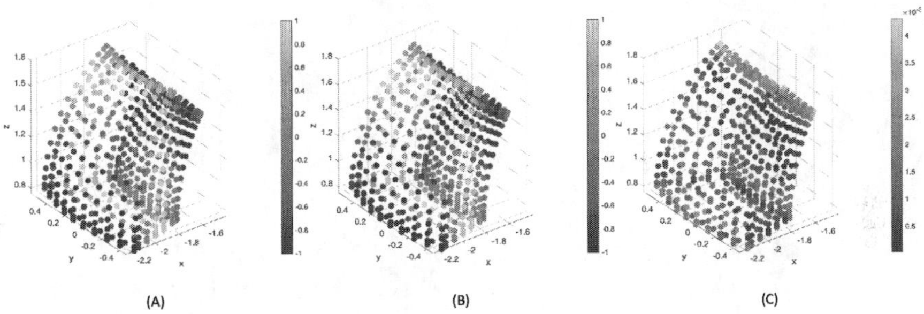

FIGURE 7. Real parts of the (A) generated and (B) prescribed fields on a shifted mesh of W, and (C) the pointwise relative error on W if the control region is one wavelength away behind the obstacle

In fact, if we move the region to about half a wavelength behind the obstacle, we observe that we can keep the error to a maximum of order 10^{-2} but $\|w_d\|_{\ell^2} = 1.1153 \times 10^6$, which is about 950% increase compared to the previous case and placing it as close as a quarter of a wavelength from the obstacle, we observe that the maximum relative error is 4.47% while the radiating effort increased by more than 200% to $\|w_d\|_{\ell^2} = 3.7357 \times 10^6$.

In Table 1, we note that we can keep the maximum pointwise relative error of the simulation and lower the total power required to be radiated if we increase the number of point sources. In fact, if we enlarge the control region in the azimuthal direction to $\theta \in \left[\frac{3\pi}{4}, \frac{4\pi}{5}\right]$, we see that there is a significant increase in radiating

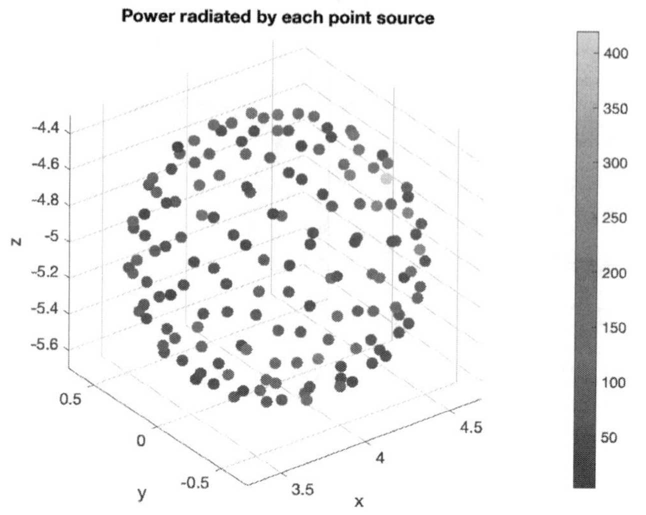

FIGURE 8. Power radiated by each point source $\mathbf{y}_j$ (magnitude)

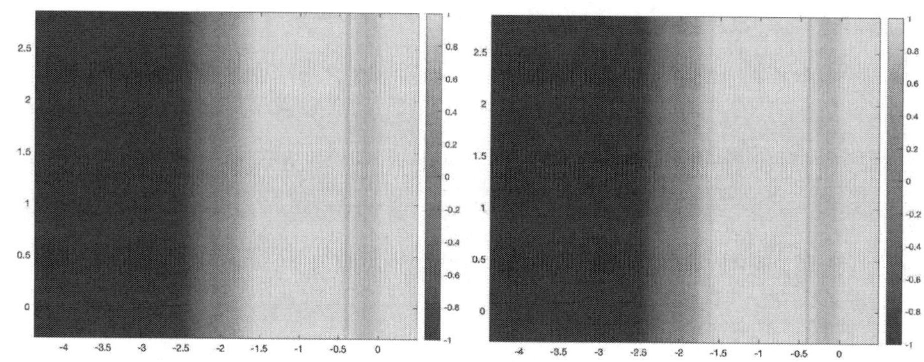

(A)

(B)

FIGURE 9. (A) Desired and (B) generated fields on a cross section of W

effort but maybe reduced if we increase the number of point sources radiating the acoustic field. If we further expand the size of the control region, with $\theta \in \left[\frac{3\pi}{4}, \frac{4\pi}{5}\right]$ and $\phi \left[\frac{5\pi}{32}, \frac{39\pi}{64}\right]$, we may observe that the power demand grow to order of 10^6 and the maximum relative error also rose to an order of 10^{-2} from 10^{-3} when computed for a smaller control region.

Thus, we take note that a constructing a larger control region requires greater power or more point sources to sustain the maximum pointwise relative error in the same order.

REMARK 3.1. We note that for all the simulations above, we use a sound-soft obstacle O, however, sound-hard obstacles may be treated similarly.

TABLE 1. Increase in the size of control region, associated power, point sources, and maximum pointwise relative error

θ	ϕ	No. of point sources	$\|w_d\|_{\ell^2}$	Max. rel. error
$\left[\frac{7\pi}{8}, \frac{11\pi}{10}\right]$	$\left[\frac{17\pi}{64}, \frac{63\pi}{128}\right]$	150	1.1724×10^3	0.33%
		250	7.5120×10^2	0.33%
		300	6.4987×10^2	0.33%
$\left[\frac{3\pi}{4}, \frac{4\pi}{5}\right]$	$\left[\frac{17\pi}{64}, \frac{63\pi}{128}\right]$	150	5.9566×10^3	0.32%
		250	3.9716×10^3	0.31%
		300	3.3282×10^3	0.31%
$\left[\frac{3\pi}{4}, \frac{4\pi}{5}\right]$	$\left[\frac{5\pi}{32}, \frac{39\pi}{64}\right]$	150	8.2482×10^6	6.87%
		250	3.5576×10^6	2.17%
		300	5.2951×10^6	3.46%

4. Conclusion

In this report, we presented a first theoretical argument and numerical support to characterize a given source such that its radiated field approximates desired patterns in prescribed exterior regions and given far field directions in environments with known impenetrable obstacles.

Our current numerical simulations suggest that in an environment with known obstacles it is possible to build an active scheme for the localization of acoustic energy while preserving desired far field behavior in given directions with applications to covert communications in the presence of obstacles, energy localization behind an obstacles with applications to imaging and interference or localized communication with dynamic quiet regions.

Our next work will be focused on considering a detailed theoretical and general sensitivity analysis for the case where general sound soft, sound hard, and penetrable obstacles are considered together with mutual coupling between obstacles as well as coupling with the antenna.

References

[1] Jens Ahrens. *Analytic Methods of Sound Field Synthesis.* T-Labs Series in Telecommunications Services. Springer, 2012.

[2] Jens Ahrens. *Analytic Methods of Sound Field Synthesis.* Springer, 2012.

[3] Jens Ahrens and Sascha Spors. Sound field reproduction using planar and linear arrays of loudspeakers. *IEEE Transactions on Audio, Speech, and Language Processing*, 18(8):2038–2050, 2010.

[4] Takuma Okamoto andAtsushi Sakaguchi. Experimental validation of spatial fourier transform-based multiple sound zone generation with a linear loudspeaker array. *The Journal of the Acoustical Society of America*, 141:1769–1780, 2017.

[5] Thomas S. Angell and Andreas Kirsch, *Optimization methods in electromagnetic radiation*, Springer Monographs in Mathematics, Springer-Verlag, New York, 2004, DOI 10.1007/b97629. MR2026856

[6] Yuri Bobrovnitskii. Impedance acoustic cloaking. *New J. Phys*, 12:043049, 2010.

[7] Charles Allan Boyles. *Acoustic waveguides: applications to oceanic science.* Wiley- Interscience, 1984.

[8] O. M. Bucci, A. Capozzoli, and G. D'Elia. Power pattern synthesis of reconfigurable conformal arrays with near-field constraints. *IEEE Trans. on Antennas and Prop.*, 52(1), 2004.

[9] James L. Buchanan, Robert P. Gilbert, Armand Wirgin, and Yongzhi S. Xu, *Marine acoustics*, Society for Industrial and Applied Mathematics, Philadelphia, PA, 2004. Direct and inverse problems, DOI 10.1137/1.9780898717983. MR2065370

[10] Jordan Cheer. Active control of scattered acoustic fields: cancellation, reproduction and cloaking. *J. Acoust. Soc. Am.*, 140(3):1502–1512, 2016.

[11] Jordan Cheer and Stephen J. Elliott. Multichannel control systems for the attenuation of interior road noise in vehicles. *Mech. Syst. Signal Process.*, 60-61:753–769, 2015.

[12] Sebastien Clauzier, Said M. Mikki, and Yahia M. M. Antar, *Design of near-field synthesis arrays through global optimization*, IEEE Trans. Antennas and Propagation **63** (2015), no. 1, 151–165, DOI 10.1109/TAP.2014.2367536. MR3301535

[13] David Colton and Rainer Kress, *Integral equation methods in scattering theory*, Classics in Applied Mathematics, vol. 72, Society for Industrial and Applied Mathematics (SIAM), Philadelphia, PA, 2013. Reprint of the 1983 original [MR0700400], DOI 10.1137/1.9781611973167.ch1. MR3397293

[14] Massimiliano Comisso, Giulia Buttazzoni, and Roberto Vescovo. Reconfigurable antenna arrays with multiple requirements: A versatile 3d approach. *International Journal of Antennas and Propagation*, 2017, 2017.

[15] Massimiliano Comisso and Roberto Vescovo. 3d power synthesis with reduction of near-field and dynamic range ratio for conformal antenna arrays. *IEEE Trans. on Antennas and Prop.*, 59(4), 2011.

[16] Anthony J. Devaney, *Mathematical foundations of imaging, tomography and wavefield inversion*, Cambridge University Press, Cambridge, 2012, DOI 10.1017/CBO9781139047838. MR2975719

[17] Neil Jerome A. Egarguin, Daniel Onofrei, and Eric Platt, *Sensitivity analysis for the active manipulation of Helmholtz fields in 3D*, Inverse Probl. Sci. Eng. **28** (2020), no. 3, 314–339, DOI 10.1080/17415977.2018.1555248. MR4062509

[18] Neil Jerome A. Egarguin, Shubin Zeng, Daniel Onofrei, and Jiefu Chen, *Active control of Helmholtz fields in 3D using an array of sources*, Wave Motion **94** (2020), 102523, 27, DOI 10.1016/j.wavemoti.2020.102523. MR4060162

[19] Neil Jerome A. Egarguin, Daniel Onofrei, Chaoxian Qi, and Jiefu Chen, *Active manipulation of Helmholtz scalar fields: near-field synthesis with directional far-field control*, Inverse Problems **36** (2020), no. 9, 095005, 28, DOI 10.1088/1361-6420/aba106. MR4149874

[20] Daniel Eggler, Hyuck Chung, Fabien Montiel, Jie Pan, and Nicole Kessissoglou, *Active noise cloaking of 2D cylindrical shells*, Wave Motion **87** (2019), 106–122, DOI 10.1016/j.wavemoti.2018.08.006. MR3944574

[21] Stephen J. Elliot, Jordan Cheer, Jung-Woo Choi, and Youngtae Kim. Robustness and regularization of personal audio systems. *IEEE Trans. of Audio, Speech, and Language process.*, 20(7):2123–2133, 2012.

[22] Stephen Elliott. *Signal Processing for Active Control.* Signal Processing and its Applications. Academic Press, 1st ed edition, 2001.

[23] Chris R. Fuller and Andreas H. von Flotow. Active control of sound and vibration. *IEEE Control Systems Magazine*, 15(6):9–19, Dec 1995.

[24] J. García-Bonito and S.J. Elliott. Active cancellation of acoustic pressure and particle velocity in the near field of a source. *Journal of Sound and Vibration*, 221(1):85 – 116, 1999.

[25] Dieter Guicking. Active control of sound and vibration history - fundamentals - state of the art. *Festschrift DPI, Universitatsverlag Gottingen*, pages 1–32, 01 2007.

[26] Zerui Han, Ming Wu, Qiaoxi Zhu, and Jun Yang. Two-dimensional multizone sound field reproduction using a wave-domain method. *The Journal of the Acoustical Society of America*, 144:1–6, 2018.

[27] Zerui Han, Ming Wu, Qiaoxi Zhu, and Jun Yang. Three-dimensional wave-domain acoustic contrast control using a circular loudspeaker array. *The Journal of the Acoustical Society of America*, 145:1–6, 2019.

[28] M. Hubenthal and D. Onofrei, *Sensitivity analysis for active control of the Helmholtz equation*, Appl. Numer. Math. **106** (2016), 1–23, DOI 10.1016/j.apnum.2016.03.003. MR3499954

[29] Finn B. Jensen, William A. Kuperman, Michael B. Porter, and Henrik Schmidt, *Computational ocean acoustics*, AIP Series in Modern Acoustics and Signal Processing, American Institute of Physics, New York, 1994. MR1259299

[30] Wenyu Jin, W. Bastiaan Kleijn, and David Virette. Multizone soundfield reproduction using orthogonal basis expansion. *2013 IEEE International Conference on Acoustics, Speech and Signal Processing*, pages 311–315, 2013.

[31] Joseph B. Keller, *Survey of wave propagation and underwater acoustics*, Wave propagation and underwater acoustics (Workshop, Mystic, Conn., 1974), Lecture Notes in Phys., Vol. 70, Springer, Berlin, 1977, pp. 1–13. MR0478970

[32] Sang-Myeong Kim, Joao A. Pereira, Antonio E. Turra, and Jun-Ho Cho. Modeling and dynamic analysis of an electrical helmholtz resonator for active control of resonant noise. *Journal of Vibration and Acoustics*, 139(5):1–9, 2017.

[33] L. Landesa, F. Obelleiro, J.L. Rodriguez, J.A. Rodriguez, F. Ares, and A.G. Pino. Pattern synthesis of array antennas with additional isolation of near field arbitrary objects. *Electronics Letters*, 34(16):1540–1541, 1998.

[34] P. Leug. Process of silencing sound oscillations. *U.S. patent no. 2043416*, 1936.

[35] Jingwei Liu, Xiaolin Wang, Ming Wu, and Jun Yang. An active control strategy for the scattered sound field control of a rigid sphere. *JASA*, 144(EL52), 2018.

[36] J. Lončarić, V. S. Ryaben′kii, and S. V. Tsynkov, *Active shielding and control of noise*, SIAM J. Appl. Math. **62** (2001), no. 2, 563–596, DOI 10.1137/S0036139900367589. MR1870707

[37] J. Lončarić and S. V. Tsynkov, *Optimization of acoustic source strength in the problems of active noise control*, SIAM J. Appl. Math. **63** (2003), no. 4, 1141–1183, DOI 10.1137/S0036139902404220. MR1989898

[38] J. Lončarić and S. V. Tsynkov, *Optimization of power in the problems of active control of sound*, Math. Comput. Simulation **65** (2004), no. 4-5, 323–335, DOI 10.1016/j.matcom.2004.01.005. MR2068701

[39] J. Lončarić and S. V. Tsynkov, *Quadratic optimization in the problems of active control of sound*, Appl. Numer. Math. **52** (2005), no. 4, 381–400, DOI 10.1016/j.apnum.2004.08.041. MR2112866

[40] M. V. Lozano and F. Ares Pena. Antenna array pattern synthesis in the presence of near zone scatterers: three-dimenssional vector case. *Jour. of Electromagnetic Waves and Appl..*, 13:1493–1507, 1999.

[41] Rajabi Majid and Mojahed Alireza. Active acoustic cloaking spherical shells. *Acta Acustica united with Acustica,*, 104(1):5–12, 2019.

[42] Qibo Mao, Shengquan Li, and Weiwei Liu. Development of a sweeping helmholtz resonator for noise control. *Applied Acoustics*, 141:348 – 354, 2018.

[43] P. A. Martin, *Acoustic scattering and radiation problems, and the null-field method*, Wave Motion **4** (1982), no. 4, 391–408, DOI 10.1016/0165-2125(82)90007-5. MR674805

[44] P. A. Martin, *Multiple scattering*, Encyclopedia of Mathematics and its Applications, vol. 107, Cambridge University Press, Cambridge, 2006. Interaction of time-harmonic waves with N obstacles, DOI 10.1017/CBO9780511735110. MR2259988

[45] P.A. Martin. On connections between boundary integralequations and t-matrix method. *Engineering Analysis with Boundary Elements*, 27:771–777, 2013.

[46] Dylan Menzies. Sound field synthesis with distributed modal constraints. *Acta Acust. Acust*, 98(1):15–27, 2012.

[47] D. A. B Miller. On perfect cloaking. *Opt. Express*, 14:12457–12466, 2006.

[48] P. A. Nelson and S. J. Elliott. *Active Control of Sound.* Academic London, 1992.

[49] H.F. Olson and E.G. May. Electronic sound absorber. *J. Acad. Soc. Am.*, 25:1130–1136, 1953.

[50] Akira Omoto, Shiro Ise, Yusuke Ikeda, Kanako Ueno, Seigo Enomoto, and Maori Kobayashi. Sound field reproduction and sharing system based on the boundary surface control principle. *Acoustical Science and Technology*, 36:1–11, 01 2015.

[51] Daniel Onofrei, *On the active manipulation of fields and applications: I. The quasistatic case*, Inverse Problems **28** (2012), no. 10, 105009, 15, DOI 10.1088/0266-5611/28/10/105009. MR2987912

[52] Daniel Onofrei, *Active manipulation of fields modeled by the Helmholtz equation*, J. Integral Equations Appl. **26** (2014), no. 4, 553–579, DOI 10.1216/JIE-2014-26-4-553. MR3299831

[53] D. Onofrei and E. Platt, *On the synthesis of acoustic sources with controllable near fields*, Wave Motion **77** (2018), 12–27, DOI 10.1016/j.wavemoti.2017.10.004. MR3754438

[54] Daniel Onofrei, Eric Platt, and Neil Jerome A. Egarguin, *Active manipulation of exterior electromagnetic fields by using surface sources*, Quart. Appl. Math. **78** (2020), no. 4, 641–670, DOI 10.1090/qam/1567. MR4148822

[55] N. Peake and D. G. Crighton, *Active control of sound*, Annual review of fluid mechanics, Vol. 32, Annu. Rev. Fluid Mech., vol. 32, Annual Reviews, Palo Alto, CA, 2000, pp. 137–164, DOI 10.1146/annurev.fluid.32.1.137. MR1744306

[56] A. W. Peterson and S. V. Tsynkov, *Active control of sound for composite regions*, SIAM J. Appl. Math. **67** (2007), no. 6, 1582–1609, DOI 10.1137/060662368. MR2349998

[57] Roy Pike and Pierre Sabatier (eds.), *Scattering. Vol. 1, 2*, Academic Press, Inc., San Diego, CA, 2002. Scattering and inverse scattering in pure and applied science. MR1878885

[58] M. A. Poletti and F. M. Fazi. An approach to generating two zones of silence with application to personal sound systems. *The Journal of the Acoust. Soc. of America*, 137:598–605, 2015.

[59] B.I. Popa, D. Shinde, A. Konneker, and S. A. Cummer. Active acoustic metamaterials reconfigurable in real time. *Phys. Rev. B*, 91, 2015.

[60] R. Kreß and G. F. Roach, *On mixed boundary value problems for the Helmholtz equation*, Proc. Roy. Soc. Edinburgh Sect. A **77** (1977), no. 1-2, 65–77, DOI 10.1017/S0308210500018047. MR473495

[61] J. A. Rodriguez, M. V. Lozano, and F. Ares Pena. Antenna array pattern synthesis in the presence of near zone scatterers: two-dimenssional scalar case. *Microwave and Opt. Tech. Lett.*., 21(4), 1999.

[62] Gálvez Marcos F. Simón, Menzies Dylan, and Fazi Filippo Maria. Dynamic audio reproduction with linear loudspeaker arrays. *JAES*, 67(4):190–200, 2019.

[63] P.A. Nelson S.J. Elliot. The active control of sound. *Electronics and Comm. Engineering Journal*, 1990.

[64] Leun Tsang, Jin Au Kong, and Kung-Hau Ding. *Scattering of electromagnetic waves.* Wiley, 1st ed edition, 2000.

[65] Fernando Guevara Vasquez, Graeme W. Milton, and Daniel Onofrei. Active exterior cloaking. *Phys. Rev. Lett.*, 103:073901, 2009.

[66] Fernando Guevara Vasquez, Graeme W. Milton, and Daniel Onofrei. Broadband exterior cloaking. *Optics Express*, 17(17):14800–14805, 2009.

[67] Fernando Guevara Vasquez, Graeme W. Milton, and Daniel Onofrei, *Exterior cloaking with active sources in two dimensional acoustics*, Wave Motion **48** (2011), no. 6, 515–524, DOI 10.1016/j.wavemoti.2011.03.005. MR2811880

[68] Junqing Zhang, Wen Zhang, Thushara D. Abhayapala, Jingli Xie, and Lijun Zhang. 2.5d multizone reproduction with active control of scattered sound fields. *IEEE Explore*, ICASSP 2019 - 2019 IEEE International Conference on Acoustics, Speech and Signal Processing (ICASSP):141–145, 12-17 May 2019.

[69] Wen Zhang, Thushara D. Abhayapala, Terence Betlehem, and Filippo Maria Fazi. Analysis and control of multi-zone sound field reproduction using modal-domain approach. *J. Acoust. Soc. Am.*, 140(3):2134–2144, 2016.

[70] Qiaoxi Zhu, Philip Coleman, Xiaojun Qiu, Ming Wu, Jun Yang, and Ian S. Burnett. Robust personal audio geometry optimization in the svd-based modal domain. *IEEE/ACM Transactions on Audio, Speech, and Language Processing*, 2018.

Department of Mathematics, University of Houston, Texas 77004
Email address: `lybesabe@uh.edu`

Department of Mathematics, University of Houston, Texas 77004
Email address: `dtonofre@central.uh.edu`

Selected Published Titles in This Series

784 **Dinh-Liem Nguyen, Loc Hoang Nguyen, and Thi-Phong Nguyen, Editors,** Advances in Inverse Problems for Partial Differential Equations, 2023

781 **Emanuel Indrei, Diego Marcon, and Levon Nurbekyan, Editors,** Geometric and Functional Inequalities and Recent Topics in Nonlinear PDEs, 2023

780 **Erik Koelink, Stefan Kolb, Nicolai Reshetikhin, and Bart Vlaar, Editors,** Hypergeometry, Integrability and Lie Theory, 2022

779 **Samuele Anni, Valentijn Karemaker, and Elisa Lorenzo García, Editors,** Arithmetic, Geometry, Cryptography, and Coding Theory 2021, 2022

778 **Carlos Galindo, Alejandro Melle Hernández, Julio José Moyano-Fernández, and Wilson A. Zúñiga-Galindo, Editors,** p-Adic Analysis, Arithmetic and Singularities, 2022

777 **Bang-Yen Chen, Nicholas D. Brubaker, Takashi Sakai, Bogdan D. Suceavă, Makiko Sumi Tanaka, Hiroshi Tamaru, and Mihaela B. Vajiac, Editors,** Differential Geometry and Global Analysis, 2022

776 **Aaron Wootton, S. Allen Broughton, and Jennifer Paulhus, Editors,** Automorphisms of Riemann Surfaces, Subgroups of Mapping Class Groups and Related Topics, 2022

775 **Fernando Galaz-García, Cecilia González-Tokman, and Juan Carlos Pardo Millán, Editors,** Mexican Mathematicians in the World, 2021

774 **Randall J. Swift, Alan Krinik, Jennifer M. Switkes, and Jason H. Park, Editors,** Stochastic Processes and Functional Analysis, 2021

773 **Nicholas R. Baeth, Thiago H. Freitas, Graham J. Leuschke, and Victor H. Jorge Pérez, Editors,** Commutative Algebra, 2021

772 **Anatoly M. Vershik, Victor M. Buchstaber, and Andrey V. Malyutin, Editors,** Topology, Geometry, and Dynamics, 2021

771 **Nicolás Andruskiewitsch, Gongxiang Liu, Susan Montgomery, and Yinhuo Zhang, Editors,** Hopf Algebras, Tensor Categories and Related Topics, 2021

770 **Stéphane Ballet, Gaetan Bisson, and Irene Bouw, Editors,** Arithmetic, Geometry, Cryptography and Coding Theory, 2021

769 **Kiyoshi Igusa, Alex Martsinkovsky, and Gordana Todorov, Editors,** Representations of Algebras, Geometry and Physics, 2021

768 **Dražen Adamović, Andrej Dujella, Antun Milas, and Pavle Pandžić, Editors,** Lie Groups, Number Theory, and Vertex Algebras, 2021

767 **Moshe Jarden and Tony Shaska, Editors,** Abelian Varieties and Number Theory, 2021

766 **Paola Comparin, Eduardo Esteves, Herbert Lange, Sebastián Reyes-Carocca, and Rubí E. Rodríguez, Editors,** Geometry at the Frontier, 2021

765 **Michael Aschbacher,** Quaternion Fusion Packets, 2021

764 **Gabriel Cunningham, Mark Mixer, and Egon Schulte, Editors,** Polytopes and Discrete Geometry, 2021

763 **Tyler J. Jarvis and Nathan Priddis, Editors,** Singularities, Mirror Symmetry, and the Gauged Linear Sigma Model, 2021

762 **Atsushi Ichino and Kartik Prasanna,** Periods of Quaternionic Shimura Varieties. I., 2021

761 **Ibrahim Assem, Christof Geiß, and Sonia Trepode, Editors,** Advances in Representation Theory of Algebras, 2021

760 **Olivier Collin, Stefan Friedl, Cameron Gordon, Stephan Tillmann, and Liam Watson, Editors,** Characters in Low-Dimensional Topology, 2020

For a complete list of titles in this series, visit the AMS Bookstore at **www.ams.org/bookstore/conmseries/**.

Handbook of Treatment Algorithms for Medical and Pharmacy Students

Dr. Santhosh Arun Kumar Komma
Mrs. Vinny Therissa Mangam
Mrs. Sravani Ratnam Arji
Dr. Sanjay Samanth Manthena
Dr. Hema Arya
Mr. Prakash Nathaniel Kumar Sarella

ISBN: 9788198765734

Scan to view online

Editors

Dr. Santhosh Arun Kumar Komma

Mrs. Vinny Therissa Mangam

Mrs. Sravani Ratnam Arji

Dr. Sanjay Samanth Manthena

Dr. Hema Arya

Mr. Prakash Nathaniel Kumar Sarella

First Edition: 2025
ISBN: 9788198765734

Maximum Retail Price: Rs. 1000/- (Inclusive of all taxes)

Published by ThinkPlus Pharma Publications,
9-15/1, Indrapalem, Kakinada, AP-533006, India.
Website: www.books.jopir.in

Cover Page Illustration: Kiran Priyam Kumar Sarella

Printed at Tarangini Printing Works, Subhadra Arcade, Bhanugudi, Kakinada, AP-533003, India.